Robust Statistics for Signal Processing

Understand the benefits of robust statistics for signal processing with this authoritative yet accessible text. The first ever book on the subject, it provides a comprehensive overview of the field, moving from fundamental theory through to important new results and recent advances. Topics covered include advanced robust methods for complex-valued data, robust covariance estimation, penalized regression models, dependent data, robust bootstrap, and tensors. Robustness issues are illustrated throughout using real-world examples and key algorithms are included in a MATLAB Robust Signal Processing Toolbox accompanying the book online, allowing the methods discussed to be easily applied and adapted to multiple practical situations. This unique resource provides a powerful tool for researchers and practitioners working in the field of signal processing.

Abdelhak M. Zoubir is Professor of Signal Processing and Head of the Signal Processing Group at Technische Universität Darmstadt, Germany. He is a fellow of the IEEE, an IEEE Distinguished Lecturer, and the coauthor of *Bootstrap Techniques for Signal Processing* (Cambridge, 2004).

Visa Koivunen is Professor of Signal Processing at Aalto University, Finland. He is also a fellow of the IEEE and an IEEE Distinguished Lecturer.

Esa Ollila is Associate Professor of Signal Processing at Aalto University, Finland.

Michael Muma is a Postdoctoral Research Fellow in the Signal Processing Group at Technische Universität Darmstadt, Germany.

Robust Statistics for Signal Processing

ABDELHAK M. ZOUBIR

Technische Universität Darmstadt

VISA KOIVUNEN

Aalto University

ESA OLLILA

Aalto University

MICHAEL MUMA

Technische Universität Darmstadt

CAMBRIDGE
UNIVERSITY PRESS

CAMBRIDGE
UNIVERSITY PRESS

University Printing House, Cambridge CB2 8BS, United Kingdom

One Liberty Plaza, 20th Floor, New York, NY 10006, USA

477 Williamstown Road, Port Melbourne, VIC 3207, Australia

314–321, 3rd Floor, Plot 3, Splendor Forum, Jasola District Centre, New Delhi – 110025, India

79 Anson Road, #06–04/06, Singapore 079906

Cambridge University Press is part of the University of Cambridge.

It furthers the University's mission by disseminating knowledge in the pursuit of education, learning, and research at the highest international levels of excellence.

www.cambridge.org
Information on this title: www.cambridge.org/9781107017412
DOI: 10.1017/9781139084291

First published 2018

Printed and bound in Great Britain by Clays Ltd, Elcograf S.p.A.

A catalogue record for this publication is available from the British Library.

Library of Congress Cataloging-in-Publication Data
Names: Zoubir, Abdelhak M., author. | Koivunen, Visa, author. | Ollila, Esa,
    1974– author. | Muma, Michael, 1981– author.
Title: Robust statistics for signal processing / Abdelhak M. Zoubir,
    Technische Universität, Darmstadt, Germany, Visa Koivunen, Aalto
    University, Finland, Esa Ollila Aalto University, Finland, Michael Muma
    Technische Universität, Darmstadt, Germany.
Description: New York, NY, USA : Cambridge University Press, 2018. | Includes
    bibliographical references and index.
Identifiers: LCCN 2018022944 | ISBN 9781107017412 (hardback : alk. paper)
Subjects: LCSH: Robust statistics. | Signal processing–Mathematics.
Classification: LCC QA276 .Z68 2018 | DDC 519.5–dc23
    LC record available at https://lccn.loc.gov/2018022944

ISBN 978-1-107-01741-2 Hardback

Additional resources for this publication at www.cambridge.org/zoubir.

Contents

Preface

With the rapid advance in signal processing driven by technological advances toward a more intelligent networked world, there is an ever-increasing need for reliable and robust information extraction and processing, and this is the domain of robust statistical signal processing. There has been a proliferation of research, applications, and results in this field and an in-depth book that provides a timely overview and a systematic account, along with foundational theory, is of importance. Collectively, the authors have significant expertise in the robust statistical signal processing field, developed over many years, and this expertise underpins the selection of material and algorithms, along with the foundational theory, that comprises the book.

Robust statistical signal processing is part of statistical signal processing that, broadly, involves making inference based on observations of signals that have been distorted or corrupted in some unknown manner (Zoubir, 2014). Often, the term *robust* is loosely used in signal processing. The focus in this book is on statistical robustness in the context of statistical uncertainty that arises in state-of-the-art statistical signal processing applications or in deviations from distributional assumptions. The emphasis is on robust statistical methods to solve problems encountered in current engineering practice. While the focus is practical, the book also presents advances in robust statistics – an established discipline of statistics which has its origins in the middle of the last century. Appropriate introductory material is provided to form a coherent development of robust statistical signal processing theory.

Classical statistical signal processing relies strongly on the normal (Gaussian) distribution, which provides, in many situations, a reasonable model for the data at hand. It also allows for closed-form derivations of optimal procedures. However, there have been deviations from Gaussianity reported in numerous measurement campaigns. An overview that covers a broad range of applications where these deviations occur is given in Zoubir et al. (2012), and in later chapters of this book. Robust statistical methods account for the fact that the postulated models for the data are fulfilled only approximately and not exactly. In contrast to classical parametric procedures, robust methods are not significantly affected by small changes in the data, such as outliers or small model departures. They also provide near-optimal performance when the assumptions hold exactly. While optimality is clearly desirable, robustness is the engineer's choice.

In Figure 0.1 the transition from classical parametric statistical signal processing to robust signal processing is illustrated. This tightrope walker metaphor, which goes back to Hampel et al. (2011), illustrates the danger of walking on a single rope, symbolizing

Figure 0.1 Tightrope walker metaphor. Artwork by Sahar Khawatmi.

a single distribution or a parametric model in a real-world situation. For some of today's problems, many robust methods and coherent results exist, as symbolized by the tight mesh that a researcher may use. However, the increasing complexity of the data found in applications necessitates new robust advanced methods. These form a broader mesh that also includes advanced robust signal processing techniques that are in demand more than ever before in today's engineering problems as a pathway to designing systems and data analysis tools.

Aim and Intended Audience

The primary aim of the book is to make robust methods accessible for everyday signal processing practice. The book has been written from a signal processing practitioner's perspective and focuses on data models and applications that we have frequently encountered, and algorithms that we have found useful. Throughout the book, examples taken from practical applications, that is, real-life examples, are used to motivate the use of robust methods and to illustrate the underlying concepts. Our objective is to give a tutorial-style treatment of fundamental concepts, as well as to provide an accessible overview of lesser-known aspects and more recent trends in robust statistics for signal processing.

The intended audience is broad and includes applied statisticians and scientists in areas such as biomedicine, data analytics, electrical and mechanical engineering, communications, and others. Some sections of the book present fundamental concepts, whereas others highlight recent developments. Therefore, beginners, for example, graduate-level students or practitioners with a basic knowledge in probability theory, linear algebra, and statistics, can use the book as a starting point to familiarize themselves with the concepts of robustness. Further, experts and more theoretically oriented readers are likely to benefit from proofs and recent results.

Organization of the Book

Chapter 1 provides the foundation for robust estimation theory that is introduced in a signal processing context and in a signal processing framework. It includes a brief historical account and provides the basic outlier and heavy-tailed distribution models. The M-estimator of location and scale parameters is explained by using the example of estimating the direct current value in independently and identically distributed noise. Important measures of robustness such as the influence function and the breakdown point are then introduced.

The important topic of parameter estimation in linear regression models is considered in Chapter 2. Linear regression models are used in many practical problems including those where the interference is impulsive or the noise has a heavy-tailed distribution. As an example, we discuss the geolocation of a user equipment in mixed line-of-sight/non-line-of-sight observations in wireless communications. Because of its practical importance, the chapter begins with a brief introduction to complex derivatives and optimization. Chapter 2 then discusses least squares estimators, least absolute deviation estimators, rank-least absolute deviation estimators, maximum likelihood estimators, M-estimators, and positive breakdown point estimators. Measures of robustness are defined for the linear regression model and simulation examples are provided to compare the estimators.

The major focus of Chapter 3 is robust and sparse estimation in linear models. Regularized robust estimators are considered and an example from the area of image denoising is given. Regularized robust estimators are also important for compressive sensing, which is a signal processing technique for efficiently acquiring and reconstructing a signal by finding solutions to underdetermined linear systems. Sparse regression methods have become increasingly important in modern data analysis due to the frequently occurring case of both the measurement and feature space being high-dimensional. In high-dimensional regression problems, the regression model is often ill-posed, that is, the number of regressors often exceeds the number of measurements. The least absolute shrinkage and selection operator (Lasso) has become a benchmark method for sparse regression, and the purpose of this chapter is to develop and review robust alternatives to the Lasso estimator and its extensions. For illustration, these methods are applied to a benchmark prostate cancer data set.

Robustness for the multichannel data case, as arises, for example, in sensor array processing, is of increasing importance in signal processing practice and is the subject of Chapter 4. The complex-valued case is considered and a brief review of complex elliptically symmetric distributions is provided before the theory of robust estimation of the multivariate location and scatter (covariance) matrix is introduced. This chapter places special attention to M-estimators for the scatter matrix as these estimators have been, by far, the most popular estimators in the signal processing literature. We illustrate the usefulness of robust covariance estimators in a signal detection application by using the normalized matched filter.

Important aspects of robust covariance estimation for sensor array processing applications are discussed in Chapter 5. First, the basic array processing signal model for ideal arrays and the underlying assumptions used in estimating the angles of arrival is presented. The uncertainties in the signal model, including modeling emitted signals, the array configuration, and the propagation environment, are considered. Some array configurations and direction-of-arrival estimation methods that provide robustness in the face of such uncertainties are briefly discussed.

Chapter 6 is concerned with tensor representations, which are a natural way to approach the modeling of high-dimensional data. First, a brief overview of tensor representations is provided and a standard notation is used. The basic ideas underpinning canonical tensor decomposition and Tucker decomposition are then described. This is followed by a discussion of a statistically robust method for finding a tensor decomposition. The case of finding a statistically robust way of decomposing tensors while promoting sparseness is also discussed.

Robust filtering constitutes an important field of research with a long and rich history, reaching back to even before the mathematical formalization of robust statistics by Huber (1964). Considering the large amount of material on robust filtering, a book on this topic is easily justified. In this context, a brief overview of selected topics is provided in Chapter 7 without any claim for completeness. In particular, we first discuss robust Wiener filtering and then describe some nonparametric nonlinear robust filters, such as the weighted median and weighted myriad filters. Real-world applicability of robust methods is exemplified by analyzing electrocardiographic data. Finally, we consider robust filtering based on state-space models, that is, the robust Kalman filter and the approximate conditional mean filter. Robust filters that can deal with outliers in the state equation and in the measurement equation are discussed. We also provide an example of tracking user equipment in a mixed line-of-sight/non-line-of-sight environment by means of robust extended Kalman filtering.

Correlated data streams are commonly measured in areas such as engineering, data analytics, economics, biomedicine, radar, or speech signal processing, to mention a few. The basic concepts of robustness introduced in the independent data case, however, cannot be straightforwardly extended to the dependent data case. Robust methods for dependent data form the most practical case for signal processing practitioners and are the subject of Chapter 8. This chapter focuses on robust parameter estimation for autoregressive moving-average (ARMA) models associated with random processes, for which the majority of the samples are appropriately modeled by a stationary and invertible ARMA model and a minority consists of outliers with respect to the ARMA model.

Robust spectral estimation is the subject of Chapter 9. Here, we consider nonparametric robust methods, as well as robust parametric spectral estimation. We discuss the robust estimation of the power spectral density of an ARMA process using methods for robust estimation of the model parameters detailed in Chapter 8. Further, we discuss parametric methods such as MUltiple SIgnal Classification for the robust estimation of line spectra by employing the robust eigendecomposition of the covariance matrix, discussed in Chapter 4.

Robust bootstrap methods are introduced in Chapter 10. The bootstrap is a powerful computational tool for statistical inference that allows for the estimation of the distribution of an estimator without distributional assumptions on the underlying data, reliance on asymptotic results, or theoretical derivations. However, the robustness properties of the bootstrap in the presence of outliers are very poor, irrespective of the robustness of the bootstrap estimator. We briefly review recent developments of robust bootstrap methods for estimation and highlight their importance. Again, the example of robust geolocation is used for illustration purposes. The chapter also features a robust bootstrap method that is scalable to very large volume and high-dimensional data sets, that is, big data. Moreover, it is compatible with distributed data storage systems and distributed and parallel computing architectures. Finally, an inference example using real-world data from the Million Song data set is considered.

Chapter 11 is devoted to real-life applications of robust methods. Here, we give several examples of how the theory detailed in preceding chapters can be applied in areas as diverse as short-term load forecasting, diabetes monitoring, heart-rate variability analysis by means of photoplethysmography, inverse atmospherical problems, and indoor localization.

To reproduce the examples that are given in the book and to allow the practitioner to directly apply the methods detailed in the book a MATLAB$^{©}$ toolbox – RobustSP – has been developed, and this is downloadable as ancillary material.

The book does not cover robustness for signal detection and robust hypothesis testing (Sion, 1958; Huber, 1965; Huber and Strassen, 1973; Österreicher, 1978; Kassam and Poor, 1985; Dabak and Johnson, 1993; Poor and Thomas, 1993; Song et al., 2002) or classification, although much progress has been made in this area recently, for example robust detection using f-divergence balls (Levy, 2009; Gül and Zoubir, 2016, 2017; Gül, 2017), the importance of density bands (Kassam, 1981; Fauß and Zoubir, 2016), or robust sequential detection (DeGroot, 1960; Brodsky and Darkhovsky, 2008; Fellouris and Tartakovsky, 2012; Fauß and Zoubir, 2015; Fauß, 2016).

Acknowledgments

I, Abdelhak Zoubir, would like to take this opportunity to express my gratitude to my esteemed doctoral advisor, Professor Johann F. Böhme, who introduced me to the fascinating world of statistical signal processing. He insisted on two ingredients for success, which accompanied me during my whole career, namely, a good and rigorous foundation in statistics and advanced real-life applications. I would also like to thank all my doctoral students, past and current ones, from whom I have learned so much. I truly had a great opportunity to work with bright minds. In particular, and without diminishing the great contributions by numerous other doctoral students, I wish to mention Michael Fauß, Gökhan Gül, and Michael Muma, who performed fundamental advancements in robust statistics for (also sequential) detection and estimation. I also had the opportunity to collaborate over several years with many colleagues from Germany and abroad. I wish to name a few: Professor Moeness Amin, Professor Fulvio Gini, Professor Fredrik Gustafsson, Professor Visa Koivunen, Professor Ali Sayed, Dr. Chong Meng Samson See, and Professor Hing Cheung So. I also wish to acknowledge the financial support of the German Research Foundation (DFG), the Hessen State Ministry for Higher Education, Research and the Arts through its LOEWE program, and industry partners. The finalization of a book with four authors would not have been possible without the timely and careful coordination by one of us, Dr. Michael Muma, without whom the project would not have been completed in a timely fashion. I wish to thank Phil Meyler from Cambridge University Press for his encouragement and support during all these years. I thank all my family (small and extended) members for their continued support and love over all these years.

I, Visa Koivunen, would like to express my gratitude to two outstanding scholars and role models, namely Professor Saleem A. Kassam from the University of Pennsylvania and Professor H. Vincent Poor from Princeton University. Their deep love for research, mentorship, and friendship as well as the opportunity to work in their research groups for several years have impacted my career in a profound way. Their 1985 article on robust techniques in signal processing inspired me to work in this area in the first place. I would also like to thank my former students Dr. Samuli Visuri and Dr. Esa Ollila for the opportunity to work with such bright minds and develop new methods and theory in this area. I would also to thank Dr. Hannu Oja for fruitful collaboration. The financial support from the Academy of Finland and Finnish Defense Forces is gratefully acknowledged. Finally, I would like to express my love and gratitude to my wife Sirpa and our sons Max and Klaus for their patience and making my life beautiful.

I, Esa Ollila, would like to express my gratitude to my thesis advisers, Professor Emeritus Hannu Oja and Professor Visa Koivunen, for introducing me to robustness as well as their generous support in my academic career. It has also been a great pleasure to collaborate with many outstanding researchers in the field of robust statistics, including Professor Emeritus Thomas P. Hettmansperger, Distinguished Professor David E. Tyler, Professor H. Vincent Poor, Professor Christophe Croux, Professor Ami Wiesel, and Professor Frédéric Pascal. The financial support from the Academy of Finland is gratefully acknowledged. I would also like to thank Elias Raninen for his help in making some of the illustrations in Chapters 2 and 3. Finally, the work put into this book was worthwhile only due to my family, my wife Hyon-Jung, our daughter Sehi and son Seha. I wish to thank them for all their love and support.

I, Michael Muma, would like to express my gratitude to Professor Zoubir for his support, mentoring, guidance, and trust. His outstanding degree of enthusiasm and motivation for research has been a great inspiration. I would like to thank Visa Koivunen and Esa Ollila for their generous support and for being role models in the area of robust signal processing. I am grateful to Yacine Chakhchoukh, Frédéric Pascal, Gonzalo Arce, Michael Fauß, Freweyni Teklehaymanot, and Jasin Machkour for sharing their valuable thoughts. Finally, I would like to express my love and gratitude to my wife Eli, my daughters Lillie and Kyra, and my parents Craig and Waltraud. Many of the smiles and happy moments I had while writing this book came from thinking about you.

We all would like to thank Roy Howard, Sara Al-Sayed, and Shahab Basiri for their careful review and constructive comments. We thank Sahar Khawatmi for the fantastic tightrope walker metaphor artwork and Leander Lenz for the help in creating ideas for the cover of our book. We would like to express our gratitude to Renate Koschella for her kindness and for always doing more than is required. We thank Martin Vetterli, Benjamín Béjar, and Marta Martinez-Camara for making us aware of the possibility to apply robust penalized methods for ill-conditioned linear inverse problems to the European Tracer Experiment data set and Bastian Alt for the implementation. We thank Nevine Demitri for sharing some of her research that applies robust methods to glucose monitoring for diabetes care and for collecting a data set that has been used in this book. We thank Yacine Chackchouck for sharing some of his research that applies robust methods to load forecasting. We thank Ulrich Hammes and Di Jin for their input toward applying robust linear regression estimators to geolocation and tracking in line-of-sight/non-line-of-sight scenarios. We are thankful to Stefan Vlaski for sharing some of his research on robust bootstrap methods. We also thank Augustin Kelava and Marlene Deja for their excellent collaboration and for their efforts in collecting data for a psychophysiological study, some of which has been used for illustration purposes in this book. We thank Tim Schäck for sharing his research on photoplethysmography and for collecting data that has been used in this book. We express our thanks to Mengling Feng and the staff at Neuro-ICU of National Neuroscience Institute, Singapore for their efforts in collecting the intracranial pressure data set.

Finally, we express our gratitude to Cambridge University Press for providing us with the opportunity to write this book and for their patience.

Abbreviations

ACE	adaptive coherence estimator
ACG	angular central Gaussian
ACM	approximate conditional mean
AIC	Akaike's information criterion
ALS	alternating least squares
AO	additive outlier
AOA	angle-of-arrival
AR	autoregressive
ARE	asymptotic relative efficiency
ARMA	autoregressive moving-average
AWGN	additive white Gaussian noise
BIC	Bayesian information criterion
BIP	bounded influence propagation
BLB	bag of little bootstraps
BLFRB	bag of little, fast, and robust bootstraps
BP	breakdown point
CANDECOMP	Canonical Decomposition
CCD	cyclic coordinatewise descent
CD	coordinate descent
cdf	cumulative distribution function
CES	complex elliptically symmetric
CFAR	constant false alarm rate
CI	confidence interval
CN	complex normal
CRLB	Cramér-Rao lower bound
CS	compressed sensing
CV	cross-validation
DC	direct current
DF	degrees of freedom
DOA	direction-of-arrival
DTFT	discrete-time Fourier transform
ECG	electrocardiogram
ECP	empirical coverage probability

EEG	electroencephalogram
EIF	empirical influence function
EKF	extended Kalman filter
EN	elastic net
ES	elliptically symmetric
ESPRIT	estimation of signal parameters via rotational invariance techniques
ETEX	European Tracer Experiment
EVD	eigenvalue decomposition
FFT	fast Fourier transform
FIR	finite impulse response
fMRI	functional magnetic resonance imaging
FNR	false negative rate
FOBI	fourth-order blind identification
FP	fixed-point
FPR	false positive rate
FRB	fast and robust bootstrap
GES	gross-error-sensitivity
GLRT	generalized likelihood ratio test
GPS	global positioning system
GUT	generalized uncorrelating transform
HOS	higher-order statistics
HQ	Hannan and Quinn
HRV	heart rate variability
IC	information criteria
ICA	independent component analysis
ICP	intracranial pressure
IF	influence function
IFB	influence function bootstrap
i.i.d.	independently and identically distributed
IO	innovations outlier
IRWLS	iterative re-weighted least squares
JADE	joint approximate diagonalization of eigen-matrices
JNR	jammer-to-noise ratio
JSR	jammer-to-signal ratio
KDE	kernel density estimator
LAD	least absolute deviations
LASSO	least absolute shrinkage and selection operator
LCD	liquid crystal display
LMS	least-median of squares
LOS	line-of-sight
LPDM	Lagrangian particle dispersion model
LSE	least squares estimator
LTI	linear time-invariant

LTS	least trimmed squares
MA	moving-average
MAP	maximum posteriori
MAPE	mean absolute percentage error
MBC	maximum-bias curve
MC	Monte Carlo
MCD	minimum covariance determinant
MDL	minimum description length
MeAD	mean absolute deviation
MHDE	minimum Hellinger distance estimator
MIMO	multiple-input multiple-output
ML	maximum likelihood
MLE	maximum likelihood estimator or estimate
MMSE	minimum mean squared error
MR	median-of-ratios
MRA	minimum redundancy arrays
MS	mean-shift
MSD	matched subspace detector
MSE	mean squared error
MSWF	multistage Wiener filter
MUSIC	MUltiple SIgnal Classification
MVDR	minimum variance distortionless response
MVE	minimum volume ellipsoid
NLOS	non-line-of-sight
NMSE	normalized mean squared error
PARAFAC	parallel factors
PCA	principal component analysis
PCI	peripheral component interconnect
PD	probability of detection
pdf	probability density function
PDH	positive definite Hermitian
PE	prediction error
PFA	probability of false alarm
PPG	photoplethysmogram
PRV	pulse rate variability
PSD	power spectral density
RA	robust autocovariance
RARE	rank-reduction
RCM	rank covariance matrix
REKF	robust extended Kalman filter
RES	real elliptically symmetric
RLAD	rank-least absolute deviations
RM	ratio-of-medians

RMSE	relative mean squared error
RMSSD	root mean square of successive differences
RO	replacement outlier
RSP	robust starting point bootstrap
RSS	residual sum-of-squares
RV	random variable
r.v.	random vector
SARIMA	seasonal integrated autoregressive moving-average
SC	sensitivity curve
SCM	sample covariance matrix
SDNN	standard deviation of N-N intervals
SML	stochastic maximum likelihood
SNR	signal-to-noise power ratio
SOI	signal of interest
SP	signal processing
SSR	sparse signal reconstruction
ST	space-time or soft-thresholding
SUT	strong uncorrelating transform
SVD	singular value decomposition
TCM	Kendall's Tau covariance matrix
TOA	time-of-arrival
UCA	uniform circular array
ULA	uniform linear array
w.l.o.g.	without loss of generality
WSN	wireless sensor network
WSS	wide-sense stationary

List of Symbols

$\mathrm{AG}_p(\boldsymbol{\mu}, \boldsymbol{\Sigma})$	p-dimensional Angular Gaussian (AG) distribution with mean vector $\boldsymbol{\mu}$ and scatter matrix $\boldsymbol{\Sigma}$
arg max	argument of the maximum
arg min	argument of the minimum
$\mathbf{a}_i$	ith column vector of a matrix $\mathbf{A} = (\mathbf{a}_1 \ \cdots \ \mathbf{a}_p)$
$\mathbf{a}_{[i]}$	ith (transposed) row vector of a matrix $\mathbf{A} = (\mathbf{a}_{[1]} \ \cdots \ \mathbf{a}_{[p]})^\top$
bias $(T(F), F)$	asymptotic bias of an estimator $T(F)$ at distribution F
$\mathbb{C}$	field of complex numbers
col$(\mathbf{X})$	column space of $\mathbf{X}$
#$\{\cdot\}$	cardinality of a set
$\mathbb{C}\mathcal{E}_p(\boldsymbol{\mu}, \boldsymbol{\Sigma}, g)$	p-dimensional complex elliptically symmetric distribution with mean vector $\boldsymbol{\mu}$, scatter matrix $\boldsymbol{\Sigma}$ and density generator g
$\mathbb{C}\mathcal{N}_p(\boldsymbol{\mu}, \boldsymbol{\Sigma})$	p-variate complex normal distribution with mean vector $\boldsymbol{\mu}$ and scatter matrix $\boldsymbol{\Sigma}$
$(\cdot)^*$	complex conjugate
$\boldsymbol{\Sigma}$	covariance matrix
$r_{xy}(\cdot)$	cross-second-order moment function of the random processes $x_t(\zeta)$ and $y_t(\zeta)$
$c_{xy}(\cdot)$	central cross-second-order moment function (cross-covariance function) of the random processes $x_t(\zeta)$, $y_t(\zeta)$
$x \overset{\mathrm{d}}{=} y$	x has the same distribution as y
$\lvert \mathbf{A} \rvert$	determinant of a matrix $\mathbf{A}$
diag$(\mathbf{a})$	diagonal matrix with $\mathbf{a} = (a_1, \ldots, a_p)^\top$ as diagonal elements
$\triangleq$	defined as
F_N	empirical distribution function
exp$(\cdot)$	exponential function
$\mathrm{E}[\cdot]$	expectation operator
$\mathrm{E}[\cdot]_F$	expectation operator at distribution F
GES $(T(F), F)$	gross error sensitivity of estimator $T(F)$ at nominal distribution F
$(\cdot)^{\mathsf{H}}$	Hermitian transpose of a matrix or a vector
$\mathbb{1}_{\mathcal{A}}$	is the indicator of the event $\mathcal{A}$
$\lfloor \cdot \rfloor$	denotes the integer part
$\mathbf{I}$	identity matrix
$\mathbf{I}_p$	$p \times p$ identity matrix

$\mathsf{IF}(y; T(F), F)$	influence function of estimator $T(F)$ at nominal distribution F
$\overset{i.i.d.}{\sim}$	independently and identically distributed
i, j, k	discrete index
$\mathsf{Im}(\cdot)$	operator extracting the imaginary part of its complex-valued argument
$(\cdot)^{-1}$	inverse of a matrix
$\langle \cdot, \cdot \rangle$	inner product
j	imaginary unit
$\mathsf{kurt}[y]$	kurtosis of a random variable y
$L_{\mathrm{ML}}(\mu, \sigma \,|\, \mathbf{y})$	negative log-likelihood function with parameters μ and σ given observations $\mathbf{y}$
$\mathbf{A}^{\dagger}$	Moore–Penrose pseudoinverse of matrix $\mathbf{A}$
$\mathsf{med}(\mathbf{x})$	median of vector $\mathbf{x}$
$\|a\|$	modulus of a complex number a
μ	location parameter (mean)
$\mathsf{MSE}\,(T(F), F)$	mean-squared error of estimator $T(F)$ of the parameter β at distribution F
$\mathcal{M}(N, \mathbf{p})$	multinomial distribution with N trials and outcome probability vector $\mathbf{p}$
$\mathcal{N}_p(\boldsymbol{\mu}, \boldsymbol{\Sigma})$	p-variate real normal distribution with mean vector $\boldsymbol{\mu}$ and scatter matrix $\boldsymbol{\Sigma}$
N	number of observations
$\|\cdot\|_1$	ℓ_1-norm
$\|\cdot\|_2$	Euclidean (ℓ_2-)norm
$\|\cdot\|_{p,q}$	mixed $\ell_{p,q}$-norm
$\|\cdot\|_{\mathrm{F}}$	Frobenius ($\ell_{2,2}$-) norm
$\mathbf{1}$	column vector of 1's
$\mathbf{A}^{\perp}$	orthogonal complement of $\mathbf{A}$
$\overset{P}{\rightarrow}$	convergence in probability
$f(x|\beta)$	pdf of random variable x given parameter β
$\mathsf{Arg}(\cdot)$	principal argument of a complex number
$\mathsf{Prob}(\cdot)$	probability
p	dimension of model or observations
$\mathbb{S}_{++}^{p \times p}$	set of all positive definite real (or complex) symmetric (or Hermitian) matrices of dimension $p \times p$
$I_{xx}(e^{j\omega})$	periodogram at frequency ω
$S_{xx}(e^{j\omega})$	power spectral density of random process x_t at frequency ω
$\psi(\cdot)$	score function of an M-estimator
$\mathcal{E}_p(\boldsymbol{\mu}, \boldsymbol{\Sigma}, g)$	p-dimensional real elliptically symmetric distribution with mean vector $\boldsymbol{\mu}$, scatter matrix $\boldsymbol{\Sigma}$ and density generator g
$\mathbb{R}$	field of real numbers
$\mathsf{Re}(\cdot)$	operator extracting the real part of its complex-valued argument
$\mathbb{R}^{+}$	$\{x \in \mathbb{R} | x > 0\}$
$\mathbb{R}_0^{+}$	$\{x \in \mathbb{R} | x \geq 0\}$

$\rho(\cdot)$	objective function of an M-estimator
$\diamond$	replica operator
$\mathsf{SC}(y; T(F_N), F_N)$	sensitivity curve of estimator $T(F)$ at empirical distribution F_N
σ	scale parameter (standard deviation)
$\boldsymbol{\Sigma}$	scatter matrix
$\mathbb{C}t_{p,\nu}(\boldsymbol{\mu}, \boldsymbol{\Sigma})$	p-dimensional complex t-distribution with mean vector $\boldsymbol{\mu}$, scatter matrix $\boldsymbol{\Sigma}$ with $\nu > 0$ degrees of freedom
$(\cdot)^{\top}$	transpose of a matrix or a vector
$\mathsf{Tr}(\cdot)$	matrix trace
$\mathcal{U}(a, b)$	uniform distribution on the interval a, b
$\sigma_y^2, \mathsf{var}(y)$	variance of a random variable y
ω	radian frequency (normalized)
χ_p^2	chi square distribution with p degrees of freedom
$\mathbb{Z}$	set of integers
$\mathbb{Z}^+$	set of positive integers

1 Introduction and Foundations

In this chapter, the basic concepts of robust estimation in a signal processing framework are introduced. After a brief historical recount, we discuss outlier and heavy-tailed distribution models. These models are common in engineering practice as is evident from numerous measurements made in different fields, for example, in digital communication. Among other heavy-tailed noise models, we introduce in this chapter the epsilon mixture model that is used extensively in subsequent chapters. We then consider the estimation of location and scale parameters in the real data case. The principles underpinning this case are demonstrated by considering the simple problem of estimating the direct current (DC) value when measurements are subject to random fluctuations that are independently and identically distributed (i.i.d.) from sample to sample. In this problem, the M-estimator is introduced and this highlights the intuitive link to maximum likelihood estimation for different noise models. Important measures of robustness then follow and these include the influence function (IF) and the breakdown point (BP). The introduction of robustness in estimation comes at the price of decreased statistical efficiency of the estimator and the trade-off between robustness and efficiency is discussed. This trade-off is demonstrated by considering location estimation based on the sample median. An understanding of this trade-off is likely to facilitate the signal processing practitioner to design robust estimators for location and scale. Because this chapter is intended to serve as an easy-to-read introduction to robustness concepts, only the real-valued case is discussed. The complex-valued case is treated in Chapter 2, where the linear regression model is introduced, which contains as a special case the location (or location-scale) model.

Several examples, along with the associated MATLAB$^{©}$ code that allows users to reproduce the results, are included in the downloadable RobustSP toolbox.

1.1   History of Robust Statistics

Statistical signal processing is an important area of research that has been successfully applied to generations of engineering problems where the extraction of useful information from empirical data is required. An effective way to incorporate knowledge from empirical data is to use parametric stochastic models. Important foundations were established in the 1920s by R. A. Fisher (Fisher, 1925) who derived many useful statistical models and methods. When applying parametric methods to real-world problems, the situation often arises that the observations do not exactly follow the assumptions made to model the problem. In these cases, the nominally excellent performance can drastically degrade.

From a practitioner's viewpoint, however, it is essential that the results associated with the parametric method used be acceptable in situations where the distributional assumptions underpinning the assumed model do not hold. One approach is to make as few assumptions as possible about the data and resort to a nonparametric statistical model. In some signal processing applications, for example in spectrum estimation, nonparametric approaches, based on the periodogram, have become widely popular. However, in today's engineering practice, parametric models continue to play an important role. This is especially the case in complex applications, where, to retrieve meaningful information, it is necessary to incorporate some knowledge about the system under consideration. So which strategy should one follow? Everyone who deals with real-world problems can relate to the famous remark by G. E. P. Box on robustness in statistical model building, "All models are wrong, but some are useful" (Box, 1979). If we acknowledge that the data model we use is at best a close approximation to the true model from which real measurements have been obtained, it is then only a small step to robust statistics.

Robustness, as treated in this book, deals with deviations from the distributional assumptions, and we mainly consider deviations from a Gaussian (normal) probability model. The word *robust* was introduced into the statistics literature by G. E. P. Box in 1953 (Box, 1953). The study of robustness, however, predates even this pioneering work. According to D. Bernoulli (Bernoulli, 1777), outlier rejection was already common practice in 1777. Mixture models and estimators that down-weight outliers were known in the 1800s and S. Newcomb even "preinvented" a kind of one-step Huber-estimator (Stigler, 1973). The question of how best to characterize uncertainties in observations has been an ongoing discussion since the early days of statistics. The first scientist to note in print that measurement errors deserve a systematic and scientific treatment was G. Galileo in 1632 (Galilei, 1632).

Since its discovery in 1733 by A. de Moivre (de Moivre, 1733), the normal distribution has played a central role in statistical modeling. It was named after C. F. Gauss, who derived it to justify his use of the least squares criterion in astronomy to locate an orbit that best fitted known observations (Gauss, 1809). He developed a theory of errors that is based on the following assumptions: (i) small errors are more likely than large errors; (ii) the likelihood of the errors being positive or negative is equal; and (iii) in the presence of several measurements of the same quantity, the most likely value of the quantity being measured is their average. On this basis, Gauss derived the formula for the normal probability density of the errors (Stahl, 2006), and this formula has since been justified in many different ways and shown to be applicable in many different contexts such that it is the default model that is used is many applications.

As H. Poincaré pointed out in 1904 (Poincaré, 1904), "Physicists believe that the Gaussian law has been proved in mathematics while mathematicians think that it was experimentally established in physics." Even today, many methods encountered in engineering practice rely on the Gaussian distribution of the data, which in many situations is well justified and motivated (Kim and Shevlyakov, 2008) by the central limit theorem. Assuming Gaussianity can be practical in many situations, for example, using the Gaussian error model can be based on the argument that it minimizes the Fisher information over the class of distributions with a bounded variance, and the Fourier transform of a Gaussian function is another Gaussian function. Assuming Gaussianity also enables a simple derivation of likelihood functions. In summary, the main justification for assuming a normal distribution is twofold. On the one hand, it provides an approximate representation for many real-world data sets. On the other hand, it is convenient from a theoretical viewpoint as it facilitates the derivation of closed-form expressions for optimal detectors or estimators. Optimality is clearly a desirable property for a detector or an estimator. Optimality, only under the assumed (nominal) distribution, however, is useless if the estimator is applied to data that does not follow this distribution. As highlighted by Tukey in 1960, even slight deviations from the assumed distribution may cause the estimator's performance to drastically degrade or to completely break down (Zoubir et al., 2012).

Robust statistics formalizes the theory of approximate parametric models. On the one hand, robust methods are able to leverage a parametric model, but on the other hand, such methods do not depend critically on the exact fulfillment of the model assumptions. In this sense, robust statistics are consistent with engineering intuition and signal processing demands. Robust methods are designed in such a way that they behave nearly optimally, if the assumed model is correct, while small deviations from the model assumptions degrade performance only slightly and larger deviations do not cause a catastrophe (Huber and Ronchetti, 2009). The theory of robust statistics was established in the middle of the twentieth century by the pioneering work of J. P. Tukey, P. J. Huber, and F. R. Hampel, who are often called the "founding fathers" of robust statistics. In 1960, J. W. Tukey (Tukey, 1960) summarized his work in the 1940s and 1950s on the effect of a small amount of contaminating data (outliers) on the sample mean and standard deviation. He introduced a contamination model and proposed some estimators that are robust against such contamination.

The first attempt toward a unified framework for robust statistics was undertaken in the seminal paper of P. J. Huber on robust location estimation in 1964 (Huber, 1964). After defining neighborhoods around a true distribution that generates the data, he proposed an estimator that yields minimax optimal performance over the entire neighborhood. This means that the estimator is optimal for the worst-case distribution within the neighborhood. For details on Huber's approach, the reader is referred to the book by P. J. Huber and E. M. Ronchetti (Huber and Ronchetti, 2009).

Further fundamental concepts of robust statistics were introduced by F. R. Hampel in 1968 (Hampel, 1968). His so-called infinitesimal approach is based on three central concepts: qualitative robustness, the IF and the BP. Intuitively, they correspond to the continuity and first derivative of a function and the distance to its nearest singularity. Interested readers are referred to the book by F. R. Hampel, E. M. Ronchetti, P. J. Rousseeuw, and W. A. Stahel (Hampel et al., 2011).

In engineering, robust estimators and detectors have been of interest since the early days of digital signal processing; see the review paper on robust methods published by Kassam and Poor in 1985 (Kassam and Poor, 1985) and references therein. Since then, many aspects of robustness have been utilized in signal processing associated with communications systems, radar and sonar, pattern recognition, biomedicine, speech, and image processing, amongst others. The increasing complexity of the data (and models) that are analyzed today have triggered new areas of research in robust statistics. Today, for example, robust methods have to deal with high-dimensional, sparse, multivariate, and/or complex-valued data. Nevertheless, much of today's work is based on the ideas that were formalized in the middle of the last century.

In many areas of engineering today, the distribution of the measurement data is far from Gaussian as it contains outliers, which cause the distribution to be heavy-tailed. In particular, measurement data from a diversity of areas (Blankenship et al., 1997; Abramovich and Turcaj, 1999; Middleton, 1999; Etter, 2003) have confirmed the presence of impulsive (heavy-tailed) noise, which can cause optimal signal processing techniques, especially the ones derived under the Gaussian probability model, to be biased or to even break down.

The occurrence of impulsive noise has been reported, for example, in outdoor mobile communication channels due to switching transients in power lines or automobile ignition (Middleton, 1999), in radar and sonar systems as a result of natural or man-made electromagnetic and acoustic interference (Abramovich and Turcaj, 1999; Etter, 2003), and in indoor wireless communication channels, owing, for example, to microwave ovens and devices with electromechanical switches, such as electric motors in elevators, printers, and copying machines (Blankenship et al., 1997). Moreover, biomedical sensor array measurements of brain activity, such as in magnetic resonance imaging (MRI) and associated with regions of the human brain where complex tissue structures are known to exist, have been found to be subject to non-Gaussian noise and interference (Alexander et al., 2002).

In geolocation position estimation and tracking, non-line-of-sight (NLOS) signal propagation, caused by obstacles such as buildings or trees, results in outliers in measurements, to which conventional position estimation methods are very sensitive (Hammes et al., 2009). In classical short-term load forecasting, the prediction accuracy is adversely influenced by outliers, which are caused by nonworking days or exceptional events such as strikes, soccer's World Cup, or natural disasters (Chakhchoukh et al., 2010). Moreover, on a computer platform, various components, such as the liquid crystal display (LCD) pixel clock and the peripheral component interconnect (PCI) express bus, cause impulsive interference that degrades the performance of the embedded wireless devices (Nassar et al., 2008). These studies show that in real-world applications, robustness against departures from Gaussianity is important. It is therefore not surprising that robustness is becoming an important area of engineering practice, and more emphasis has been given in recent years to the design and development of robust systems. The complexity of new engineering systems and the high robustness requirements in many applications suggest the urgent need to further revisit robust estimation techniques and present them in an accessible manner.

1.2   Robust *M*-estimators for Single-Channel Data

In this section, robust M-estimators for single-channel data are introduced. M-estimation is easily accessible in the single-channel context; in later chapters, we will show how this concept can be applied to other areas such as multichannel data and linear regression.

1.2.1   Location and Scale Estimation

The robust estimation of the location and scale parameters of a univariate random variable is considered to be the origin of what we know today as robust statistics. In the 1940s and 1950s, J. W. Tukey, one of the pioneers of statistics of the twentieth century, investigated the effect of small amounts of contaminating data (outliers) on the sample mean and standard deviation. Tukey also proposed robust estimators that are not severely affected by outliers (Tukey, 1960). In 1964, P. J. Huber formalized robust statistical theory and introduced M-estimation in his seminal paper on robust location estimation (Huber, 1964).

Consider the Thevenin equivalent model of a DC electrical system, as illustrated in Figure 1.1, with a Thevenin equivalent voltage of μ and a Thevenin equivalent resistance of R Ohm. Thermal noise in the resistance leads to a time-varying noise signal, denoted $v(t)$, in series with the DC voltage source. The voltage at the system output is denoted $y(t)$.

The noise signal arising from a resistor has a uniform power spectral density over a band that usually well exceeds the bandwidth of a measurement system, and time samples from such a signal are consistent with samples from a Gaussian probability density function (pdf). From a random process perspective, the noise arising from a resistor has a white power spectral density and a Gaussian amplitude pdf, that is, the noise is additive white Gaussian noise (AWGN). Consistent with this, and in an electrical context, the measurement of a DC level is modeled according to

$$y_i = \mu + v_i, \quad i = 1, \ldots, N, \tag{1.1}$$

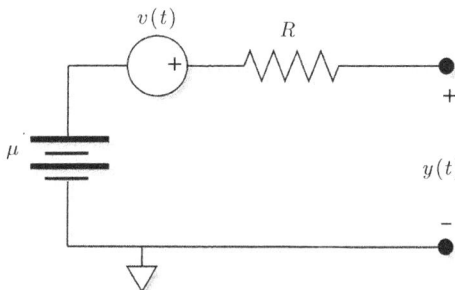

Figure 1.1 Thevenin equivalent model for a DC electrical system.

where y_i are random variables[1] that model measurements taken at time instances $t_i, i \in \{1, \ldots, N\}$ and v_i are identically and independently distributed (i.i.d.) variables for $i = 1, \ldots, N$. Because the DC voltage is constant over time, it is represented by the deterministic scalar quantity μ. A common assumption is that the random variable v_i follows the zero-mean Gaussian distribution, whose pdf is given by

$$f\left(v_i \middle| \mu_v, \sigma_v^2\right) = \frac{1}{\sqrt{2\pi \sigma_v^2}} e^{-\frac{(v_i - \mu_v)^2}{2\sigma_v^2}}, \tag{1.2}$$

with $\mu_v = 0$. Under these assumptions, the pdf of the measurements is

$$f\left(y_i \middle| \mu, \sigma^2\right) = \frac{1}{\sqrt{2\pi \sigma^2}} e^{-\frac{(y_i - \mu)^2}{2\sigma^2}}, \tag{1.3}$$

where $\sigma^2 = \sigma_v^2$.

For many of today's engineering applications, however, the AWGN model is not an adequate representation. When considering the measurement of a DC voltage, impulsive non-Gaussian noise can be injected, for example, by DC–DC converters, switching mode power supplies found in light dimmers, or switching thermostats in fridges or cookers.

Maximum Likelihood Estimation of Location and Scale

The goals of location and scale estimation are to determine the values μ and σ, which best model the observations/measurements $\mathbf{y} = (y_1, \ldots, y_N)^\top$ from (1.1). Provided that the Gaussian noise assumption is fulfilled, that is, that the pdf of v_i is given by (1.2), the maximum likelihood estimators (MLEs) are the sample mean and the sample standard deviation.

Assuming statistical independence of y_i, $i = 1, \ldots, N$, this directly follows from (1.1) and (1.3) by taking the partial derivatives of the Gaussian negative log-likelihood function

$$L_{\mathrm{ML}}(\mu, \sigma | \mathbf{y}) = \frac{N}{2} \ln(2\pi\sigma^2) + \frac{\sum_{i=1}^{N}(y_i - \mu)^2}{2\sigma^2} \tag{1.4}$$

with respect to the unknown parameters μ, σ

$$\frac{\partial}{\partial \mu} L_{\mathrm{ML}}(\mu, \sigma | \mathbf{y}) = -\frac{2\sum_{i=1}^{N}(y_i - \mu)}{2\sigma^2}$$

$$\frac{\partial}{\partial \sigma} L_{\mathrm{ML}}(\mu, \sigma | \mathbf{y}) = \frac{N}{\sigma} - \frac{\sum_{i=1}^{N}(y_i - \mu)^2}{\sigma^3}$$

and setting them equal to zero. Thus, the sample mean is such that

$$\sum_{i=1}^{N}(y_i - \hat{\mu}) = 0$$

---

[1] Throughout the book, we will not explicitly differentiate between a random variable X and its realization x. This should be understood from the context.

and the sample standard deviation is such that

$$N - \frac{\sum_{i=1}^{N}(y_i - \hat{\mu})^2}{\hat{\sigma}^2} = 0$$

$$\Leftrightarrow \frac{1}{N}\frac{\sum_{i=1}^{N}(y_i - \hat{\mu})^2}{\hat{\sigma}^2} = 1.$$

Solving yields the well-known estimators of location

$$\hat{\mu} = \frac{1}{N}\sum_{i=1}^{N} y_i \tag{1.5}$$

and scale

$$\hat{\sigma} = \sqrt{\frac{1}{N}\sum_{i=1}^{N}(y_i - \hat{\mu})^2}. \tag{1.6}$$

Because the objective function that is defined in (1.4) is a jointly convex function in $(\mu, 1/\sigma^2)$, a global minimizer can be found. Provided that the Gaussian assumption is fulfilled, the sample mean and sample standard deviation that are defined, respectively, in (1.5) and (1.6) are optimal in the sense that they attain the Cramér–Rao lower bound (CRLB). This means that the distributions of $\sqrt{N}(\hat{\mu} - \mu)$ and $\sqrt{N}(\hat{\sigma} - \sigma)$ for $N \to \infty$ tend to Gaussian distributions whose mean values are the true values (consistency) and whose covariance is equal to the inverse of the Fisher information matrix (efficiency).

For the Gaussian distribution, and consistent with (1.5) and (1.6), it is optimal to give all observations equal importance in the objective function. However, if the noise pdf $f\left(v_i | \mu_v, \sigma_v^2\right)$ is that of a non-Gaussian random variable, and the measured data contains outliers, our intuition dictates that we weight the observations $y_i, i = 1, \ldots, N$, in a manner to give more importance to observations that are close to the measurement model as compared to the ones that are unlikely to occur. This is precisely what robust location and scale estimation is about.

A frequently used approach to derive robust estimators is to compute the MLE for a heavy-tailed noise model, for example, the Laplace or the Cauchy noise distribution (see Figure 1.2). The Laplace distribution has the pdf

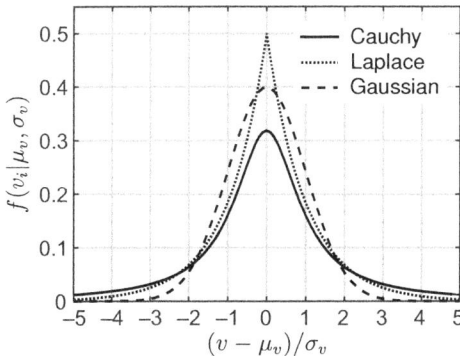

Figure 1.2 The Gaussian, Laplace, and Cauchy pdfs.

Table 1.1 The probability that a random variable takes on a value that is more than a few σ away from μ.

	Gaussian	Laplace	Cauchy
$\mathrm{Prob}(\lvert y - \mu\rvert > 3\sigma)$	0.0027	0.0498	0.2048
$\mathrm{Prob}(\lvert y - \mu\rvert > 4\sigma)$	$6.3342 \cdot 10^{-5}$	0.0183	0.1560
$\mathrm{Prob}(\lvert y - \mu\rvert > 5\sigma)$	$5.7330 \cdot 10^{-7}$	0.0067	0.1257
$\mathrm{Prob}(\lvert y - \mu\rvert > 10\sigma)$	0	$4.5400 \cdot 10^{-5}$	0.0635

$$f(v_i|\mu_v,\sigma_v) = \frac{1}{\sqrt{2}\sigma_v} e^{-\frac{\sqrt{2}|v_i - \mu_v|}{\sigma_v}}, \tag{1.7}$$

where μ_v and σ_v are the location and scale parameters, respectively. The Cauchy distribution has the pdf

$$f(v_i|\mu_v,\sigma_v) = \frac{1}{\pi\sigma_v} \cdot \frac{\sigma_v^2}{(v_i - \mu_v)^2 + \sigma_v^2}. \tag{1.8}$$

As shown in Table 1.1, the probability that a Laplace, or Cauchy, distributed random variable takes on a value that is more than three standard deviations away from μ is significantly different from zero. For the Cauchy-distributed random variable, even the probability of taking on a value that is more than ten σ away from μ is 0.0635.

If the model for the noise is the Laplace distribution, that is, (1.7) holds with $\mu_v = 0$, the pdf of y_i becomes

$$f(y_i|\mu,\sigma) = \frac{1}{\sqrt{2}\sigma} e^{-\frac{\sqrt{2}|y_i - \mu|}{\sigma}},$$

where $\sigma = \sigma_v$. For N observations, the negative log-likelihood function is then given by

$$L_{\mathrm{ML}}(\mu,\sigma|\mathbf{y}) = N\ln(\sqrt{2}\sigma) + \frac{\sqrt{2}}{\sigma}\sum_{i=1}^{N}|y_i - \mu|. \tag{1.9}$$

Taking the partial derivative with respect to μ yields

$$\frac{\partial}{\partial\mu}L_{\mathrm{ML}}(\mu,\sigma|\mathbf{y}) = \frac{\sqrt{2}}{\sigma}\sum_{i=1}^{N}\frac{\partial\,|y_i - \mu|}{\partial\mu}$$

$$= -\frac{\sqrt{2}}{\sigma}\sum_{i=1}^{N}\mathrm{sign}(y_i - \mu) \tag{1.10}$$

where the identity

$$\frac{\partial|x|}{\partial x} = \frac{x}{|x|} = \mathrm{sign}(x).$$

has been used and the sign function is defined as

$$\text{sign}(x) = \begin{cases} +1, & \text{if } x > 0, \\ 0, & \text{if } x = 0, \\ -1, & \text{if } x < 0. \end{cases} \tag{1.11}$$

The MLE of the location parameter μ, is the solution of

$$\frac{\sqrt{2}}{\sigma} \sum_{i=1}^{N} \text{sign}(y_i - \mu) = 0$$

and the sample median, that is,

$$\text{med}(\mathbf{y}) = \begin{cases} y_{\left(\frac{N+1}{2}\right)}, & \text{if } N \text{ is odd}, \\ \frac{1}{2}(y_{\left(\frac{N}{2}\right)} + y_{\left(\frac{N}{2}+1\right)}), & \text{if } N \text{ is even}, \end{cases} \tag{1.12}$$

given ordered samples $\{y_{(1)} \leq \ldots \leq y_{(N-1)} \leq y_{(N)}\}$. From

$$\frac{\partial}{\partial \sigma} L_{\text{ML}}(\mu, \sigma \,|\mathbf{y}) = 0$$

the MLE of the scale parameter turns out to be the mean of the absolute deviations from the median, that is,

$$\hat{\sigma} = \frac{1}{N} \sum_{i=1}^{N} |y_i - \text{med}(\mathbf{y})|. \tag{1.13}$$

Weighted medians are addressed in Section 2.4.1 in the context of linear regression. Median and weighted median filters are discussed in Section 7.2.

The negative log-likelihood function for the Cauchy distribution given a sample size N is

$$L_{\text{ML}}(\mu, \sigma \,|\mathbf{y}) = N \ln(\sigma \pi) + \sum_{i=1}^{N} \ln \left(1 + \left(\frac{y_i - \mu}{\sigma} \right)^2 \right). \tag{1.14}$$

Taking the partial derivatives of (1.14) with respect to the unknown parameters μ and σ yields

$$\frac{\partial}{\partial \mu} L_{\text{ML}}(\mu, \sigma \,|\mathbf{y}) = -2\sigma \sum_{i=1}^{N} \frac{y_i - \mu}{\sigma^2 + (y_i - \mu)^2} \tag{1.15}$$

and

$$\frac{\partial}{\partial \sigma} L_{\text{ML}}(\mu, \sigma \,|\mathbf{y}) = \frac{N}{\sigma} - \frac{2}{\sigma} \sum_{i=1}^{N} \frac{1}{\sigma^2 + (y_i - \mu)^2}. \tag{1.16}$$

To find the Cauchy location and scale estimates, a numerical solution to

$$\sum_{i=1}^{N} \frac{y_i - \mu}{\sigma^2 + (y_i - \mu)^2} = 0$$

$$\sum_{i=1}^{N} \frac{1}{\sigma^2 + (y_i - \mu)^2} = \frac{N}{2}$$

is required. Kalluri and Arce (2000), for example, provide an algorithm that employs a fixed-point (FP) search that is guaranteed to converge to a local minimum. However, one needs to be careful with the choice of the local minimum because the Cauchy likelihood can have multiple spurious roots (Reeds, 1985). Finding a global optimum for the location and scale of non-Gaussian ML functions is still an open problem.

M-estimation of Location and Scale

An important class of robust estimators are *M*-estimators (Huber and Ronchetti, 2009), which are a generalization of MLEs. Because this chapter is intended to serve as an easy-to-read introduction to robustness concepts, only the real-valued case is discussed. *M*-estimation of location and scale is extended to the complex-valued case in Section 2.5 of the next chapter, where linear regression is discussed.[2]

M-estimators replace the negative log-likelihood function $L_{\text{ML}}(\mu, \sigma | \mathbf{y})$ with a different objective function $L_{\text{M}}(\mu, \sigma | \mathbf{y}) = \rho(\mu, \sigma | \mathbf{y})$. If $\rho(\cdot)$ is differentiable, with

$$\psi(x) = \frac{d\rho(x)}{dx}, \tag{1.17}$$

then the *M*-estimating equations follow:

$$\sum_{i=1}^{N} \psi\left(\frac{y_i - \hat{\mu}}{\hat{\sigma}}\right) = 0 \tag{1.18}$$

and

$$\frac{1}{N} \sum_{i=1}^{N} \psi\left(\frac{y_i - \hat{\mu}}{\hat{\sigma}}\right) \cdot \left(\frac{y_i - \hat{\mu}}{\hat{\sigma}}\right) = b. \tag{1.19}$$

Here, b is a positive constant that must satisfy $0 < b < \rho(\infty)$. If $f(y_i | \mu, \sigma)$ is symmetric, then $\rho(\mu, \sigma | \mathbf{y})$ is even and, hence, $\psi(\mu, \sigma | \mathbf{y})$ is odd. MLEs are included within the class of *M*-estimators by setting $\rho(\mu, \sigma | \mathbf{y}) = L_{\text{ML}}(\mu, \sigma | \mathbf{y})$. For example, in the Gaussian noise case, the MLE is obtained by letting $\psi(x) = x$ and $b = 1$ in (1.18) and (1.19).

M-estimators are classified into two categories depending on the shape of $\psi(x)$, namely the *monotone* and the *redescending* *M*-estimators. Within the redescending class, the *M*-estimators for which $\psi(x)$ returns exactly to zero, that is, $\psi(x_0) = 0$ for some value x_0, are called *strongly redescending*. For a detailed discussion of different ψ functions, the interested reader is referred, for example, to Huber and Ronchetti (2009, chapter 4).

---

[2] The linear regression model contains as a special case the location (or location-scale) model when the design matrix is a column vector of ones.

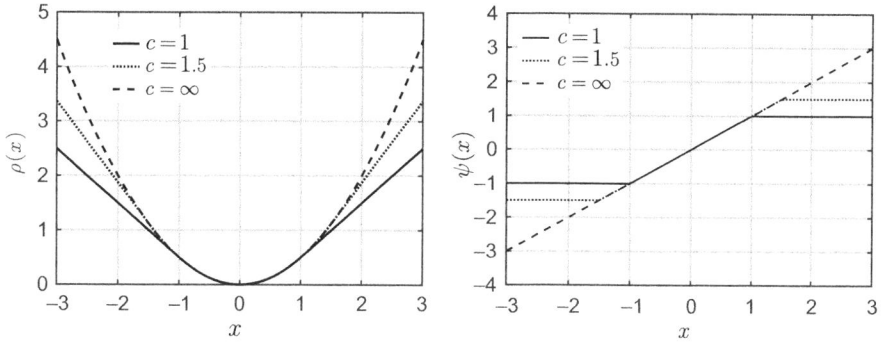

Figure 1.3 Graphs of $\rho(x)$ and $\psi(x)$ that define Huber's estimator, which belongs to the category of monotone *M*-estimators.

Perhaps the most commonly used monotone *M*-estimator is Huber's, which is also known as the soft limiter and is defined according to

$$\rho(x) = \begin{cases} \frac{1}{2}x^2 & |x| \le c \\ c|x| - \frac{1}{2}c^2 & |x| > c \end{cases} \tag{1.20}$$

or, equivalently, according to

$$\psi(x) = \begin{cases} x & |x| \le c \\ c\,\text{sign}(x) & |x| > c. \end{cases} \tag{1.21}$$

These functions are shown in Figure 1.3.

Tukey's biweight (or bisquare) *M*-estimator is a popular member of the (strongly) redescending category and is defined by

$$\rho(x) = \begin{cases} \frac{c^2}{6}\left(1 - \left(1 - \frac{x^2}{c^2}\right)^3\right) & |x| \le c \\ \frac{c^2}{6} & |x| > c \end{cases} \tag{1.22}$$

and

$$\psi(x) = \begin{cases} x\left(1 - \frac{x^2}{c^2}\right)^2 & |x| \le c \\ 0 & |x| > c. \end{cases} \tag{1.23}$$

These functions are shown in Figure 1.4. For $c \to \infty$, Huber's and Tukey's *M*-estimators of location converge to the sample mean. By contrast, for $c \to 0$, they converge to the sample median.

The main advantage of monotone estimators is the fact that they are unique by virtue of the convexity of the optimization problems (1.18) and (1.19). This is not the case for redescending *M*-estimators. For this reason, estimates that were obtained from a monotone *M*-estimator are often used as a starting point for computing redescending *M*-estimators.

In principle, based on (1.18) and (1.19), any equation-solving method can be used, for example, the Newton–Raphson method, to compute *M*-estimates. However, methods

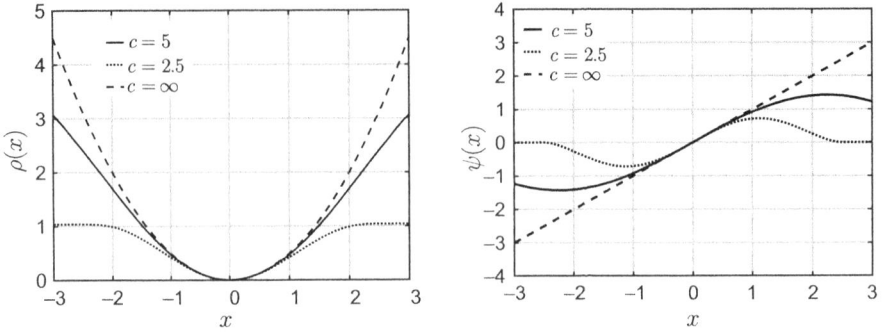

Figure 1.4 Graphs of $\rho(x)$ and $\psi(x)$ that define Tukey's biweight (bisquare) estimator, which belongs to the category of (strongly) redescending M-estimators.

based on derivatives may be unsafe because if ψ is bounded, which is the case for robust M-estimators, its derivative tends to zero at infinity (see, e.g., Maronna et al. [2006] or Huber and Ronchetti [2009] for a detailed discussion). M-estimators of location and scale can conveniently be computed using an iterative re-weighting algorithm. In practice, the location and scale parameters are usually computed separately. This is due to the nonconvexity of the joint objective function $L_M(\mu, \sigma | \mathbf{y})$ and the lack of robustness of the joint solution for monotone M-estimators (Maronna et al., 2006; Huber and Ronchetti, 2009).

An M-estimate of location with a previously computed robust scale estimate $\hat{\sigma}$ is obtained using an iterative reweighting algorithm as detailed in Algorithm 1: Mloc. In this algorithm, $\xi \in \mathbb{R}$ is a small positive constant. Alternatively, Mloc can also be terminated after a fixed number of iterations consistent with the stopping criterion being $n < N_{\text{iter}}$. For location estimation, the weights are determined by

$$
W(x) = \begin{cases} \psi(x)/x, & \text{if } x \neq 0, \\ \left. \dfrac{d\psi(x)}{dx} \right|_{x=0}, & \text{if } x = 0. \end{cases} \tag{1.24}
$$

Note that in most cases, $\frac{d\psi(x)}{dx}|_{x=0}$ exists and because $\psi(x)$ is approximately linear at the origin, that is, $\psi(x)|_{x=0} \approx x$, it usually follows that $W(x = 0) = 1$. MATLAB$^{©}$ implementations of Huber's and Tukey's M-estimates of location with previously computed scale are provided in the RobustSP toolbox.

For monotone ψ functions, the starting point $\hat{\mu}^{(0)}$ only influences the number of iterations required until convergence. For redescending M-estimators, a simple robust initialization of location, that is, the median

$$
\hat{\mu}^{(0)} = \text{med}(\mathbf{y}),
$$

is recommended. Additionally, using a precomputed scale estimate for $\hat{\sigma}$, for example, the normalized median absolute deviation

$$
\text{madn}(\mathbf{y}) = 1.4826 \cdot \text{med}(|\mathbf{y} - \text{med}(\mathbf{y})|) \tag{1.25}
$$

Algorithm 1: Mloc: computes an *M*-estimate of location with previously computed scale.

---

  input : observations $\mathbf{y} \in \mathbb{R}^N$, tuning parameter $c \in \mathbb{R}$, scale estimate $\hat{\sigma}$
  output : $\hat{\mu}$ that solves (1.18)
  initialize: initial location estimate $\hat{\mu}^{(0)}$, iteration index $n = 1$, tolerance level $\xi \in \mathbb{R}$

1 **while** $\dfrac{|\hat{\mu}^{(n+1)} - \hat{\mu}^{(n)}|}{\hat{\sigma}} > \xi$ **do**

2   compute weights

$$w_i^{(n)} = W\left(\frac{y_i - \hat{\mu}^{(n)}}{\hat{\sigma}}\right)$$

     compute location estimates

$$\hat{\mu}^{(n+1)} = \frac{\sum_{i=1}^{N} w_i^{(n)} y_i}{\sum_{i=1}^{N} w_i^{(n)}}$$

     increment iteration index

$$n \leftarrow n + 1$$

3 **return** $\hat{\mu} \leftarrow \hat{\mu}^{(n+1)}$

---

is necessary and sufficient for convergence to a "good" solution, that is, a normally distributed and highly efficient estimator $\hat{\mu}$ for symmetrically distributed data (Maronna et al., 2006). The normalization factor 1.4826 has been introduced in (1.25) to make the estimator consistent with the standard deviation of a normal distribution. As discussed in Section 2.5.3, a different normalization factor is required for consistency for the case of a complex normal distribution.

Intuitively, *M*-estimates of scale can be represented as a weighted root-mean-square estimate. Let $W(x)$ be defined as in (1.24). Then, the equation for the *M*-estimation of scale (1.19) becomes

$$\frac{1}{b}\sum_{i=1}^{N} W\left(\frac{y_i - \hat{\mu}}{\hat{\sigma}}\right) \cdot \left(\frac{y_i - \hat{\mu}}{\hat{\sigma}}\right)^2 = N$$

$$\Leftrightarrow \hat{\sigma} = \sqrt{\frac{1}{Nb}\sum_{i=1}^{N} W\left(\frac{y_i - \hat{\mu}}{\hat{\sigma}}\right) \cdot (y_i - \hat{\mu})^2}. \tag{1.26}$$

The weights are chosen such that (1.26) is quadratic near the origin and then increases less rapidly, therewith down-weighting unusually large-valued data (outliers). Just as for location estimation, the recommended computation is done by iterative reweighting. Based on (1.26), *M*-estimates of scale, with previously computed robust location estimate $\hat{\mu}$, are computed through use of Algorithm 2: Mscale. The recommended location estimate is

$$\hat{\mu} = \text{med}(\mathbf{y})$$

and for the initial scale estimate

$$\hat{\sigma}^{(0)} = \text{madn}(\mathbf{y}).$$

Just as in Algorithm 1, $\xi \in \mathbb{R}$ is a small positive constant, and Mscale can also be terminated after a fixed number of iterations. The positive constant b must satisfy $0 < b < \rho(\infty)$ and is chosen as $\mathsf{E}[\rho(u)]$, where u is a standard normal random variable to achieve consistency with the Gaussian distribution.

Algorithm 2: Mscale: computes an M-estimate of scale with previously computed location.

---

    input   : observations $\mathbf{y} \in \mathbb{R}^N$, tuning parameter $c \in \mathbb{R}$, location estimate $\hat{\mu}$
    output : $\hat{\sigma}$ that solves (1.26)
    initialize: initial scale estimate $\hat{\sigma}^{(0)}$, iteration index $n = 1$, tolerance level $\xi \in \mathbb{R}$
1 **while** $|\hat{\sigma}^{(n+1)}/\hat{\sigma}^{(n)} - 1| > \xi$ **do**

2   compute weights

$$w_i^{(n)} = W\left(\frac{y_i - \hat{\mu}}{\hat{\sigma}^{(n)}}\right)$$

       compute scale estimates

$$\hat{\sigma}^{(n+1)} = \sqrt{\frac{1}{Nb} \sum_{i=1}^{N} w_i^{(n)} (y_i - \hat{\mu})^2}$$

       increment iteration index

$$n \leftarrow n + 1$$

3 **return** $\hat{\sigma} \leftarrow \hat{\sigma}^{(n+1)}$

MATLAB© implementations of Huber's and Tukey's M-estimators of scale with previously computed location are provided in the program Mscale that is included in the RobustSP toolbox.

1.3   Measures of Robustness

The following section provides the background that is necessary to assess the robustness of a given method. From a practical engineering viewpoint, robustness measures help in understanding the "response" of a method to departures or variations from the exact assumptions made.

Ideally, we would like the assumptions made with respect to the data, which we have incorporated into our parametric statistical model, for example the AWGN model in (1.1), to hold exactly. However, an engineering (signal processing) practitioner cannot rely on optimism when proposing a solution to a real-world problem. Instead, the practitioner must always be able to answer the question "What happens in a more pessimistic case?" The answer to this question is the goal of this section and to this end, appropriate tools, such as the sensitivity curve, the IF, the maximum-bias curve (MBC), and the BP, are introduced. We use simple signal and contamination models to explain and illustrate the main ideas and fundamental concepts. In later chapters, we will show how robustness measures can be adapted to more complex problems that are consistent with what a practitioner is likely to encounter.

Robustness measures also facilitate the construction of new methods that respond in a desirable fashion to departures from the data assumptions made. For example, a new method should not behave in an erratic manner if we were to replace a small portion of the data with outliers. And if the portion of outliers in the data set increases, this should not cause a catastrophe, that is, a breakdown of the method. At the same time, robust methods should behave reasonably well, that is, nearly optimally, when the data is consistent with the assumptions made. For example, estimators should be consistent for the ideal case in which the data conforms to the assumptions made. *Additionally*, estimators should not drift too far away when the data does not exactly coincide with our optimistic view of it. In summary, robustness measures allow us to assess whether our method possesses the following desirable features (Huber and Ronchetti, 2009):

1.   Near-optimal behavior at the assumed model.
2.   Small deviations from the model should only lead to small performance loss.
3.   Large deviations from the model should not cause a breakdown.

It is important to note that the use of robust methods does not remove the requirement to select an appropriate statistical model. On the contrary, we should take great care to find the best possible model. However, even the best model remains only a useful representation that cannot hold exactly. Intuitively speaking, there is always a difference between the representation (model) and the data/measurement. Robust methods can tolerate these differences to some extent, because they do not require the model assumptions to hold exactly. It suffices that the assumptions hold approximately.

1.3.1   The Influence Function and Qualitative Robustness

As a first measure of robustness, we discuss the IF, which has been considered by many to be the most useful heuristic tool of robust statistics (Huber and Ronchetti, 2009). It was introduced by F. R. Hampel (1968, 1974) under the name of *influence curve*, and was later renamed to *influence function*, after applying it to more general problems, for example, vector-valued parameter estimation.

Using the famous bridge analogy that was introduced by (Huber, 1972), robustness describes the stability of the bridge in the face of perturbations. Qualitative robustness requires, intuitively, that small perturbations have small effect. Small perturbations, for example, refer to replacing a small proportion of the data by arbitrary values: The effect refers to a bias in the distribution of the estimator. The IF measures the effects of infinitesimal perturbations (Hampel et al., 2011).

Sensitivity Curve

To understand the IF, we begin by introducing a closely related finite sample measure of robustness, Tukey's sensitivity curve (Tukey, 1977). It is also referred to as the *empirical influence function* and is defined as follows:

DEFINITION 1   *The sensitivity curve of an estimator $\hat{\beta}_N$ is defined according to*

$$\mathrm{SC}\left(y, \hat{\beta}_N\right) \triangleq N \cdot \left(\hat{\beta}_N(y_1, y_2, ..., y_{N-1}, y) - \hat{\beta}_{N-1}(y_1, y_2, ..., y_{N-1})\right) \qquad (1.27)$$

The sensitivity curve displays the bias of an estimator $\hat{\beta}_N$ when an additional observation, that takes on the value y, is added to a sample $\mathbf{y}_{N-1} = (y_1, y_2, \ldots, y_{N-1})^\top$. Here, $\hat{\beta}_N(\cdot)$ represents an estimator of a deterministic parameter β based on N samples. For example, when estimating the location parameter, the sensitivity curve becomes

$$\mathrm{SC}(y, \hat{\mu}_N(y)) \triangleq N \cdot \left(\hat{\mu}_N(y_1, y_2, ..., y_{N-1}, y) - \hat{\mu}_{N-1}(y_1, y_2, ..., y_{N-1})\right). \qquad (1.28)$$

The sensitivity curve displays the bias of $\hat{\mu}_N$ that is introduced by the additional observation y

$$\mathrm{bias}(y, \hat{\mu}_N(y)) = \hat{\mu}_N(y_1, y_2, ..., y_{N-1}, y) - \hat{\mu}_{N-1}(y_1, y_2, ..., y_{N-1})$$

and

$$\mathrm{SC}(y, \hat{\mu}_N(y)) = N \cdot \mathrm{bias}(y, \hat{\mu}_N(y)).$$

To understand the sensitivity curve, we revisit the example of measuring a DC voltage in noise. For this example, we assess the sensitivity of the previously introduced location estimators, that is, the sample mean, sample median, Huber's M-estimator, and Tukey's M-estimator by means of $\mathrm{SC}(y, \hat{\mu}_N)$. Figure 1.5 displays the results of evaluating (1.28), where it is assumed that $\mathbf{y}_{N-1}$ follows (1.1), that is,

$$y_i = \mu + v_i, \quad i = 1, \ldots, N - 1,$$

with $N = 100$, and y_i being Gaussian with $\mu = 5$ V and $\sigma = 1$ V. The additional observation y in (1.28) takes values $0 \le y \le 10$. For Huber's M-estimate, $c = 1.28$, while for Tukey's M-estimate $c = 4.68$. A MATLAB$^{\copyright}$ code to reproduce this example is provided in the RobustSP toolbox.

These sensitivity curves provide the practitioner with valuable intuitive information about the estimators. Observe the following from Figure 1.5:

1.   At $y = 0$ all estimators provide estimates close to μ, which is a first hint of the unbiasedness of the estimator for Gaussian data.
2.   The bias that is introduced for the sample mean depends linearly on the value of y, which confirms its sensitivity against even a single outlier.

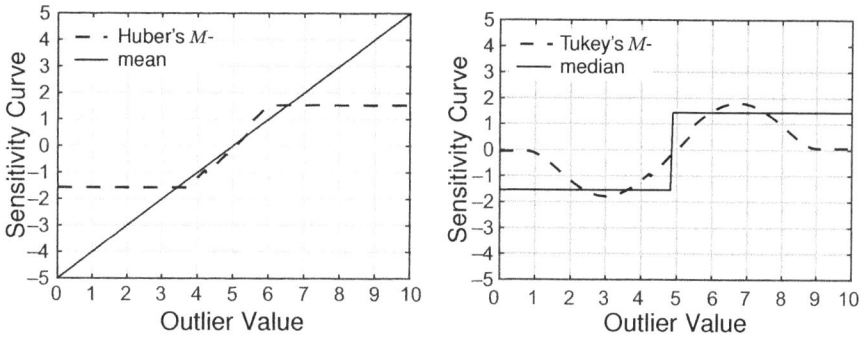

Figure 1.5 Sensitivity curves of the sample mean, Huber's M-estimator, the sample median, and Tukey's M-estimator for the estimation of a DC value in noise.

3. The bias that is introduced by an additional observation that takes on the value y is finite for all values of y for the sample median, Huber's M-estimator, and Tukey's M-estimator, which confirms their robustness against a single outlier in the data set.

4. The sensitivity curves of the sample mean, Huber's M-estimator, and Tukey's M-estimator are smooth, which means that small changes in the data lead to small changes in the estimates.

5. The sensitivity curve of the sample median has a discontinuity at $\mu = 5$ V. Here, a small change in the data may lead to a considerable change in the estimate.

6. The maximal value of the sensitivity curve of the sample median is the smallest of all considered location estimators, which means that it is maximally bias robust.

7. Within the 3σ-region around μ, the sensitivity curves of Huber's M-estimator and Tukey's M-estimator are nearly linear. The behavior of these estimates is similar to that of the sample mean that corresponds to the MLE when the data is Gaussian. This is a first indicator that Huber's and Tukey's M-estimators are highly efficient.

Similarly to location estimation, sensitivity curves can also be computed for scale estimation by evaluating

$$\text{SC}(y, \hat{\sigma}_N) \triangleq N \cdot \left(\hat{\sigma}_N (y_1, y_2, \ldots, y_{N-1}, y) - \hat{\sigma}_{N-1} (y_1, y_2, \ldots, y_{N-1}) \right). \quad (1.29)$$

Figure 1.6 displays the sensitivity curves of the Gaussian MLE (1.6), the Laplace MLE (1.13), Huber's M-estimator, and Tukey's M-estimator of the scale (1.19) of the noise distribution for the DC estimation example. The MATLAB$^{\copyright}$ code to reproduce this example is provided in the RobustSP toolbox.

Interestingly, for the estimation of the scale parameter, only Tukey's redescending M-estimator has a bounded sensitivity curve. This difference to location estimation occurs because (1.19) contains the product

$$\psi \left(\frac{y_i - \hat{\mu}}{\hat{\sigma}} \right) \cdot \left(\frac{y_i - \hat{\mu}}{\hat{\sigma}} \right),$$

which for monotone M-estimators is an unbounded function of y_i.

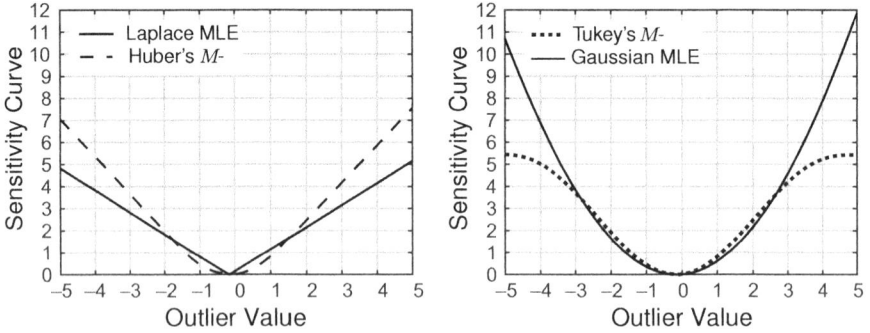

Figure 1.6 Sensitivity curves for the Laplace MLE, Huber's M-estimator, the Gaussian MLE, and Tukey's M-estimator of the scale of the noise distribution for the DC estimation example.

All the empirical information about the estimators, as summarized in Figures 1.5 and 1.6, has been obtained without any tedious derivations and only by typing a few lines of code to evaluate (1.28) and (1.29). Before formally defining the IF, we will show its connection to the sensitivity curve by considering the case when the estimator is given as a functional, that is,

$$\hat{\beta}_N = T(F_N), \tag{1.30}$$

where F_N denotes the empirical distribution of the sample $(y_1, y_2, \ldots, y_N)$, which for any value of N, is defined as

$$F_N(t) = \frac{1}{N} \sum_{i=1}^{N} \mathbb{1}_{\{y_i < t\}}. \tag{1.31}$$

Here, $\mathbb{1}_A$ is the indicator of the event A, and the functional representation of an estimator is written as $T(F_N)$. The word *functional* is used to indicate that the domain of $T(F_N)$ is not a subset of $\mathbb{R}$ (or $\mathbb{C}$), but a subset of functions, that is, the empirical distribution functions, in this case. Most of the commonly used estimators depend on the sample only through the empirical distribution, so that

$$\hat{\beta}_N(y_1, y_2, \ldots, y_N) = T(F_N)$$

holds. In all cases of practical interest, there exists a value for N, such that

$$T(F_N) \xrightarrow{P} T(F). \tag{1.32}$$

This means that the sequence $\{T(F_1), T(F_2), \ldots, T(F_N)\}$ converges in probability toward $T(F)$ in the sense that

$$\lim_{N \to \infty} \text{Prob}(|T(F_N) - T(F)| > \xi) = 0$$

for any $\xi > 0$.

The following functional representations of some previously introduced estimators are useful examples. In these examples, the link between the distribution and the sample values is noted.

Example 1 The sample mean can be written in a functional form as a Lebesque–Stieltjes integral

$$T(F_N) = \frac{1}{N} \sum_{i=1}^{N} y_i = \int_{-\infty}^{\infty} t \, dF_N(t), \tag{1.33}$$

where $F_N(x)$ is defined in (1.31). As $N \to \infty$, the law of large numbers (1.32) holds, and the so-called asymptotic estimator becomes

$$T(F) = \mathsf{E}_F[t] = \int_{-\infty}^{\infty} t \, dF(t). \tag{1.34}$$

Here, $\mathsf{E}_F[\cdot]$ denotes the expectation operator at distribution F.

Example 2 The α-trimmed mean can be written in a functional form as

$$T(F) = \frac{1}{1 - 2\alpha} \int_{\alpha}^{1-\alpha} F^{-1}(t) dt, \tag{1.35}$$

and the corresponding sample statistic is

$$T(F_N) = \frac{1}{1 - 2\alpha} \int_{\alpha}^{1-\alpha} F_N^{-1}(t) dt. \tag{1.36}$$

Example 3 The median functional $T(F)$ and the corresponding sample statistic $T(F_N)$ can be expressed in several ways (Becker et al., 2013; Oja, 2013), one possible definition is

$$T(F) = \inf\{x : F(x) \geq 0.5\},$$

that is, it is the 0.5 quantile of F. If it is not unique, $T(F)$ is defined as the midpoint of the 0.5 quantile. If $(y_1, y_2, \ldots, y_N)$ is a sample from F and $T(F)$ is unique then (1.32) holds.

Example 4 An M-estimator of location with known scale σ can be represented in functional form as

$$\mathsf{E}_F\left[\psi\left(\frac{t - T(F)}{\sigma}\right)\right] = 0,$$

i.e.,

$$\int_{-\infty}^{\infty} \psi\left(\frac{t - T(F)}{\sigma}\right) dF(t) = 0,$$

whereas the joint M-estimation of location and scale is determined by two equations of the form

$$\int_{-\infty}^{\infty} \psi \left(\frac{t - T(F)}{S(F)} \right) dF(t) = 0,$$

$$\int_{-\infty}^{\infty} \psi \left(\frac{t - T(F)}{S(F)} \right) \left(\frac{t - T(F)}{S(F)} \right) dF(t) = b,$$

where $S(F)$ is the functional representation of the M-estimator of scale.

The functional form of an estimator allows the sensitivity curve to be written as

$$\mathsf{SC}(y, T(F_N), F_N) = \frac{T\left(\left(1 - \frac{1}{N}\right) F_{N-1} + \frac{1}{N}\delta_y\right) - T(F_{N-1})}{1/N}, \qquad (1.37)$$

where δ_y is the point mass 1 at y.

$\mathsf{SC}(y, T(F_N), F_N)$ in (1.37) is a useful heuristic tool as it tells us in which way the estimator $\hat{\beta}_N = T(F_N)$ is affected by an additional sample that takes on the value y. Consider, for example, the case when the additional sample $y_N = y$ is an outlier. The sensitivity curve shows the difference between the value taken on by the estimator when evaluated based on an outlier contaminated distribution

$$\left(1 - \frac{1}{N}\right) F_{N-1} + \frac{1}{N}\delta_y$$

and the value based on the distribution F_{N-1}. In this case, δ_y is the contaminating (outlier generating) distribution. The value of y that maximizes $\mathsf{SC}(y, T(F_N), F_N)$, that is,

$$\arg\max_{y} \mathsf{SC}(y, T(F_N), F_N),$$

shows the contaminating distribution (in this case, corresponding to the outlier amplitude) to which our estimator $\hat{\beta}_N = T(F_N)$ is most sensitive.

The Influence Function

The influence function (IF) (Hampel, 1968, 1974) of an estimator $\hat{\beta}$ is the counterpart to its sensitivity curve, as N tends to infinity. Let $\hat{\beta}_\infty$ denote the estimator as N tends to infinity and let $T(F)$ be its functional representation. Then, the IF shows the approximate behavior of the asymptotic estimator when the sample contains a small fraction ε of identical outliers. The IF is defined as follows:

DEFINITION 2 *Influence function*

$$\mathsf{IF}(y; T(F), F) = \lim_{\epsilon \downarrow 0} \frac{T(F_\varepsilon) - T(F)}{\epsilon} = \left[\frac{\partial T(F_\varepsilon)}{\partial \epsilon} \right]_{\epsilon=0} \qquad (1.38)$$

where $T(F)$ and $T(F_\varepsilon)$ are the functional representations of the estimator when the data is distributed following, respectively, F and the contaminated distribution

$$F_\varepsilon = (1 - \epsilon)F + \epsilon\delta_y \qquad (1.39)$$

with δ_y being the point-mass probability on y and ϵ the fraction of contamination. The
main argument of the IF is y, the position of the infinitesimal contamination.

The IF is the first derivative of an estimator $T(F)$ evaluated based on the underlying
distribution function F. The parameter y is a coordinate in the space of probability
distributions (Hampel et al., 2011). Consistent with (1.39), the assumption is made
that the true distribution lies in an ε-neighborhood of the nominal distribution F. By
evaluating the influence at F, the goal is to gain insight into the behavior of $T(F)$ within
the entire neighborhood. For increasing deviations from the model, the information that
we inferred, when evaluating the IF at F, becomes less accurate. However, as long as
we remain sufficiently far away from the BP (to be introduced in the next section), the
IF may still be a useful approximation for the actual behavior.

By comparing (1.37) and (1.38) and setting $\varepsilon = 1/N$, it becomes clear that the IF is
just the limit version of the sensitivity curve. Mathematically speaking, for example,
Croux (1998) and Maronna et al. (2006) proved the almost sure convergence (convergence with probability 1) of the sensitivity curve to the IF for trimmed means and
general M-estimators. The sensitivity curve of the median, however, is not a consistent
estimator for the IF (Croux, 1998).

Desirable properties of the IF are boundedness and continuity. Boundedness ensures
that a small fraction of contaminated data, or outliers, can have only a limited effect
on the estimate, whereas continuity means that small changes in the data lead to small
changes in the estimate. If both properties are satisfied, an estimator is considered to be
qualitatively robust against infinitesimal contamination.

For the previously considered DC estimation example, when using location M-
estimates with previously computed scale, as discussed in Section 1.2.1, under the
condition that F is symmetric, and assuming that $\hat{\sigma}$ is an equivariant scale estimator,
(1.38) simplifies to

$$\mathsf{IF}(y; T(F), F) = \hat{\sigma} \frac{\psi\left(\frac{y-T(F)}{\hat{\sigma}}\right)}{\mathsf{E}\left[\psi'\left(\frac{x-T(F)}{\hat{\sigma}}\right)\right]}, \tag{1.40}$$

where $\psi'(x) \triangleq \frac{d\psi(x)}{dx}$, and $T(F)$ is the functional representation of an estimator of the
location μ. The expectation $\mathsf{E}\left[\cdot\right]$ in the denominator of (1.40) is evaluated at the nominal
distribution F.

Figure 1.7 shows the IFs of the median, the mean and Huber's M-estimator of location
in the case of F being the standard Gaussian distribution. It is assumed that $\hat{\sigma} = 1$,
that is, the scale estimator is consistent assuming an underlying standard Gaussian
distribution. These IFs are similar to the sensitivity curves shown in Figure 1.5.

Many properties of an estimator can be derived from its IF (see Hampel et al. [2011]
for a comprehensive discussion). In following sections, the properties, which we consider to be of the most importance for the engineering practitioner, are listed. The formal
proofs including all underlying assumptions are found, for example, in Hampel et al.
(2011); Maronna et al. (2006); and Huber and Ronchetti (2009).

Assuming small values of ε, the *asymptotic bias* of an estimator $T(F)$ at a nominal distribution F that is contaminated by a fraction ε of equal outliers of amplitude y, can be approximated by

$$\mathsf{bias}(T(F), F) \approx \varepsilon \cdot \mathsf{IF}(y; T(F), F).$$

The *asymptotic variance* of an estimator $T(F)$, at a distribution F, is related to the IF by

$$\mathsf{var}\,(T(F), F) = \mathsf{E}\left[\mathsf{IF}(y; T(F), F)^2\right].$$

For example, for the location M-estimator with previously computed scale $\hat{\sigma}$, under the condition that F is symmetric, and assuming that $\hat{\sigma}$ is an equivariant scale estimator, the asymptotic variance follows from (1.40) according to

$$\mathsf{var}\,(T(F), F) = \mathsf{E}\left[\left(\hat{\sigma}\,\frac{\psi\left((y - T(F))/\hat{\sigma}\right)}{\mathsf{E}\left[\psi'\left((y - T(F))/\hat{\sigma}\right)\right]}\right)^2\right].$$

From the asymptotic variance, we can derive the *asymptotic relative efficiency* (ARE), which is an asymptotic measure of the performance loss with respect to the optimal MLE. The ARE of an estimator $T(F)$ is defined as follows:

DEFINITION 3   *The ARE of $T(F)$ at F is defined by*

$$\mathsf{ARE}\,(T(F), F) \triangleq \frac{\mathsf{var}\,(T_{\mathrm{ML}}(F), F)}{\mathsf{var}\,(T(F), F)}. \tag{1.41}$$

Here, $\mathsf{var}\,(T_{\mathrm{ML}}(F), F)$ *is the asymptotic variance of the MLE,* $T_{\mathrm{ML}}(F)$, *under the nominal distribution F.*

The ARE takes values

$$0 \leq \mathsf{ARE}\,(T(F), F) \leq 1,$$

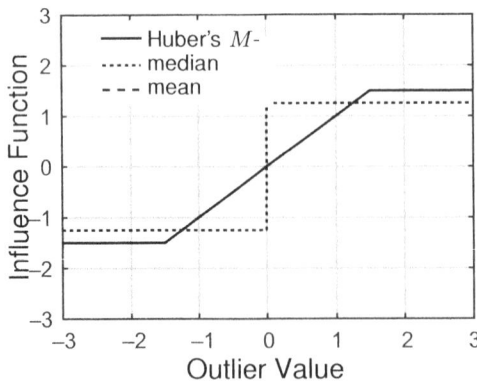

Figure 1.7 IFs of the median, the mean, and Huber's M-estimator of location.

Table 1.2 Relative efficiency of the sample mean, the sample median, Huber's M-estimator with $c_{.95} = 1.345$, and Tukey's M-estimator with $c_{.95} = 4.685$, and with F being the standard normal distribution.

	mean	median	Huber's M $c_{.95} = 1.345$	Tukey's M $c_{.95} = 4.685$
ARE $(T(F))$	1	$2/\pi$	0.95	0.95

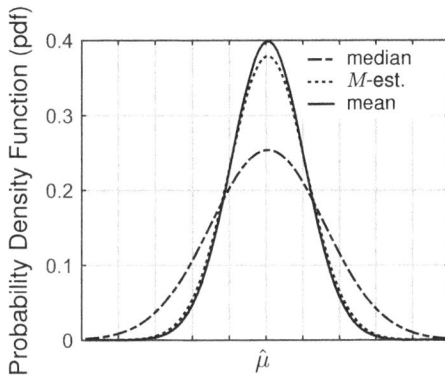

Figure 1.8 Distributions, based on F being the standard normal distribution, of the sample mean, the sample median, and Huber's and Tukey's M-estimators with $c = c_{.95}$.

where a higher value corresponds to a smaller performance loss. For unbiased estimators, $\mathsf{ARE}\,(T(F), F)$ can be used as a measure for the mean-squared error (MSE) performance loss because

$$\mathsf{MSE}\,(T(F), F) = (\mathsf{bias}\,(T(F), F))^2 + \mathsf{var}\,(T(F), F)$$

and for an unbiased estimator $\mathsf{bias}\,(T(F), F) = 0$, which yields

$$\mathsf{MSE}\,(T(F), F) = \mathsf{var}\,(T(F), F)\,.$$

Figure 1.8 displays the distributions, based on F being the standard normal distribution, of the sample mean, the sample median and Huber's M-estimator for $c_{.95} = 1.345$. Here, $c_{.95}$ means that the tuning parameter of Huber's M-estimator is chosen such that $\mathsf{ARE}\,(T(F), F) = 0.95$. Clearly, all the estimators are unbiased; however, their asymptotic variances differ. As shown in Table 1.2, the sample median has a low efficiency at the Gaussian distribution, while, for example, Huber's M-estimator can be tuned to be highly efficient by choosing $c_{.95} = 1.345$.

A further measure, that can be derived from the IF, is the *gross error sensitivity (GES)*, which is the supremum of the IF.

DEFINITION 4 *Gross-error sensitivity*
GES of $T(F)$ at F is defined by

$$\mathsf{GES}\,(T(F), F) \triangleq \sup_{y} |\mathsf{IF}(y; T(F), F)|\,,$$

where the supremum is taken over all y where $\mathsf{IF}(y; T(F), F)$ exists.

The GES measures the influence of a small amount of contamination, of fixed size, on the estimator $T(F)$. Thus, it can be regarded as an upper bound on the (standardized) asymptotic bias of the estimator (Hampel et al., 2011). A small value of $\mathsf{GES}\,(T(F), F)$ is preferred.

Unfortunately, maximizing the ARE conflicts with maximizing the GES. As an illustrating example, consider Huber's M-estimator. From (1.40), it becomes clear that

$$\mathsf{GES}\,(T(F), F) \sim c,$$

that is, a smaller value of c in (1.21), is preferred. By contrast, the maximal ARE is obtained when $T(F) = T_{\mathrm{ML}}(F)$. For F being Gaussian, this means that $c \to \infty$. Therefore, there is always a trade-off between maximizing robustness against deviations from the model, for example, minimizing the bias caused by some percentage of outliers and achieving maximum efficiency at the assumed model. Estimators that provide a good trade-off are sought.

Qualitative Robustness of an Estimator

Until now, we have mainly considered small departures from the model. These are concepts that are related to the IF and to qualitative robustness. Qualitative robustness is based on the intuition that small changes in the data should lead to only small changes in the estimates. More formally, qualitative robustness complements the notion of differentiability of an estimator, as measured by the continuity of the IF with respect to the Prokhorov distance (see Definition 3 in Hampel et al., 2011).

Using the bridge analogy, as stated in the beginning of this section, the IF measures the effects of infinitesimal perturbations. We next proceed to robustness measures that reveal the impact of larger perturbations. It is important to know how the performance degrades with increasing amounts, or intensities, of perturbations. Moreover, it is important to be able to answer the question of how big the perturbation can become before a breakdown occurs.

1.3.2   The Breakdown Point and Quantitative Robustness

The Breakdown Point

Beyond its so-called BP, an estimator $\hat{\beta}$ no longer provides any information about the true parameter β. This section will give a simple formalization for this intuitive statement. In its original formulation, as introduced by Hampel (1968), the BP was defined based on the Prokhorov distance (Prokhorov, 1956) of two probability distributions. Extensions to other distances, such as the Lévi or Kolmogorov distance have been given (Huber and Ronchetti, 2009).

However, from a practical viewpoint, it often suffices to consider what is referred to in Hampel (1968) as the *gross-error breakdown point* or in Maronna et al. (2006) as the *contamination breakdown point*. This variant of the BP of an estimator $\hat{\beta}$ is the largest proportion of outliers that the data may contain such that the resulting estimate $\hat{\beta}$ still provides some information about the true parameter β that we are interested in. This means that it still must contain information about the "typical data,"

If we assume that the majority of points belong to the nominal distribution F, and the outliers from some other distribution $G \in \mathcal{G}$, where $\mathcal{G}$ is a family of distributions, the data is distributed according to the so-called *ε-contamination model*

$$F_\varepsilon = (1 - \varepsilon)F + \varepsilon G. \tag{1.42}$$

Here, ε is the probability that a data point is an outlier. Based on this contamination model, the asymptotic BP of an estimator $T(F)$ at F is the largest value ε^* such that for every $\varepsilon < \varepsilon^*$, $T((1 - \varepsilon)F + \varepsilon G)$ as a function of G remains bounded, and also bounded away from the boundary of the parameter space.

When estimating the location parameter of a random variable, for example representing the DC value of a voltage in noise, $\beta = \mu$ in (1.1), the parameter space is $-\infty < \beta < \infty$, and a breakdown means that the bias of the estimator goes to ∞. For scale estimation, breakdown can refer to explosion with $\hat{\sigma} \to \infty$ or implosion with $\hat{\sigma} \to 0$. Similarly, as discussed in Chapter 5, for scatter matrices we can define the BP as the percentage of contamination ε^* for which the eigenvalues are bounded away from 0 and ∞. For penalized regression estimators, the boundary of the parameter space, and therewith the BP, depends on the penalty term. In the dependent data setting, for example in ARMA parameter estimation, the situation is even more involved, as the type of contamination determines the point to which the estimators are biased. As discussed in Chapter 8, this point may not necessarily lie at the boundary of the parameter space.

In case the estimate is not uniquely defined, that is, when an equation has multiple roots, the boundedness of all solutions is required. The maximally possible value that ε^* can take on is 50 percent. This is because its definition is based on "typical data." If $\varepsilon > 50\%$, the outliers would be treated as typical and the estimate would provide us instead with information on G. This is highly undesirable when we are in fact interested in the part of the data that carries information on β that is distributed according to F.

In many practical cases, we can consider the finite-sample BP ε^* of an estimator $\hat{\beta}_N$ based on a data set $\mathbf{y} = (y_1, \ldots, y_N)^\top$.

DEFINITION 5 *The finite-sample BP of an estimator $\hat{\beta}_N$ based on a data set $\mathbf{y} = (y_1, \ldots, y_N)^\top$ is defined as*

$$\varepsilon^*(\mathbf{y}, \hat{\beta}_N) = M^*/N, \tag{1.43}$$

where M^* is the maximal value of $M \in \{0, \ldots, N\}$, such that $\hat{\beta}_N(\mathbf{y})$ is bounded and also bounded away from the boundary of the parameter space $\forall \mathbf{y} \in \mathcal{X}_M$. Here, $\mathcal{X}_M$ refers to the set of all data sets $\mathbf{x}$ of size N having $N - M$ elements in common with $\mathbf{y}$.

The finite-sample BP, also referred to in Maronna et al. (2006) as the *replacement finite-sample breakdown point*, does not contain probability distributions and is easy to evaluate. Further possible definitions of the finite-sample BP are discussed in Huber and Ronchetti 2009 (p. 281).

To better understand the finite-sample BP, consider the data values illustrated in Figure 1.9 and the finite-sample BP of three different estimators of location, namely, the mean that is computed by evaluating (1.5), the median as defined in (1.12), and the

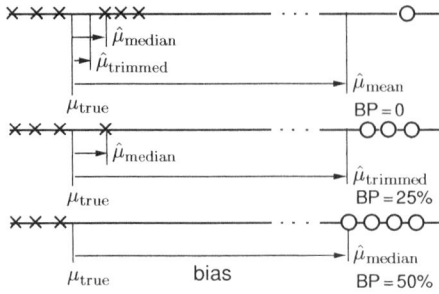

Figure 1.9 BPs of the mean, the α-trimmed mean, and the median estimators of location. Outliers are marked by "○," whereas normal data points are marked by "×."

α-trimmed mean, which computes the mean after ignoring the largest and smallest α % of the data. "Clean" observations are depicted as crosses and outliers as circles.

As illustrated in Figure 1.9, the finite-sample BP of the mean is 0, which means that a single outlier has an unbounded effect on the bias of its sample mean. The α-trimmed mean, with $\alpha = 0.25$ in this example, resists one outlier by ignoring the largest and smallest 25 percent of the data. The BP of the sample median equals the highest possible value of 50 percent, which means that its bias remains bounded even in situations when up to half of the observations, in this case three, are replaced by arbitrarily large values.

The Maximum-Bias Curve

The IF considers infinitesimal amounts of contamination ($\varepsilon \to 0$) and the BP describes the largest tolerable fraction of contamination ($\varepsilon = \varepsilon^*$). The MBC, in general, shows the worst possible bias of the estimator as a function of ε.

Assume that the true distribution F lies within some ε-*neighborhood*

$$\mathcal{F}(F, \varepsilon) = \{(1 - \varepsilon)F + \varepsilon\mathcal{G}\} \tag{1.44}$$

where $\mathcal{G}$ is an arbitrary set of distributions. It then follows that the asymptotic bias of an estimator $T(F)$ at any distribution $F \in \mathcal{F}(F, \varepsilon)$ is

$$\text{bias}\,(T(F), \beta) = T(F) - \beta$$

and the maximum asymptotic bias within the ε-neighborhood is displayed by the MBC.

DEFINITION 6   *The MBC of an estimator $T(F)$ of the parameter β is defined as*

$$\text{MB}\,(T(F), \varepsilon, \beta) = \max\{|\text{bias}\,(T(F))| : F \in \mathcal{F}(F, \varepsilon)\}. \tag{1.45}$$

Here, $\mathcal{F}(F, \varepsilon)$ is a ε-neighborhood as given in (1.44).

To illustrate the MBC, consider the location estimation example defined by (1.1); MBCs are shown in Figure 1.10. Equation (1.45) is evaluated to obtain $\text{MB}\,(T(F), \varepsilon, \mu)$ as a function of ε for the sample mean, the sample median, and Huber's M-estimator with $c_{.95} = 1.345$. The parameter space has been assumed to be the whole set of real numbers and the BP is the value $\varepsilon < \varepsilon^*$ for which $\text{MB}\,(T(F), \varepsilon, \mu)$ remains bounded.

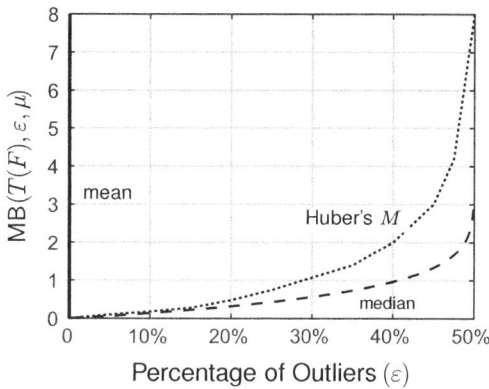

Figure 1.10 MBC of the sample mean, the sample median, and Huber's M-estimator with $c_{.95} = 1.345$. Note that the MBC for the sample mean tends to infinity for $\varepsilon > 0$.

It is evident from Figure 1.10 that estimators with equal BP have different MBCs. While Huber's M-estimator and the median both have a BP of 50 percent, their maximum-bias for $\varepsilon < 50\%$ differs. This difference is the reason why the MBC is useful in comparing different estimators.

1.4 Concluding Remarks

The objective of this chapter has been to give a tutorial-style introduction to the fundamental concepts of robust statistics for signal processing. In this chapter, we have defined measures of robustness, such as the BP, the IF and the MBC. Using these concepts, we have motivated the use of robust estimators, such as the M-estimator, for location and scale parameters instead of the conventionally used sample mean and sample standard deviation.

In the next chapters, we will introduce, in an accessible manner, the most important aspects, some of the less-known aspects, and most recent trends of robust statistical signal processing. Throughout the book, examples and real-life applications are used to illustrate difficult concepts. Robustness has become an important area of engineering practice and more emphasis has been given to the design and development of robust systems in recent years. Optimal systems are desirable; however, optimality is harder to achieve with the type of signal disturbance, such as impulsive interference, encountered in today's applications.

2 Robust Estimation: The Linear Regression Model

In this chapter, robust estimation for the linear regression model is discussed. Many of today's engineering problems can be formulated as problems of linear regression for which the parameters of interest are sought. These problems appear in areas as diverse as wireless communication (Wang and Poor, 1999; Zoubir and Brcich, 2002; Hammes et al., 2008; Kumar and Rao, 2009; Guvenc and Chong, 2009), ultrasonic systems (Prieto et al., 2009), computer vision (Stewart, 1999; Ye et al., 2003), electric power systems (Mili et al., 2002), automated detection of defects (Thomas and Mili, 2007), biomedical signal analysis (Bénar et al., 2007), genomics (Stranger et al., 2007), and so forth. There exists a multitude of robust procedures, and the collection of estimators to choose from has become excessively large. This chapter introduces a subset of approaches, based on various considerations such as robustness measures, as well as conceptual and computational complexity that are interesting and useful in practice. This chapter considers the case when the number of observations N exceeds the number of parameters p to be estimated, that is, $N > p$. The next chapter addresses the constrained or penalized regression model that allows the number of predictors to exceed the number of observations ($p > N$ case), a problem that frequently occurs in many modern data analysis problems, for example, in genomewide association analysis in genetics; see for example, Wu et al. (2009).

2.1 Complex Derivatives and Optimization

Recall that the set of complex numbers, denoted by $\mathbb{C}$, is the plane $\mathbb{R} \times \mathbb{R} = \mathbb{R}^2$ equipped with complex addition and complex multiplication making it a complex field. The complex *conjugate* of $x = (a, b) = a + \jmath b \in \mathbb{C}$ is defined as $x^* = (a, -b) = a - \jmath b$, where $\jmath = \sqrt{-1}$ denotes the *imaginary unit* and $a = \mathrm{Re}(x) \in \mathbb{R}$ and $b = \mathrm{Im}(x) \in \mathbb{R}$ are the real and imaginary part of x. Then, $|x| = \sqrt{xx^*} = \sqrt{a^2 + b^2}$ denotes the *modulus* of x. The notation $\langle \mathbf{a}, \mathbf{b} \rangle = \mathbf{a}^H \mathbf{b}$ is used to denote the Hermitian inner product defined in the complex vector space $\mathbb{C}^p$.

Here and in the forthcoming chapters, we will need to optimize a real-valued function over a complex-valued (vector or matrix) parameter. Therefore, we need to review basic definitions of complex derivatives. Consider a complex function $f : \mathbb{C} \to \mathbb{C}$ of a complex variable $x = (a, b) = a + \jmath b$:

$$f(x) = f(a, b) = u(a, b) + \jmath v(a, b),$$

where $u : \mathbb{C} \to \mathbb{R}$ and $v : \mathbb{C} \to \mathbb{R}$. Define

$$\frac{\partial f}{\partial a}(c) = \frac{\partial u}{\partial a}(c) + J\frac{\partial v}{\partial a}(c) \text{ and } \frac{\partial f}{\partial b}(c) = \frac{\partial u}{\partial b}(c) + J\frac{\partial v}{\partial b}(c),$$

where we assume that the functions u and v possess first real partial derivatives w.r.t. a and b at a point $c \in \mathbb{C}$. Observe that these are directional derivatives of f at c in the direction $t = 1 = (1, 0) \in \mathbb{C}$ and $t = J = (0, 1) \in \mathbb{C}$; that is, they describe the rates of change of $f(x)$ as x moves toward c along the real axis and imaginary axis, respectively. We then define

$$\frac{\partial f}{\partial x}(c) = \frac{1}{2}\left(\frac{\partial f}{\partial a}(c) - J\frac{\partial f}{\partial b}(c)\right), \quad \frac{\partial f}{\partial x^*}(c) = \frac{1}{2}\left(\frac{\partial f}{\partial a}(c) + J\frac{\partial f}{\partial b}(c)\right)$$

and call them the *complex derivative* and the *complex conjugate derivative* of f at c, respectively.

In Kreutz-Delgado (2007), the complex derivatives are called the $\mathbb{R}$-derivative and the conjugate $\mathbb{R}$-derivative, respectively. The differential calculus based on these operators is known as Wirtinger calculus (Remmert, 1991) or $\mathbb{CR}$-calculus (Kreutz-Delgado, 2007). For rules of calculus related to complex derivatives, see Kreutz-Delgado (2007), Eriksson et al. (2009, 2010), Adali et al. (2011), Ollila et al. (2011b), Ollila (2010, chapter 3), and Schreier and Scharf (2010, appendix 2).

For a complex-valued parameter $\mathbf{x} = (x_1, \dots, x_p)^\top = \mathbf{a} + J\mathbf{b} \in \mathbb{C}^p$, $\mathbf{a}, \mathbf{b} \in \mathbb{R}^p$, the complex derivative and conjugate derivative operators, with respect to $\mathbf{x}$ and $\mathbf{x}^*$, are defined similarly according to

$$\frac{\partial}{\partial \mathbf{x}} = \left(\frac{\partial}{\partial x_1}, \dots, \frac{\partial}{\partial x_p}\right)^\top \quad \text{and} \quad \frac{\partial}{\partial \mathbf{x}^*} = \left(\frac{\partial}{\partial x_1^*}, \dots, \frac{\partial}{\partial x_p^*}\right)^\top.$$

It is known (Brandwood, 1983; van den Bos, 1994) that both complex derivatives (and their multivariate extensions) vanish at stationary points of a real-valued function of a complex variable, but the (conjugate) derivative $\nabla = \partial/\partial \mathbf{x}^*$ defines the direction of the maximum rate of change, that is, it defines the *complex gradient*. Thus, a necessary and sufficient condition for $\mathbf{c} \in \mathbb{C}^p$ to be a stationary point of an $\mathbb{R}$-differentiable function $f : \mathbb{C}^p \to \mathbb{R}$ is that

$$\nabla_{\mathbf{x}^*} f(\mathbf{c}) = \frac{\partial f}{\partial \mathbf{x}^*}(\mathbf{c}) = \mathbf{0}, \tag{2.1}$$

which is the Karush-Kuhn-Tucker (KKT) condition.

When computing complex derivatives of a function $f(x)$, it is more convenient to express the function as $f(x, x^*)$ as if x and x^* are independent variables. Then, the complex derivatives can be evaluated by treating x (equivalently x^*) as a constant in f. For example, the partial derivative of a function $f(x) = |x|^2$ can be computed by writing it as $f(x, x^*) = xx^*$ and then $\partial f/\partial x = x^*$ and $\partial f/\partial x^* = x$. Indeed, the usefulness of the complex derivatives stems from an easily verifiable fact that they follow formally the same sum, product, and quotient rules as the ordinary partial derivatives. However, the *chain rule* for the composition function $(f \circ g)(x) = f(g(x))$ is different and is then given by

$$\frac{\partial f \circ g}{\partial x^*} = \frac{\partial f}{\partial x}(g) \cdot \frac{\partial g}{\partial x^*} + \frac{\partial f}{\partial x^*}(g) \cdot \frac{\partial g^*}{\partial x^*}.$$

Furthermore, if $f : \mathbb{R} \to \mathbb{R}$ and $g : \mathbb{C}^p \to \mathbb{R}$, then the complex conjugate derivative of $(f \circ g)(\mathbf{x}) = f(g(\mathbf{x}))$ at $\mathbf{c} \in \mathbb{C}^p$ is

$$\frac{\partial f \circ g}{\partial \mathbf{x}^*}(\mathbf{c}) = f'(g(\mathbf{c})) \frac{\partial g}{\partial \mathbf{x}^*}(\mathbf{c}), \tag{2.2}$$

where $f'(t) = \frac{\mathrm{d}}{\mathrm{d}t} f(t)$ is the real derivative of f. For example, suppose $f(t) = \sqrt{t}$ and $g(x) = |x|^2$, where $t > 0$ and $x \in \mathbb{C}$. Then, the conjugate derivative of $(f \circ g)(x) = |x|$ at $c \in \mathbb{C}, c \neq 0$, according to the rule that is specified by (2.2) is

$$\frac{\partial |x|}{\partial x^*}(c) = \frac{1}{2} \frac{c}{|c|}. \tag{2.3}$$

However, at a point $c = 0$, the function $|x|$ is not $\mathbb{R}$-differentiable. Some useful differentiation results on linear and quadratic forms are:

$$\nabla_{\mathbf{x}^*} \mathbf{x}^{\mathsf{H}} \mathbf{a} = \mathbf{a}, \quad \nabla_{\mathbf{x}^*} \mathbf{x}^{\top} \mathbf{a} = \mathbf{0}, \quad \nabla_{\mathbf{x}^*} \mathbf{x}^{\mathsf{H}} \mathbf{A} \mathbf{x} = \mathbf{A} \mathbf{x}, \tag{2.4}$$

which are valid for all $\mathbf{a} \in \mathbb{C}^p$ and $\mathbf{A} \in \mathbb{C}^{p \times p}$. Using the rules specified in (2.4), yields

$$\nabla_{\boldsymbol{\beta}^*} \|\mathbf{y} - \mathbf{X}\boldsymbol{\beta}\|_2^2 = \nabla_{\boldsymbol{\beta}^*} (\mathbf{y} - \mathbf{X}\boldsymbol{\beta})^{\mathsf{H}} (\mathbf{y} - \mathbf{X}\boldsymbol{\beta}) = -\mathbf{X}^{\mathsf{H}} (\mathbf{y} - \mathbf{X}\boldsymbol{\beta}),$$

where $\|\cdot\|_2^2$ is the squared ℓ_2-norm. This will be used in the next section as $\|\mathbf{y} - \mathbf{X}\boldsymbol{\beta}\|_2^2$ is the residual sum of squares criterion for regression.

Similarly, for a complex-valued function of a matrix variable $\mathbf{X} = (x_{ij}) \in \mathbb{C}^{n \times m}$, $f : \mathbb{C}^{n \times m} \to \mathbb{C}$, we define $\partial f / \partial \mathbf{X}$ and $\partial f / \partial \mathbf{X}^*$ as an $n \times m$ complex matrix whose (i, j)th entry is $\partial f / \partial x_{ij}$ and $\partial f / \partial x_{ij}^*$, respectively. For example, suppose the function $f : \mathbb{C}^{n \times n} \to \mathbb{C}$ is of the form $\mathrm{Tr}(\mathbf{A} \mathbf{X}^{-1})$ or determinant $|\mathbf{X}|$. The corresponding matrix differentials are

$$\frac{\partial \mathrm{Tr}(\mathbf{A} \mathbf{X}^{-1})}{\partial \mathbf{X}} = -(\mathbf{X}^{-1} \mathbf{A} \mathbf{X}^{-1})^{\top}, \quad \frac{\partial \mathrm{Tr}(\mathbf{A} \mathbf{X}^{-1})}{\partial \mathbf{X}^*} = \mathbf{0}, \quad \frac{\partial |\mathbf{X}|}{\partial \mathbf{X}} = |\mathbf{X}| \mathbf{X}^{-\top}, \quad \frac{\partial |\mathbf{X}|}{\partial \mathbf{X}^*} = \mathbf{0}.$$

Additional matrix differential results can be found in Hjorungnes and Gesbert (2007).

The subdifferential (Boyd and Vandenberghe, 2004) generalizes the concept of a derivative for convex functions that are not everywhere differentiable. For example, the function $f(x) = |x|$ is not differentiable at a point $c = 0$. However, because the function is convex, we can exploit the definition of a subgradient. A vector $\mathbf{g} \in \mathbb{R}^p$ is called a *subgradient* of a convex function $f : \mathbb{R}^p \to \mathbb{R}$ at $\boldsymbol{\beta}$ if

$$f(\boldsymbol{\beta}') \geq f(\boldsymbol{\beta}) + \langle \mathbf{g}, \boldsymbol{\beta}' - \boldsymbol{\beta} \rangle, \quad \forall \boldsymbol{\beta}' \in \mathbb{R}^p.$$

The set of all subgradients of f at $\boldsymbol{\beta}$ is called the *subdifferential* of f at $\boldsymbol{\beta}$, denoted $\partial f(\boldsymbol{\beta})$. In other words,

$$\partial f(\boldsymbol{\beta}) = \{\mathbf{g} \in \mathbb{R}^p : f(\boldsymbol{\beta}') \geq f(\boldsymbol{\beta}) + \langle \mathbf{g}, \boldsymbol{\beta}' - \boldsymbol{\beta} \rangle, \forall \boldsymbol{\beta}' \in \mathbb{R}^p\}.$$

At points where the function is differentiable, the subdifferential reduces to the gradient, that is, $\partial f(\boldsymbol{\beta}) = \{\nabla f(\boldsymbol{\beta})\}$. For a complex-valued convex function $f : \mathbb{C}^p \to \mathbb{R}$, we define the subdifferential at a point $\boldsymbol{\beta}$ as the set

$$\partial f(\boldsymbol{\beta}) = \{\mathbf{g} \in \mathbb{C}^p : f(\boldsymbol{\beta}') \geq f(\boldsymbol{\beta}) + 2\mathrm{Re}(\langle \mathbf{g}, \boldsymbol{\beta}' - \boldsymbol{\beta} \rangle), \ \forall \boldsymbol{\beta}' \in \mathbb{C}^p\}.$$

Any element $\mathbf{g} \in \partial f(\boldsymbol{\beta})$ is then called a *subgradient* of f at $\boldsymbol{\beta} \in \mathbb{C}^p$.

The generalized KKT theory states that a vector $\hat{\boldsymbol{\beta}}$ is a global optimum of a convex function $f : \mathbb{C}^p \to \mathbb{R}$ if

$$\mathbf{0} \in \partial f(\hat{\boldsymbol{\beta}}).$$

This means that the gradient in KKT condition (2.1) is replaced by the subdifferential. The difference is that the subdifferential is a set, so a vector of zeros needs to belong to this set, that is, $\mathbf{0}$ is a subgradient. For example, the least absolute deviation (LAD) or rank-LAD criterion functions defined in Section 2.4 are not differentiable at all points. Similarly, all penalized criterion functions that are based on the Lasso penalty $\|\boldsymbol{\beta}\|_1$ (see Chapter 3) are not differentiable at all points and, therefore, the theory of subdifferentials is needed for their analysis.

The subdifferential of the modulus $|\beta|$ of a complex variable $\beta \in \mathbb{C}$ is

$$\partial |\beta| = \begin{cases} \frac{1}{2}\mathrm{sign}(\beta), & \text{for } \beta \neq 0 \\ \frac{1}{2}s & \text{for } \beta = 0 \end{cases} \tag{2.5}$$

where s is some complex number satisfying $|s| \leq 1$. Thus, the subdifferential of $|\beta|$ is the usual complex conjugate derivative in (2.3) except at a point $\beta = 0$. In (2.5), $\mathrm{sign}(\cdot)$ denotes the complex *signum function*,

$$\mathrm{sign}(x) = \begin{cases} x/|x|, & \text{for } x \neq 0 \\ 0, & \text{for } x = 0 \end{cases} \tag{2.6}$$

which is a natural extension of the sign function to the complex-valued case. Indeed, if $\mathrm{Im}(x) = 0$, then (2.6) reduces to (1.11), as expected.

2.2 The Linear Model and Organization of the Chapter

Consider the case in which we have N measurements or *outputs* (also referred to as responses) $y_i \in \mathbb{F}\ (= \mathbb{C}$ or $\mathbb{R})$ and each output is associated with a p-dimensional vector of *inputs* (also referred to as predictors, covariates, etc.) $\mathbf{x}_{[i]}^\top = (x_{i1}, \ldots, x_{ip}) \in \mathbb{F}^p$. Then, given N observations $\mathcal{Z} = \{\mathbf{z}_i = (y_i, \mathbf{x}_{[i]}); i = 1, \ldots, N\}$, called the training data in machine learning, the aim is to approximate the output y_i using a *linear* combination of the inputs

$$\hat{y}_i = h(\mathbf{x}_{[i]}) = \sum_{j=1}^{p} x_{ij}\beta_j = \mathbf{x}_{[i]}^\top \boldsymbol{\beta}.$$

A linear model is widely used and is the subject of this chapter. The model used is

$$y_i = \mathbf{x}_{[i]}^\top \boldsymbol{\beta} + v_i, \quad i = 1, \ldots, N, \tag{2.7}$$

where the random error terms v_i, $i = 1, \ldots, N$, account for both the modeling and measurement errors. The goal is to estimate the vector $\boldsymbol{\beta} = (\beta_1, \ldots, \beta_p)^\top \in \mathbb{F}^p$ of *regression coefficients* given the data $\mathcal{Z}$ or, equivalently, to find a model $\hat{h}(\mathbf{x}) = \mathbf{x}^\top \hat{\boldsymbol{\beta}}$ that provides a good fit to the data $\mathcal{Z}$.

The linear model is usually written using matrix–vector notation and according to

$$\mathbf{X} = \begin{pmatrix} x_{11} & x_{12} & \cdots & x_{1p} \\ x_{21} & x_{22} & \cdots & x_{2p} \\ \vdots & \vdots & \ddots & \vdots \\ x_{N1} & x_{N2} & \cdots & x_{Np} \end{pmatrix} = \begin{pmatrix} \mathbf{x}_{[1]}^\top \\ \mathbf{x}_{[2]}^\top \\ \vdots \\ \mathbf{x}_{[N]}^\top \end{pmatrix}$$

$$= \begin{pmatrix} \mathbf{x}_1 & \mathbf{x}_2 & \cdots & \mathbf{x}_p \end{pmatrix}$$

that is, each row of $\mathbf{X}$ is an input vector. The column vectors $\{\mathbf{x}_j\}$ are often referred to as *feature vectors*. With this notation, the linear model defined in (2.7) can be written as

$$\mathbf{y} = \mathbf{X}\boldsymbol{\beta} + \mathbf{v} \tag{2.8}$$

$$= \mathbf{x}_1 \beta_1 + \cdots + \mathbf{x}_p \beta_p + \mathbf{v} \tag{2.9}$$

where $\mathbf{v} = (v_1, \ldots, v_N)^\top$ and $\mathbf{y} = (y_1, \ldots, y_N)^\top$ are N-dimensional vectors of error terms and outputs, respectively.

Often the regression model also includes a constant term and, for this case, the model is

$$y_i = \beta_0 + \mathbf{x}_{[i]}^\top \boldsymbol{\beta} + v_i, \quad i = 1, \ldots, N,$$

where $\beta_0 \in \mathbb{F}$ is called the *intercept*. The model can be written in the form (2.7) by letting $\mathbf{x}_{[i]} = (1, x_{i1}, \ldots, x_{ip})^\top$, $i = 1, \ldots, N$, be the vector of predictors for the ith response and $\boldsymbol{\beta} = (\beta_0, \beta_1, \ldots, \beta_p)^\top$ be the regression vector. When the vector notation of (2.8) is used, $\mathbf{X}$ has size $N \times (p + 1)$, the first column being the N-vector of ones, denoted by $\mathbf{1}$. In this case, we let the column index of $\mathbf{X}$ run from 0 to p, so that $\mathbf{x}_0 = \mathbf{1}$ and $\mathbf{X} = \begin{pmatrix} \mathbf{x}_0 & \mathbf{x}_1 & \cdots & \mathbf{x}_p \end{pmatrix}$. From now on, we assume a linear model with an intercept, unless mentioned otherwise.

Most criterion functions in regression seek to minimize the distance between $\mathbf{y}$ and the fit $\hat{\mathbf{y}}$, that is, the residual vector

$$\hat{\mathbf{r}} = \mathbf{y} - \hat{\mathbf{y}} = \mathbf{y} - \mathbf{X}\hat{\boldsymbol{\beta}}$$

should be as small as possible when a suitable metric is used. The most commonly used metric is the squared ℓ_2-norm that leads to minimizing the residual sum-of-squares (RSS)

$$L_{\text{RSS}}(\boldsymbol{\beta}) = \|\mathbf{y} - \mathbf{X}\boldsymbol{\beta}\|_2^2 = \sum_{i=1}^{N} \left| y_i - \mathbf{x}_{[i]}^\top \boldsymbol{\beta} \right|^2. \tag{2.10}$$

Minimization of $L_{\mathrm{RSS}}(\boldsymbol{\beta})$ is achieved by using the least squares estimate (LSE) $\hat{\boldsymbol{\beta}}_{\mathrm{LS}}$. The associated estimate for the scale parameter σ of the error terms is the sample standard deviation,

$$\hat{\sigma}_{\mathrm{SD}}^2 = \frac{1}{N}\|\hat{\mathbf{r}}\|_2^2 = \frac{1}{N}\sum_{i=1}^{N}|y_i - \mathbf{x}_{[i]}^{\top}\hat{\boldsymbol{\beta}}_{\mathrm{LS}}|^2, \tag{2.11}$$

or its unbiased version, $s^2 = [N/(N-p-1)]\hat{\sigma}_{\mathrm{SD}}^2$. The LSE is the MLE for the Gaussian noise case (see Chapter 1). However, it is also highly sensitive to outliers and is inefficient when the error terms are non-Gaussian. This chapter introduces several robust criterion functions $L(\boldsymbol{\beta})$ that provide robust alternatives to the LSE discussed in Section 2.3. In Section 2.4, we discuss LAD and rank-LAD regression methods. M-estimates of regression based on general (robust) loss functions and an auxiliary scale estimate of error terms are developed in Section 2.5. Most robust loss functions require an estimate of scale and therefore joint estimation of the unknown regression vector $\boldsymbol{\beta}$ and scale σ is often desirable. In Section 2.6, we describe an elegant approach to estimate the unknown regression vector and scale using Huber's criterion, which finds minimizers of a (jointly) convex criterion function $L_{\mathrm{H}}(\boldsymbol{\beta}, \sigma)$. Section 2.7 introduces measures of robustness, such as the IF and the BP. The different types of outliers that can occur in linear regression, as well as their effect on the estimators, is also discussed in this section. Because this section is intended to serve as an easy-to-read introduction to robustness concepts, only the real-valued case is discussed. Section 2.8 briefly discusses positive breakdown estimators, while Section 2.9 provides examples and simulation results that facilitate the comparison of the methods discussed in the chapter.

To convince the reader of the need for robust regression approaches, we provide a simple illustration. Figure 2.1 depicts a scatter plot of $\{(y_i, x_i)\}_{i=1}^{N}$ of $N = 10$ real-valued data points that are consistent with the linear model of a single predictor with no intercept. It also shows the obtained least squares (LS), LAD, and rank-LAD estimates of regression, as well as the M-estimates of regression using an auxiliary scale estimate based on Tukey's loss function. An outlier is created by setting $y_{10} = -20$. The LS regression line is strongly pulled toward the outlier whereas the regression lines obtained by robust estimators appear less affected by the outlier and thus provide an accurate fit to the majority of the data.

2.3 The Least Squares Estimator

The LSE $\hat{\boldsymbol{\beta}}_{\mathrm{LS}}$, or simply $\hat{\boldsymbol{\beta}}$, is a vector in $\mathbb{F}^{p+1}$ ($\mathbb{F} = \mathbb{C}$ or $\mathbb{R}$) for which the fit $\hat{\mathbf{y}} = \mathbf{X}\hat{\boldsymbol{\beta}}$ is closest (assuming the Euclidean metric) to $\mathbf{y}$ compared to any other point in the column space of $\mathbf{X}$, denoted $\mathrm{col}(\mathbf{X})$. In other words, $\|\hat{\mathbf{r}}\|_2^2 = \|\mathbf{y} - \hat{\mathbf{y}}\|_2^2$ is the minimum value of the RSS criterion in (2.10). The geometry behind the LS estimation is illustrated in Figure 2.2 for the real-valued case of two predictors ($p = 2$).

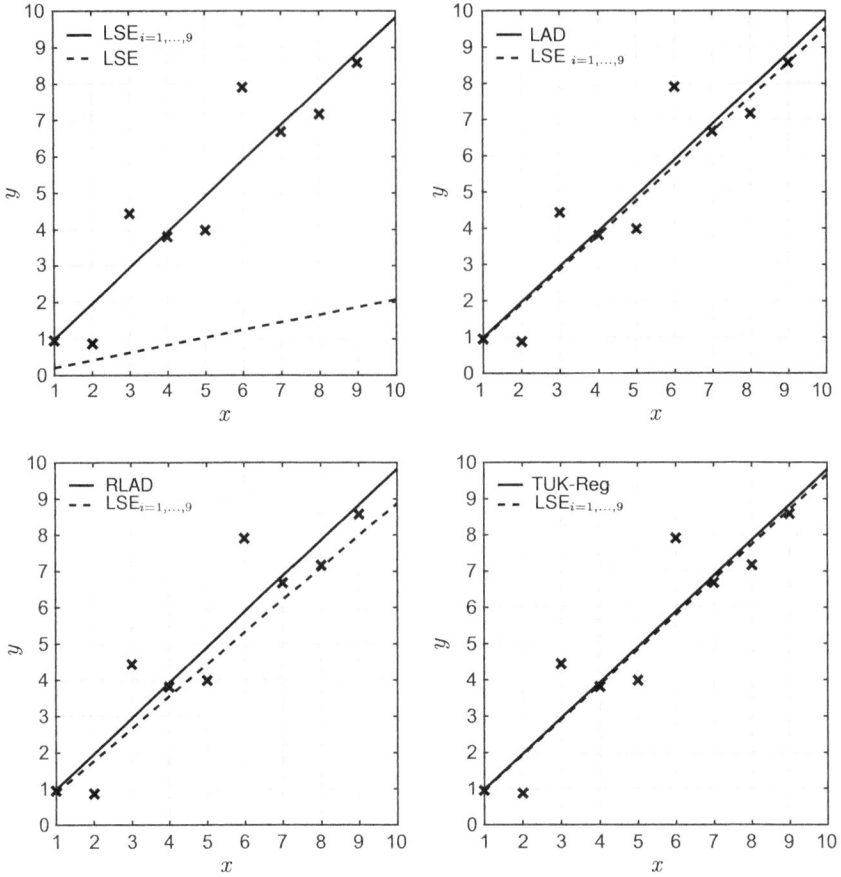

Figure 2.1 Scatter plots of a data set $\mathcal{Z} = \{\mathbf{z}_i = (y_i, x_i)\}_{i=1}^{N} \subset \mathbb{R}^2$ consisting of $N = 10$ data points along with regression lines $h(x) = x\hat{\beta}$ produced by LS, LAD, rank-LAD, and Tukey's M-estimators (based on auxiliary scale estimate). For the data point $\mathbf{z}_{10}$, an outlier is created by setting $y_{10} = -20$. $\text{LSE}_{i=1,\dots,9}$ denotes the LS solution based on the nonoutlying points.

The LSE is then a solution to the zero-gradient equation

$$\nabla_{\beta^*} L_{\text{RSS}}(\hat{\beta}) = \mathbf{0} \Leftrightarrow \sum_{i=1}^{N} (y_i - \mathbf{x}_{[i]}^{\top} \hat{\beta}) \mathbf{x}_{[i]}^* = \mathbf{0},$$

which can be represented more compactly in matrix form as

$$\mathbf{X}^{\mathsf{H}}(\mathbf{y} - \mathbf{X}\hat{\beta}) = \mathbf{0} \Leftrightarrow \mathbf{X}^{\mathsf{H}}\mathbf{X}\hat{\beta} = \mathbf{X}^{\mathsf{H}}\mathbf{y}. \qquad (2.12)$$

This equation is referred to as the *normal equation*. The normal equation (2.12) can also be expressed as

$$\langle \mathbf{x}_0, \hat{\mathbf{r}} \rangle = \sum_{i=1}^{N} \hat{r}_i = 0 \quad \text{and} \quad \langle \mathbf{x}_j, \hat{\mathbf{r}} \rangle = 0, \quad j = 1, \dots, p, \qquad (2.13)$$

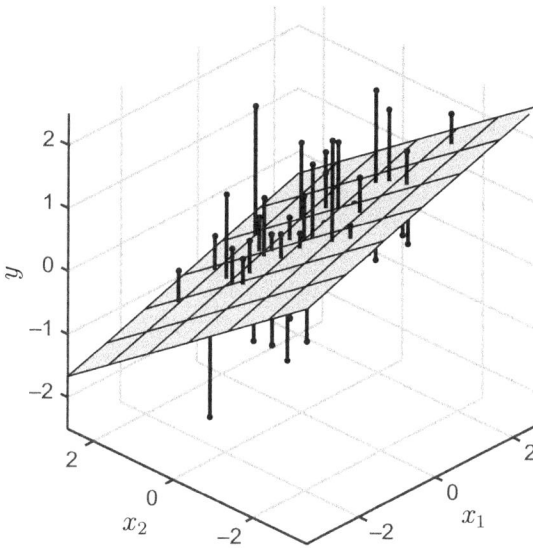

Figure 2.2 The LS plane $\hat{y} = \hat{h}(\mathbf{x}) = \mathbf{x}^\top \hat{\boldsymbol{\beta}} = \beta_0 + \beta_1 x_1 + \beta_2 x_2$ is such that the sum of the squared residuals $\sum_{i=1}^{N} |y_i - \hat{y}_i|^2$ is smallest among any other affine hyperplanes $h(\mathbf{x}) = \mathbf{x}^\top \boldsymbol{\beta}$.

where $\langle \mathbf{a}, \mathbf{b} \rangle = \mathbf{a}^H \mathbf{b}$. The normal equations state that the sample mean of the residuals is zero and that the residual vector is orthogonal to the column space of $\mathbf{X}$. Because we assume that $\mathbf{X}$ is of full rank, that is, $\text{rank}(\mathbf{X}) = p + 1$, the LSE is the unique solution of the normal equation. Furthermore, from (2.12), we observe that the LSE can be expressed as a linear function of $\mathbf{y}$ as

$$\hat{\boldsymbol{\beta}} = (\mathbf{X}^H \mathbf{X})^{-1} \mathbf{X}^H \mathbf{y}. \tag{2.14}$$

The LS fit is thus

$$\hat{\mathbf{y}} = \mathbf{X}\hat{\boldsymbol{\beta}} = \mathbf{X}(\mathbf{X}^H \mathbf{X})^{-1} \mathbf{X}^H \mathbf{y} = \mathbf{H}\mathbf{y},$$

where

$$\mathbf{H} = \mathbf{X}(\mathbf{X}^H \mathbf{X})^{-1} \mathbf{X}^H$$

is called the *hat matrix*. The hat matrix is a projection matrix ($\mathbf{H}^H = \mathbf{H}$ and $\mathbf{H}^2 = \mathbf{H}$) onto $\text{col}(\mathbf{X})$. The residual vector is

$$\hat{\mathbf{r}} = \mathbf{y} - \hat{\mathbf{y}} = (\mathbf{I} - \mathbf{H})\mathbf{y} = \mathbf{H}^\perp \mathbf{y}$$

where $\mathbf{H}^\perp = \mathbf{I} - \mathbf{H}$ is a projection matrix onto the orthogonal complement of $\text{col}(\mathbf{X})$. This geometry of LS estimation is illustrated in Figure 2.3 for the case of two predictors ($p = 2$).

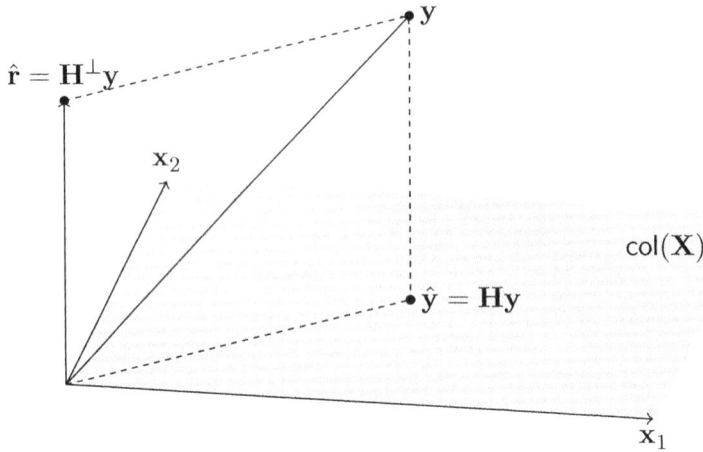

Figure 2.3 The LS fit $\hat{\mathbf{y}}$ is a projection of the response vector $\mathbf{y}$ onto the column space of $\mathbf{X}$. The fit $\hat{\mathbf{y}}$ is closest to $\mathbf{y}$ compared to any other point in $\mathrm{col}(\mathbf{X})$, that is, the LS residual $\hat{\mathbf{r}} = \mathbf{y} - \mathbf{X}\hat{\boldsymbol{\beta}} = \mathbf{H}^{\perp}\mathbf{y}$ has a smaller ℓ_2-norm than any other residual vector $\mathbf{r} = \mathbf{r}(\boldsymbol{\beta}) = \mathbf{y} - \mathbf{X}\boldsymbol{\beta}$ for any candidate $\boldsymbol{\beta} \in \mathbb{F}^{p+1}$.

2.4   Least Absolute Deviation and Rank-Least Absolute Deviation Regression

The LAD estimate $\hat{\boldsymbol{\beta}}_{\mathrm{LAD}}$, or simply $\hat{\boldsymbol{\beta}}$, is a solution to the optimization problem

$$\underset{\boldsymbol{\beta} \in \mathbb{F}^{p+1}}{\text{minimize}} \left\{ L_{\mathrm{LAD}}(\boldsymbol{\beta}) = \|\mathbf{y} - \mathbf{X}\boldsymbol{\beta}\|_1 = \sum_{i=1}^{N} |y_i - \mathbf{x}_{[i]}^{\top}\boldsymbol{\beta}| \right\}. \qquad (2.15)$$

When a residual $r_i \equiv r_i(\boldsymbol{\beta}) = y_i - \mathbf{x}_{[i]}^{\top}\boldsymbol{\beta}$ is very large, it will have a smaller impact on the LAD criterion compared to the RSS criterion where the magnitudes of the residuals are squared. This explains why the resulting solution is more resilient to outliers. An overview of the history of LAD regression can be found in Koenker (2005).

The LAD criterion (2.15) is not a smooth function due to the use of the ℓ_1-norm, which results in it being nondifferentiable at points $\boldsymbol{\beta} \in \mathbb{F}^{p+1}$ for which the residual vanishes, $r_i(\boldsymbol{\beta}) = 0$. The LAD criterion equals the negative log-likelihood function when the error terms have a double exponential (Laplace) distribution, that is, the LAD estimator is the MLE in this case. The estimator has a bounded IF, but the IF is non-smooth (discontinuous) due to the nondifferentiability of the ℓ_1-loss function at the origin.

We also consider an estimator that minimizes the LAD criterion of pairwise differences of the residuals, that is,

$$\underset{\boldsymbol{\beta} \in \mathbb{F}^{p}}{\text{minimize}} \left\{ L_{\mathrm{R}}(\boldsymbol{\beta}) = \sum_{i<j} |r_i - r_j| = \sum_{i<j} |(y_i - y_j) - (\mathbf{x}_{[i]} - \mathbf{x}_{[j]})^{\top}\boldsymbol{\beta}| \right\}. \qquad (2.16)$$

The number of pairwise differences is $\binom{N}{2} = \frac{N(N-1)}{2}$ and, hence, this criterion does not lend itself easily to large-scale problems. Nevertheless, randomized algorithms can be utilized to compute an approximate solution. The intercept β_0, however, cannot be estimated using the criterion function (2.16) as it cancels out naturally due to the pairwise differences used,

$$r_i - r_j = y_i - \beta_0 - \mathbf{x}_{[i]}^\top \boldsymbol{\beta} - (y_j - \beta_0 - \mathbf{x}_{[j]}^\top \boldsymbol{\beta})$$
$$= (y_i - y_j) - (\mathbf{x}_{[i]} - \mathbf{x}_{[j]})^\top \boldsymbol{\beta}.$$

We defer the problem of estimating the intercept to Section 2.4.3. Thus, in (2.15) we estimate a $(p + 1)$-dimensional regression parameter having as its first element the intercept β_0, whereas in (2.16) we estimate only a p-dimensional regression vector.

In the real-valued case, it is easy to verify that $L_R(\boldsymbol{\beta})$ can be expressed in an equivalent form based on the ranks of the residuals. Namely,

$$L_R(\boldsymbol{\beta}) = \sum_{i<j} |r_i - r_j| = 2 \sum_{i=1}^{N} R(r_{(i)}) r_{(i)}, \qquad (2.17)$$

$$R(r_{(i)}) = i - \frac{N+1}{2} := centered\ rank\ \text{(Wilcoxon score)} \qquad (2.18)$$

where $r_{(i)}$ denotes ordered residuals, $r_{(1)} \le r_{(2)} \le \ldots \le r_{(N)}$. The latter equation (based on ranks) is called the *rank regression* dispersion function (Jaeckel, 1972). Hence, we refer to the solution $\hat{\boldsymbol{\beta}}_R$ of (2.16) as the *rank-LAD estimate* (or the *RLAD estimate*) and $L_R(\boldsymbol{\beta})$ as the *rank-LAD criterion*.

The LAD and rank-LAD loss functions are convex, which makes them suitable candidates for penalized regression as well. The penalized estimates, the LAD-Lasso and the rank-Lasso estimators, are discussed in Chapter 3. Due to convexity, the estimates can be computed using linear programming techniques or specialized algorithms such as the iteratively reweighted least squares (IRWLS) method. The benefit of the IRWLS algorithm is that it is simple and easy to implement. It allows an initial guess (a warm start) for each iteration, which is a useful feature in the penalized case in which a regularization path needs to be computed.

Example 5 We compare the optimization landscapes of LS, LAD, and rank-LAD criteria in the real-valued linear model with $p = 2$ predictors and no intercept,

$$y_i = x_{i1}\beta_1 + x_{i2}\beta_2 + v_i, \quad \beta_1 = \beta_2 = 1,$$

where the predictors $x_{i1}, x_{i2} \overset{i.i.d.}{\sim} \mathcal{N}(0,1)$, $v_i \overset{i.i.d.}{\sim} \mathcal{N}(0, \sigma^2)$ with $\sigma = 0.2$. Figure 2.4 depicts the surface and contour plots of the criterion functions for a data set generated from the preceding model. The number of data points is $N = 20$. We observe that the RSS criterion is smooth, strictly convex (because $N > p$ and $\mathbf{X}$ is of full rank), and has elliptical equidensity contours. The LAD criterion, however, is clearly piecewise linear. We also observe that the rank-LAD criterion appears as a smoothed version of the LAD criterion. In the upper panel, the minimizers of the criterion functions, the LSE, the LAD, and the rank-LAD estimates, are marked with an asterisk.

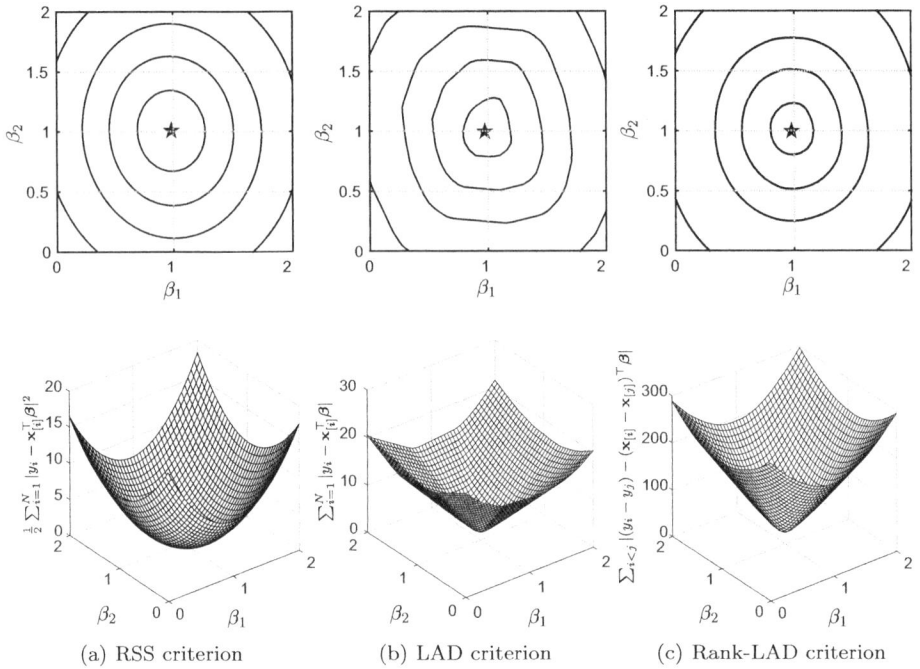

(a) RSS criterion   (b) LAD criterion   (c) Rank-LAD criterion

Figure 2.4 The figures depict the contour and surface plots of the RSS, LAD, and rank-LAD criterion function for the simulated data based on $N = 20$ observations and $p = 2$ predictors. The LSE, LAD, and rank-LAD estimates, respectively, are marked by an asterisk in the respective contour plots.

2.4.1   Simple Linear Regression without an Intercept

It is instructive to start with the single predictor ($p = 1$) case with no intercept, $\beta_0 = 0$. We show that the LAD estimate $\hat{\beta}_{\text{LAD}} \in \mathbb{F}$ (or the rank-LAD estimate $\hat{\beta}_{\text{R}} \in \mathbb{F}$) is a weighted median of the slopes of the lines passing through the origin $(0,0)$ and the points $(y_j, x_j), j = 1, \ldots, N$, or alternatively, the pairs of observations (y_i, x_i) and (y_j, x_j), $i, j = 1, \ldots, N$.

In the real-valued case, we show than an explicit expression for the solution is found using ranking of the slopes. In the complex case, however, one requires an iterative gradient descent type algorithm.

Assume that $x_i \neq 0$ for $i = 1, \ldots, N$. The LAD criterion can then be written in the equivalent form

$$L_{\text{LAD}}(\beta) = \sum_{i=1}^{N} |b_i - \beta| \, w_i, \tag{2.19}$$

where $b_i = y_i / x_i$ are the slopes of the regression lines passing through the origin and a data point (y_i, x_i) and $w_i = |x_i|$ denotes the respective weight, $i = 1, \ldots, N$. Thus, the LAD estimate $\hat{\beta}_{\text{LAD}}$ is a *weighted median* of slopes b_i with weights w_i. We use the notation $\hat{\beta} = \text{med}(b_i \mid w_i)$ to denote that $\hat{\beta}$ is a minimizer of (2.19).

If an output y_i is associated with a large input x_i, then this means that the respective slope b_i is associated with a large weight w_i. Consequently, such a measurement has a greater effect on the solution. This feature is meaningful because all slopes are unbiased, that is, $\mathsf{E}[b_i] = \beta$, but have largely varying variances, $\mathsf{var}(b_i) = \mathsf{var}(y_i/x_i) = \sigma^2/|x_i|^2$, where $\sigma^2 = \mathsf{var}(v_i)$. Thus, a slope b_i with a small weight $|x_i|$ estimates β with a large error.

Weighted Median Regression: The Real-Valued Case

Let $\{b_{(i)}\}_{i=1}^N \in \mathbb{R}$ denote the ordered slopes in decreasing order $b_{(1)} \leq b_{(2)} \leq \ldots \leq b_{(N)}$ and let $w_{(i)}$ denote the respective weights. We can write

$$L_{\text{LAD}}(\beta) = \sum_{i=1}^N |b_i - \beta| w_i = \sum_{b_i > \beta} (b_i - \beta) w_i - \sum_{b_i \leq \beta} (b_i - \beta) w_i$$

$$= \begin{cases} \sum_{i=1}^N b_{(i)} w_{(i)} - \beta \sum_{i=1}^N w_{(i)}, & \beta < b_{(1)} \\ \sum_{i=2}^N b_{(i)} w_{(i)} - b_{(1)} w_{(1)} + \beta \left\{ w_{(1)} - \sum_{i=2}^N w_{(i)} \right\}, & b_{(1)} \leq \beta < b_{(2)} \\ \sum_{i=3}^N b_{(i)} w_{(i)} - b_{(1)} w_{(1)} - b_{(2)} w_{(2)} \\ \quad + \beta \left\{ w_{(1)} + w_{(2)} - \sum_{i=3}^N w_{(i)} \right\}, & b_{(2)} \leq \beta < b_{(3)} \\ \text{etc.} \end{cases}$$

The preceding expression illustrates that the LAD criterion is a piecewise linear function with respect to β. Furthermore, the slope of the LAD criterion in the interval $b_{(k-1)} \leq \beta < b_{(k)}$ is

$$w_{(1)} + \cdots + w_{(k-1)} - \sum_{i=k}^N w_{(i)} = 2 \sum_{i=1}^{k-1} w_{(i)} - \sum_{i=1}^N w_{(i)}. \tag{2.20}$$

If the slope changes its sign in the next interval from negative to positive, then the LAD solution is the end point $\hat{\beta}_{\text{LAD}} = b_{(k)}$, where k is an index that satisfies

$$\sum_{i=1}^{k-1} w_{(i)} < \frac{1}{2} \sum_{i=1}^N w_{(i)}, \quad \sum_{i=1}^k w_{(i)} \geq \frac{1}{2} \sum_{i=1}^N w_{(i)}. \tag{2.21}$$

Thus, the LAD solution, although not given in closed form, can easily be computed using (2.21). Note that the criterion (2.19) is differentiable everywhere except at $\beta = b_i = y_i/x_i, i = 1, \ldots, N$. Because a minimum occurs at one of the slopes, that is, at the point of nondifferentiability, $L_{\text{LAD}}(\beta)$ has no root at the minimum.

Recall (cf. Section 2.1) that $\hat{\beta}$ is a minimizer of $L_{\text{LAD}}(\beta)$ if and only if $0 \in \partial L_{\text{LAD}}(\hat{\beta})$. The LAD criterion can be written using the ordered slopes and weights as $L_{\text{LAD}}(\beta) = \sum_{i=1}^N |b_{(i)} - \beta| w_{(i)}$. Then, note that the subdifferential of $|b_{(i)} - \beta|$ is

$$\partial |b_{(i)} - \beta| = \begin{cases} \text{sign}(\beta - b_{(i)}), & \text{for } \beta \neq b_{(i)} \\ [-1, +1] & \text{for } \beta = b_{(i)} \end{cases}.$$

The subdifferential of $L_{\mathrm{LAD}}(\beta)$ at a point $b_{(k)}$ is

$$\partial L_{\mathrm{LAD}}(b_{(k)}) = \sum_{i \neq k} \mathrm{sign}(b_{(k)} - b_{(i)})w_{(i)} + sw_{(k)}$$

$$= \sum_{i=1}^{k-1} w_{(i)} + sw_{(k)} - \sum_{i=k+1}^{N} w_{(i)}$$

$$= 2\sum_{i=1}^{k-1} w_{(i)} + (1+s)w_{(k)} - \sum_{i=1}^{N} w_{(i)} \qquad (2.22)$$

for some number $s \in [-1, 1]$. Note that $\hat{\beta} = b_{(k)}$ if and only if $0 \in \partial L_{\mathrm{LAD}}(b_{(k)})$, which, due to (2.22) is equivalent to the condition

$$\frac{1}{w_{(k)}}\left\{2\sum_{i=1}^{k-1} w_{(i)} - \sum_{i=1}^{N} w_{(i)}\right\} \in [-2, 0]. \qquad (2.23)$$

Thus, if condition (2.3) is satisfied, then $\hat{\beta} = b_{(k)}$ must be the LAD estimate. Note that (2.23) is a reformulation of the condition (2.21).

In a similar fashion, we may write the rank-LAD criterion function $L_{\mathrm{R}}(\beta)$ as

$$L_{\mathrm{R}}(\beta) = \sum_{i<j} |(y_i - y_j) - \beta(x_i - x_j)| = \sum_{i<j} |b_{ij} - \beta|w_{ij} \qquad (2.24)$$

where

$$b_{ij} = \frac{y_i - y_j}{x_i - x_j} \qquad (2.25)$$

is the slope of the line that passes through the data points (y_i, x_i) and (y_j, x_j) and $w_{ij} = |x_i - x_j|$ is the associated weight. Thus, the rank-LAD estimate $\hat{\beta}_{\mathrm{R}}$ can be found in the same way as the LAD estimate, but now based on the slopes b_{ij} with associated weights $w_{ij}, i, j = 1, \ldots, N, i < j$.

Example 6   Consider the data set defined in Figure 2.1. The graphs of the LAD and rank-LAD criterion functions defined in (2.19) and (2.24) are shown, respectively, in Figure 2.5. The estimates $\hat{\beta}_{\mathrm{LAD}}$ and $\hat{\beta}_{\mathrm{R}}$ can be computed as the weighted median of the slopes b_i and b_{ij}, respectively. The values of the estimates are denoted by vertical dashed lines in the graphs. The points of discontinuity, occurring at $\beta = y_i/x_i = b_i$ for the LAD case and $\beta = (y_i - y_j)/(x_i - x_j) = b_{ij}$ for the rank-LAD case, respectively, are plotted as circles ($\circ$) in the graphs. Note that the rank-LAD criterion, though also piecewise linear, appears to be much smoother. A similar effect can be observed in Figure 2.4b and 2.4c.

Weighted Median Regression: The Complex-Valued Case

Finding the solution to (2.19) requires a different approach in the complex-valued case as there is a lack of natural ordering of complex-valued slopes $b_i \in \mathbb{C}$ or $b_{ij} \in \mathbb{C}$.

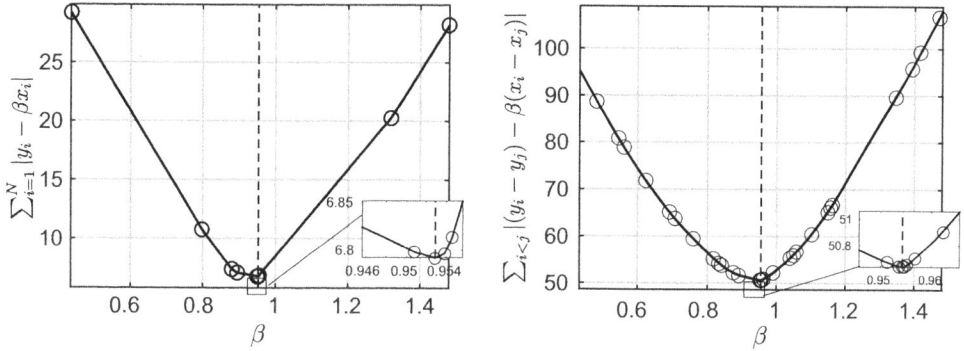

Figure 2.5 Left: graph of $L_{\mathrm{LAD}}(\beta)$. Right: graph of $L_{\mathrm{R}}(\beta)$. The vertical dashed line shows the locations of the estimates $\hat{\beta}_{\mathrm{LAD}}$ and $\hat{\beta}_{\mathrm{R}}$. The points of nondifferentiablity, appearing at $\beta = b_i = y_i/x_i$ and $\beta = b_{ij} = (y_i - y_j)/(x_i - x_j)$ for the LAD and rank-LAD criteria, respectively, are denoted with circles ($\circ$), $i, j = 1, \ldots, N$, $i < j$.

The theory of convex optimization states that $\hat{\beta}_{\mathrm{LAD}}$ is a solution to (2.19) if $0 \in \partial L_{\mathrm{LAD}}(\hat{\beta})$. Recall that $L_{\mathrm{LAD}}(\hat{\beta})$, as defined by (2.19), is not differentiable with respect to the slopes b_i. Hence, b_k is a solution if $0 \in \partial L_{\mathrm{R}}(b_k)$. This is equivalent to the condition that

$$\left| \frac{1}{w_k} \sum_{i \neq k} \mathrm{sign}(b_k - b_i) w_i \right| \leq 1. \tag{2.26}$$

If condition (2.26) is not met, then $\hat{\beta} \neq b_k \; \forall k$ and the solution may be found by setting the complex conjugate derivative of $L_{\mathrm{LAD}}(\beta) \equiv L_{\mathrm{LAD}}(\beta, \beta^*)$ to zero, that is, by solving the estimating equation

$$0 = \sum_{i=1}^{N} \mathrm{sign}(b_i - \hat{\beta}) w_i, \tag{2.27}$$

where $\mathrm{sign}(x)$ is defined in (2.6). Writing $\mathrm{sign}(x) = x \cdot (1/|x|)$ allows us to rewrite (2.27) as

$$0 = \sum_{i=1}^{N} (b_i - \hat{\beta}) \frac{w_i}{|b_i - \hat{\beta}|} \quad \Leftrightarrow \quad \hat{\beta} = \left(\sum_{i=1}^{N} \frac{w_i}{|b_i - \hat{\beta}|} \right)^{-1} \sum_{i=1}^{N} \frac{b_i w_i}{|b_i - \hat{\beta}|}.$$

Thus, $\hat{\beta}$ solves a FP equation, which suggests an iterative algorithm of the form

$$\beta^{(n+1)} = \left(\sum_{i=1}^{N} \frac{w_i}{|b_i - \beta^{(n)}|} \right)^{-1} \sum_{i=1}^{N} \frac{b_i w_i}{|b_i - \beta^{(n)}|}, \quad n = 0, 1, \ldots \tag{2.28}$$

which can start by using an arbitrary initial start value $\beta^{(0)}$. We can rewrite equation (2.28) in the more intuitive form

$$\beta^{(n+1)} = \beta^{(n)} + \left(\sum_{i=1}^{N} \frac{w_i}{|b_i - \beta^{(n)}|} \right)^{-1} \sum_{i=1}^{N} w_i \, \mathrm{sign}(b_i - \beta^{(n)}). \tag{2.29}$$

This follows by writing

$$\sum_{i=1}^{N} \frac{b_i w_i}{|b_i - \beta^{(n)}|} = \sum_{i=1}^{N} \frac{b_i w_i - \beta^{(n)} w_i + \beta^{(n)} w_i}{|b_i - \beta^{(n)}|}$$

$$= \sum_{i=1}^{N} \text{sign}(b_i - \beta^{(n)}) w_i + \beta^{(n)} \sum_{i=1}^{N} \frac{w_i}{|b_i - \beta^{(n)}|},$$

which can then be substituted into equation (2.28). The form of (2.29) is consistent with a gradient descent type algorithm, with an update of the form $\beta^{(n+1)} = \beta^{(n)} + \mu \delta^{(n)}$. Note that the rank-LAD estimate $\hat{\beta}_{\text{R}}$ is computed similarly, but by using slopes b_{ij} with weights w_{ij}, $i < j$, $i, j \in \{1, \dots, N\}$.

The iterative algorithm in (2.28) was developed by (Weiszfeld, 1937) for solving the weighted spatial median (or Fermat–Weber problem) of real-valued observations. The connection to the weighted spatial median in the real-valued case follows from the fact that (2.19) can be expressed using a real-valued notation, as

$$L_{\text{LAD}}(\beta_{\text{r}}, \beta_{\text{i}}) = \sum_{i=1}^{N} w_i \left\| \begin{pmatrix} b_{\text{r},i} \\ b_{\text{i},i} \end{pmatrix} - \begin{pmatrix} \beta_{\text{r}} \\ \beta_{\text{i}} \end{pmatrix} \right\|_2, \tag{2.30}$$

where $b_i = b_{\text{r},i} + jb_{\text{i},i}$ and $\beta = \beta_{\text{r}} + j\beta_{\text{i}}$. Thus, if $(\hat{\beta}_{\text{r}}, \hat{\beta}_{\text{i}})$ solves (2.30), then $\hat{\beta} = \hat{\beta}_{\text{r}} + j\hat{\beta}_{\text{i}}$ is a solution of (2.19), and vice versa. Thanks to this connection, it is known that the iterative algorithm (2.29) always converges to the true solution provided that $\beta^{(n)}$ does not take on a value defined by the slopes $b_1, \dots, b_N$ at any time during an iteration. Vardi and Zhang (2000) improved the behavior of Weiszfeld's algorithm by introducing a simple modification to the iterative update in the event the algorithm takes on a value equal to a data value. This modification of Vardi and Zhang (2000) can easily be extended to the complex-valued case.

Our implementation wmed in the RobustSP toolbox first checks if the condition (2.26) holds and if it does not, then it runs the iterative algorithm (2.29) until convergence.

2.4.2   Simple Linear Regression with Intercept

If the intercept is included in the linear regression model, then the LAD criterion is

$$L_{\text{LAD}}(\beta_0, \beta) = \sum_{i=1}^{N} |y_i - \beta_0 - x_i \beta| \tag{2.31}$$

and a cyclic coordinatewise optimization algorithm can be used to compute the estimates of the slope and the intercept. A cyclic coordinatewise algorithm (Tseng, 2001) minimizes the objective function in one parameter while keeping the other fixed at its current iterate value. If β is fixed at its current iterate, $\beta = \beta^{(n-1)}$, then the next update

$\beta_0^{(n)}$ of the intercept β_0 is

$$\beta_0^{(n)} = \arg\min_{\beta_0 \in \mathbb{F}} \sum_{i=1}^{N} |(y_i - x_i\beta^{(n-1)}) - \beta_0| = \text{med}(y_i - x_i\beta^{(n-1)}). \tag{2.32}$$

Then, β_0 is held fixed at its current value $\beta_0^{(n)}$ and an update $\beta^{(n)}$ of the slope β is

$$\beta^{(n)} = \arg\min_{\beta \in \mathbb{F}} \sum_{i=1}^{N} |(y_i - \beta_0^{(n)}) - x_i\beta| = \text{med}\left(\frac{y_i - \beta_0^{(n)}}{x_i} \mid |x_i|\right). \tag{2.33}$$

The algorithm alternates between median and weighted median calculations of (2.32) and (2.33) until convergence. The algorithm always decreases the objective function (2.31) at each update but may not always converge to a global minimum as it may get trapped at nonoptimal points. To avoid this problem, Li and Arce (2004) proposed a refinement of the preceding cyclic approach for LAD regression in the real-valued case.

In the real-valued case, there also exists an explicit (noniterative) way to compute the LAD estimates of slope $\hat{\beta}_{\text{LAD}}$ and intercept $\hat{\beta}_{0,\text{LAD}}$. Consider a line that passes through two data points (x_i, y_i) and (x_j, y_j), $i \neq j$. The slope of this line is b_{ij} in (2.25) and the intercept is

$$b_{0,ij} = \frac{x_j y_i - x_i y_j}{x_j - x_i}.$$

The pairs $(b_{0,ij}, b_{ij})$ are called *elemental estimates* (Mayo and Gray, 1997) of regression, and there are $\binom{N}{2} = N(N-1)/2$ such estimates. The LAD estimate $(\hat{\beta}_{0,\text{LAD}}, \hat{\beta}_{\text{LAD}})$ is then one of the elemental estimates. This result is easy to show by noting that one can first solve for β_0 by considering β_1 fixed, which yields the conditional solution $\hat{\beta}_0(\beta_1)$ by using (2.32), and then solving for β_1 by using (2.33) given the conditional solution for β_0.

REMARK 1 *Elemental estimates have been used for constructing robust regression estimates for more than half a century. Their basic properties are easy to establish. When the error terms are i.i.d. with* $\text{var}(v_i) = \text{var}(y_i) = \sigma^2$, *the elemental estimates are unbiased, that is,* $\text{E}[b_{0,ij}] = \beta_0$ *and* $\text{E}[b_{ij}] = \beta$. *Their variances are*

$$\text{var}(b_{0,ij}) = \frac{\text{var}(x_j y_i) + \text{var}(x_i y_j) - 2\text{Re}\{\text{cov}(x_j y_i, x_i y_j)\}}{|x_j - x_i|^2}$$

$$= \frac{|x_j|^2 + |x_i|^2}{|x_j - x_i|^2} \sigma^2,$$

and

$$\text{var}(b_{ij}) = \frac{2}{|x_j - x_i|^2} \sigma^2.$$

Here,

$$\text{cov}(a, b) = \text{E}[(x - \text{E}[x])(y - \text{E}[y])^*] \tag{2.34}$$

denotes the covariance of two random variables x and y and we used the fact that
$\text{cov}(x_j y_i, x_i y_j) = x_j x_i^* \text{cov}(y_i, y_j) = 0$ *as* x_i *and* x_j *are nonrandom and* y_i *and* y_j *are independent. Among the first robust high-breakdown estimates of regression of intercept and slope are those by Theil (1950) who defined (in the real-valued case) regression estimates as median values of elemental fits, that is,*

$$\hat{\beta}_0 = \text{med}_{i<j}(b_{0,ij}) \quad and \quad \hat{\beta} = \text{med}_{i<j}(b_{ij}).$$

Because the elemental estimates are unbiased, it follows that Theil's estimates are unbiased and, due to their construction, they enjoy a high BP. Perhaps a less known result (Ollila et al., 2003b, theorem 4) is that the LS estimate can be expressed as weighted means of elemental estimates for the real case according to:

$$\hat{\beta}_{\text{LS},0} = \left[\sum_{i<j} w_{ij}^2\right]^{-1} \sum_{i<j} w_{ij}^2 b_{0,ij} \tag{2.35}$$

$$\hat{\beta}_{\text{LS}} = \left[\sum_{i<j} w_{ij}^2\right]^{-1} \sum_{i<j} w_{ij}^2 b_{ij}, \tag{2.36}$$

where $w_{ij} = |x_j - x_i|$.

2.4.3   Computation of Least Absolute Deviation and Rank-Least Absolute Deviation Estimates

In the previous section, we addressed the computation of LAD and rank-LAD estimates when only a single predictor ($p = 1$) is available. Next, we consider the general case of $p > 1$. A global solution to (2.15) can be found by linear programming such as detailed in Barrodale and Roberts (1973). The IRWLS algorithm (see also Section 2.5.3) can also be used as described in the following text. Note that $-2\times$ the complex conjugate gradient of $L_{\text{LAD}}(\boldsymbol{\beta}) = \|\mathbf{y} - \mathbf{X}\boldsymbol{\beta}\|_1$ is

$$\mathbf{s}(\boldsymbol{\beta}) = \mathbf{X}^H \text{sign}(\mathbf{y} - \mathbf{X}\boldsymbol{\beta}) = \sum_{i=1}^N \text{sign}(y_i - \mathbf{x}_{[i]}^\top \boldsymbol{\beta}) \mathbf{x}_{[i]}^* \tag{2.37}$$

which has discontinuities at all points where the criterion is not differentiable. It then follows that $\hat{\boldsymbol{\beta}}$ is a global minimizer of $L_{\text{LAD}}(\boldsymbol{\beta})$ if and only if $\mathbf{0} \in \partial L_{\text{LAD}}(\hat{\boldsymbol{\beta}})$. This means that $\hat{\boldsymbol{\beta}}$ is either a point where $\mathbf{s}(\hat{\boldsymbol{\beta}}) = \mathbf{0}$ or a point where the criterion is not differentiable, that is, a point that satisfies $r_i(\hat{\boldsymbol{\beta}}) = 0$ for at least one i. The gradient function (2.37) can also be expressed as

$$\mathbf{s}(\boldsymbol{\beta}) = \sum_{i=1}^N w_i(\boldsymbol{\beta})(y_i - \mathbf{x}_{[i]}^\top \boldsymbol{\beta}) \mathbf{x}_{[i]}^*,$$

where $w_i = w_i(\boldsymbol{\beta}) = |y_i - \mathbf{x}_{[i]}^\top \boldsymbol{\beta}|^{-1}$. It then follows that $\mathbf{s}(\hat{\boldsymbol{\beta}}) = \mathbf{0}$ when

$$\left(\sum_{i=1}^N w_i(\hat{\boldsymbol{\beta}}) \mathbf{x}_{[i]}^* \mathbf{x}_{[i]}^\top\right) \hat{\boldsymbol{\beta}} = \sum_{i=1}^N w_i(\hat{\boldsymbol{\beta}}) y_i \mathbf{x}_{[i]}^*,$$

which yields the *weighted normal equations*

$$(\mathbf{X}^H \mathbf{W} \mathbf{X}) \hat{\boldsymbol{\beta}} = \mathbf{X}^H \mathbf{W} \mathbf{y} \quad \text{or} \quad \hat{\boldsymbol{\beta}} = (\mathbf{X}^H \mathbf{W} \mathbf{X})^{-1} \mathbf{X}^H \mathbf{W} \mathbf{y}$$

where $\mathbf{W} = \operatorname{diag}(w_1(\hat{\boldsymbol{\beta}}), \ldots, w_N(\hat{\boldsymbol{\beta}}))$. This resembles a weighted LS solution, but here the weights are not fixed but are functions of the unknown $\hat{\boldsymbol{\beta}}$.

Given an initial start $\boldsymbol{\beta}^{(0)}$, the IRWLS algorithm proceeds by updating the weights $w_i^{(n)}$ using the current solution $\boldsymbol{\beta}^{(n)}$ and then by computing an update of the regression vector

$$\boldsymbol{\beta}^{(n+1)} = (\mathbf{X}^H \mathbf{W}^{(n)} \mathbf{X})^{-1} \mathbf{X}^H \mathbf{W}^{(n)} \mathbf{y}, \quad n = 0, 1, \ldots \quad (2.38)$$

It is relatively easy to show that one can again rewrite the update (2.38) as

$$\boldsymbol{\beta}^{(n+1)} = \boldsymbol{\beta}^{(n)} + (\mathbf{X}^H \mathbf{W}^{(n)} \mathbf{X})^{-1} \mathbf{s}(\boldsymbol{\beta}^{(n)}) \quad (2.39)$$

which is of the form $\boldsymbol{\beta}^{(n+1)} = \boldsymbol{\beta}^{(n)} + \mu \boldsymbol{\delta}^{(n)}$ and closely resembles the gradient descent update rule. In particular, we see that the iterations terminate when $\boldsymbol{\delta}^{(n)} \approx \mathbf{0}$, that is, when the gradient function vanishes, $\mathbf{s}(\boldsymbol{\beta}^{(n)}) \approx \mathbf{0}$.

The pseudo-code of the IRWLS algorithm using the update rule (2.39) (and assuming $p > 1$) is specified in Algorithm 3. The initial start $\boldsymbol{\beta}^{(0)}$ can be, for example, the LSE. It is important to monitor if any one of the residuals $r_i^{(n)}$ become zero or close to zero during iterations. This is obviously problematic because the weight $w_i^{(n)} = 1/|r_i^{(n)}|$ becomes ill-defined. For this case, one can *replace* zero (and close to zero) residuals with a small positive quantity: $|r_i^{(n)}| \leftarrow \Delta$ when $|r_i^{(n)}| \leq \Delta$. Note that in Step 3 of Algorithm 3, the residuals (and hence weights) that fall below a threshold ($\Delta = 10^{-6}$) in absolute value are regularized in this manner. The function ladreg in the RobustSP toolbox uses the pseudo-code of Algorithm 3 when $p > 1$, and the weighted median approach in the single predictor case ($p = 1$) explained in previous subsections.

Let us now discuss how the rank-LAD estimate can be computed. Define

$$\tilde{\mathbf{y}} = \begin{pmatrix} y_1 - y_2 \\ y_2 - y_3 \\ \vdots \\ y_{N-1} - y_N \end{pmatrix} \quad \text{and} \quad \tilde{\mathbf{X}} = \begin{pmatrix} (\mathbf{x}_{[1]} - \mathbf{x}_{[2]})^\top \\ (\mathbf{x}_{[2]} - \mathbf{x}_{[3]})^\top \\ \vdots \\ (\mathbf{x}_{[N-1]} - \mathbf{x}_{[N]})^\top \end{pmatrix}, \quad (2.40)$$

where $\tilde{\mathbf{y}}$ (respectively $\tilde{\mathbf{X}}$) contains all pairwise differences $y_i - y_j$, $i < j$, $i, j \in \{1, \ldots, N\}$ of the responses (respectively predictors $\mathbf{x}_{[i]} - \mathbf{x}_{[j]}$, $i < j$). Then, we may write the rank-LAD criterion, as

$$L_R(\boldsymbol{\beta}) = \sum_{i<j} \left| (y_i - y_j) - (\mathbf{x}_{[i]} - \mathbf{x}_{[j]})^\top \boldsymbol{\beta} \right| = \| \tilde{\mathbf{y}} - \tilde{\mathbf{X}} \boldsymbol{\beta} \|_1,$$

which is simply the LAD criterion for regressing $\tilde{\mathbf{X}}$ on $\tilde{\mathbf{y}}$. This means that any algorithm for LAD regression, for example, Algorithm 3, can be used to compute the rank-LAD

Algorithm 3: ladreg: computes the LAD estimate of regression using the IRWLS algorithm.

---

input : $\mathbf{y} \in \mathbb{F}^N$, $\mathbf{X} \in \mathbb{F}^{N \times (p+1)}$, initial start $\boldsymbol{\beta}^{(0)} \in \mathbb{F}^{p+1}$
output : $\hat{\boldsymbol{\beta}}_{\text{LAD}} \in \mathbb{F}^{p+1}$, minimizer of the LAD criterion $L_{\text{LAD}}(\boldsymbol{\beta})$
initialize: $N_{\text{iter}} \in \mathbb{N}$ (max. number of iterations), tolerance level $\delta > 0$

1 **for** $n = 0, 1, \ldots, N_{\text{iter}}$ **do**
2   update residual vector
$$\mathbf{r}^{(n)} \leftarrow \mathbf{y} - \mathbf{X}\boldsymbol{\beta}^{(n)}$$

3   update weights
$$w_i^{(n)} \leftarrow \begin{cases} 1/|r_i^{(n)}|, & \text{if } |r_i^{(n)}| > 10^{-6} \\ 1/10^{-6}, & \text{if } |r_i^{(n)}| \leq 10^{-6} \end{cases}$$

     for $i = 1, \ldots, N$ and set $\mathbf{W}^{(n)} = \text{diag}(w_1^{(n)}, \ldots, w_N^{(n)})$.
4   update regression coefficients vector
$$\boldsymbol{\beta}^{(n+1)} \leftarrow \boldsymbol{\beta}^{(n)} + \boldsymbol{\delta}^{(n)},$$

     where $\boldsymbol{\delta}^{(n)}$ solves
$$(\mathbf{X}^H \mathbf{W}^{(n)} \mathbf{X})\boldsymbol{\delta} = \mathbf{X}^H \text{sign}(\mathbf{r}^{(n)}).$$

5   **if** $\|\boldsymbol{\delta}^{(n)}\|_2 / \|\boldsymbol{\beta}^{(n)}\|_2 < \delta$ **then**
6   **return** $\hat{\boldsymbol{\beta}}_{\text{LAD}} \leftarrow \boldsymbol{\beta}^{(n+1)}$

---

estimate, but now using $\tilde{\mathbf{y}}$ and $\tilde{\mathbf{X}}$ as the input data. The intercept cannot be estimated using the rank-LAD criterion, but a natural estimate is

$$\hat{\beta}_{0,\text{R}} = \arg\min_{\beta_0 \in \mathbb{F}} \sum_{i<j} \left| \frac{\hat{r}_i + \hat{r}_j}{2} - \beta_0 \right|, \tag{2.41}$$

where $\hat{r}_i = y_i - \mathbf{x}_{[i]}^{\top} \hat{\boldsymbol{\beta}}_{\text{R}}$ are the obtained residuals of the rank-LAD fit and $\hat{\boldsymbol{\beta}}_{\text{R}}$ is the rank-LAD regression estimate. In the real-valued case ($\mathbb{F} = \mathbb{R}$), the preceding intercept estimate $\hat{\beta}_{0,\text{R}}$ is called the Hodges–Lehmann (HL) median (Lehmann and D'Abrera, 1975). In the complex-valued case, the solution to (2.41) is best described as the spatial HL median of the residuals. In the RobustSP toolbox, the function rladreg computes the rank-LAD regression estimate $\hat{\boldsymbol{\beta}}_{\text{R}}$.

2.5 ML- and *M*-estimates of Regression with an Auxiliary Scale Estimate

This section discusses ML- and *M*-estimation of regression with an auxiliary scale estimate, both for the real-valued case and for the complex-valued case. The location model (or the location-scale model) was discussed in Chapter 1 only for the real-valued case so as to provide an accessible introduction to robustness concepts. However, the regression model (2.8) contains, as a special case, the location model (or the location-scale model) when the design matrix is $\mathbf{X} = \mathbf{1}$. In fact, $\mathbf{X} = \mathbf{1}$ yields

$$y_i = \mu + v_i, \quad i = 1, \dots, N,$$

with $\mu = \beta_0$. We now continue with the general regression case.

2.5.1 Objective Function Approach vs. Estimating Equation Approach

In the real-valued case, the error (noise) terms v_i are assumed to be i.i.d. continuous random variables from a symmetric distribution. A distribution is symmetric (around zero) if $x \stackrel{\mathrm{d}}{=} -x$, where $\stackrel{\mathrm{d}}{=}$ is notation for "has the same distribution as." In the complex-valued case, a natural analogy is to assume that the errors v_i follow a circular distribution ($e^{J\theta}x \stackrel{\mathrm{d}}{=} x \ \forall \theta \in \mathbb{R}$, see also Section 4.1 for more discussion). In both cases, the pdf of the error terms can be expressed in the form

$$f(v) = C \cdot \frac{1}{\sigma^{2/\gamma}} g\left(\frac{|v|^2}{\sigma^2}\right), \quad v \in \mathbb{F}(=\mathbb{R} \text{ or } \mathbb{C}), \tag{2.42}$$

where $\sigma > 0$ is the *scale parameter*; the function $g : \mathbb{R}_0^+ \to \mathbb{R}^+$, called the density generator, satisfies $\int_0^\infty g(t)\mathrm{d}t < \infty$; C is a normalizing constant (not dependent on σ) that guarantees that $f(v)$ integrates to 1; and γ is a constant, defined as

$$\gamma = \begin{cases} 1, & \text{complex-valued case, } \mathbb{F} = \mathbb{C} \\ 2, & \text{real-valued case, } \mathbb{F} = \mathbb{R}. \end{cases} \tag{2.43}$$

In the definition of g, the following notation is used: $\mathbb{R}^+ = \{x \in \mathbb{R} | x > 0\}$ and $\mathbb{R}_0^+ = \{x \in \mathbb{R} | x \geq 0\}$. The pdf of $y_i = \mathbf{x}_{[i]}^\top \boldsymbol{\beta} + v_i$ is then

$$f_i(y) = C \cdot \frac{1}{\sigma^{2/\gamma}} g\left(\frac{|y_i - \mathbf{x}_{[i]}^\top \boldsymbol{\beta}|^2}{\sigma^2}\right), \quad i = 1, \dots, N.$$

For example, if we assume (real- or complex-valued) Gaussian errors, $v_i \sim \mathcal{N}(0, \sigma^2)$, then the pdf of v_i is of the form (2.42) with the density generator being $g(t) = \exp(-t/\gamma)$.

Suppose for a moment that the scale parameter σ is known. In this case, the negative log-likelihood function (ignoring the scaling $C \cdot 1/\sigma^{2/\gamma}$ of the pdf that is not dependent on $\boldsymbol{\beta}$) is

$$L_{\mathrm{ML}}(\boldsymbol{\beta}) = \sum_{i=1}^N \rho_{\mathrm{ML}}\left(\frac{y_i - \mathbf{x}_{[i]}^\top \boldsymbol{\beta}}{\sigma}\right) \tag{2.44}$$

where the ML loss function $\rho_{\mathrm{ML}} : \mathbb{F} \to \mathbb{R}^+$ is defined as

$$\rho_{\mathrm{ML}}(x) = -\ln g(|x|^2), \quad x \in \mathbb{F}. \tag{2.45}$$

The MLE $\hat{\boldsymbol{\beta}}_{\mathrm{ML}}$ of $\boldsymbol{\beta} \in \mathbb{F}^{p+1}$ is defined as the minimizer of $L_{\mathrm{ML}}(\boldsymbol{\beta})$. When the *ML loss function* is ($\mathbb{R}$-)differentiable, the MLE is a critical point of $L_{\mathrm{ML}}(\boldsymbol{\beta})$ and, hence, a solution to

$$\sum_{i=1}^{N} \psi_{\mathrm{ML}}\left(\frac{y_i - \mathbf{x}_{[i]}^{\top}\hat{\boldsymbol{\beta}}}{\sigma}\right)\mathbf{x}_{[i]}^* = \mathbf{0}, \qquad (2.46)$$

where

$$\psi_{\mathrm{ML}}(x) = \frac{\partial}{\partial x^*}\rho_{\mathrm{ML}}(x) \qquad (2.47)$$

is the complex conjugate derivative in the complex case ($\mathbb{F} = \mathbb{C}$) and the conventional derivative, $\psi_{\mathrm{ML}}(x) = \rho_{\mathrm{ML}}'(x)$, in the real case ($\mathbb{F} = \mathbb{R}$). We refer to $\psi_{\mathrm{ML}}(x)$ as the *ML score function*.

The nature of *M*-estimation is such as to disentangle the estimators obtained from minimizing (2.44) from the distributions that generated the negative log-likelihood function. When using the LSE, for example, one does not need to assume that it is based on an independent sample from a Gaussian distribution. *M*-estimates of regression are defined as generalized ML estimates that allow more general loss functions or score functions that are not necessarily related to any (symmetric or circularly symmetric) density and, hence, are not necessarily ML estimators of such a class. Robust loss functions (cf. [Huber and Ronchetti, 2009] or [Maronna et al., 2006]) are defined in the real-valued case and some care must be given with respect to the properties of complex-valued loss functions. We start here by defining a loss function ρ. The following discussion is consistent with Ollila (2015b).

DEFINITION 7   *A function $\rho : \mathbb{F} \to \mathbb{R}_0^+$ for $\mathbb{F} = \mathbb{C}$ (or $\mathbb{F} = \mathbb{R}$) is called a loss function if it satisfies:*

(L1) ρ *is circularly symmetric, $\rho(e^{J\theta}x) = \rho(x)$, $\forall\theta \in \mathbb{R}$ (or symmetric, $\rho(x) = \rho(-x)$).*

(L2) ρ *is continuous, increasing for $|x| > 0$ and with $\rho(0) = 0$.*

Definition 7 permits more general loss functions than ML loss functions. When the loss function is differentiable, the derivative of ρ, $\psi(x) = \frac{\partial}{\partial x^*}\rho(x)$ in the complex-valued case (respectively, $\psi(x) = \rho'(x)$, in the real-valued case) is referred to as a *score function*. Due to (L1), a score function has the property

$$\psi(e^{J\theta}x) = e^{J\theta}\psi(x), \; \forall\theta \in \mathbb{R} \quad \text{and} \quad \psi(-x) = -\psi(x) \qquad (2.48)$$

in the complex-valued and real-valued cases, respectively. If a loss function ρ is convex/strictly convex, then the corresponding score function also satisfies the condition

$$\psi(x) \text{ is monotonically nondecreasing/strictly increasing.} \qquad (2.49)$$

REMARK 2   *The condition (L1) is equivalent to the statement*

$$\rho(x) = \rho_0(|x|^2), \quad \text{for some } \rho_0 : \mathbb{R}_0^+ \to \mathbb{R}_0^+ \qquad (2.50)$$

for $x \in \mathbb{F}$. Thus, we can write a score function ψ in the form

$$\psi(x) = \gamma \cdot \rho_0'(|x|^2)x, \tag{2.51}$$

where ρ_0' denotes the (real) derivative of the real-valued function $\rho_0(t)$ and γ is as in (2.43).

Consistent with Definition 7 for a loss function $\rho : \mathbb{F} \to \mathbb{R}_0^+$, an objective function approach for *M*-estimation of regression leads to the *M*-estimate of regression as being a solution of

$$\underset{\boldsymbol{\beta} \in \mathbb{F}^{p+1}}{\text{minimize}} \left\{ L_\rho(\boldsymbol{\beta}) = \sum_{i=1}^{N} \rho\left(\frac{y_i - \mathbf{x}_{[i]}^\top \boldsymbol{\beta}}{\sigma}\right) \right\}. \tag{2.52}$$

Such solutions coincide with minimizing (2.44) when ρ is the ML loss function as then $L_\rho(\boldsymbol{\beta}) = L_{\text{ML}}(\boldsymbol{\beta})$. If the loss function is not convex, finding the global minimizer of (2.52) becomes challenging. An estimating equation approach for *M*-estimation leads to the *M*-estimate of regression as being a solution to the estimating equation

$$\sum_{i=1}^{N} \psi\left(\frac{y_i - \mathbf{x}_{[i]}^\top \hat{\boldsymbol{\beta}}}{\sigma}\right)\mathbf{x}_{[i]}^* = \mathbf{0}, \tag{2.53}$$

where $\psi : \mathbb{F} \to \mathbb{R}_0^+$ is a score function. If ρ is differentiable and convex, then (2.52) and (2.53) are equivalent characterizations for the problem. The latter form is more commonly used when the loss function ρ is nonconvex and, hence, ψ is not monotonically nondecreasing/strictly increasing, but redescending. For example, Tukey's loss function is nonconvex and the estimating equation (2.53) can have multiple roots or equivalently stated, the corresponding objective function in (2.52) has multiple local minima. Algorithms, such as the IRWLS described in the following text, only seek a local solution, that is, an estimate satisfying (2.53). Thus, the computation of an *M*-estimate based on a redescending ψ-function requires a robust initial start $\hat{\boldsymbol{\beta}}^{(0)}$ for the algorithm not to converge to a spurious (inaccurate) local solution. The initial start, however, is usually found by computing an *M*-estimate with a convex ρ function that has a corresponding monotone ψ function.

2.5.2   Examples of Loss Functions

In this section, we provide examples for loss functions that will be used extensively in the book. We refer the reader to Chapter 1, where we introduced *M*-estimators.

Gaussian, that is, LS (ℓ_2-) loss function: If $v_i, i = 1, \ldots, N$, follows the real (or complex) Gaussian distribution, $v_i \sim \mathcal{N}(0, \sigma^2)$, then its pdf is of the form (2.42) with a density generator $g(t) = \exp(-(1/\gamma)t)$ and, hence, the Gaussian (ML) loss function and the associated score function are

$$\rho_{\text{G}}(x) = (1/\gamma) \cdot |x|^2, \quad \psi_{\text{G}}(x) = x, \quad x \in \mathbb{F}.$$

The Gaussian loss function is strictly convex and the unique solution to (2.52) is the LSE.

Laplace, that is, LAD (ℓ_1-) loss function: Consider the case in which the errors v_i $i = 1, \ldots, N$, follow the generalized Gaussian distribution with exponent $s = 1/2$. In the real-valued case, this distribution is also referred to as the Laplace or the double exponential distribution. The pdf of v_i is of the form (2.42) with density generator $g(t) = \exp(-\sqrt{t})$ and, hence, the Laplace (ML) loss is

$$\rho_{\ell_1}(x) = |x|, \quad x \in \mathbb{F},$$

which is the LAD loss function and, hence, the obtained MLE is the LAD estimator. The loss function is also convex but not differentiable at $x = 0$.

Huber's loss function: In his seminal work and for the ε-contaminated Gaussian model (1.42), Huber (1964) derived a family of univariate heavy-tailed distributions, which he called the least favorable distributions (LFDs). For the ε-contaminated Gaussian model, the LFD corresponds to a symmetric unimodal distribution that follows a Gaussian distribution in the middle and a double exponential distribution in the tails. The corresponding MLE is referred to as Huber's M-estimator. Huber's loss function is thus a hybrid of the ℓ_2- and the ℓ_1-loss functions defined previously, using the ℓ_2-loss function for relatively small errors and ℓ_1-loss function for relatively large errors. The pdf of the LFD has the form $f(x) = C \cdot (1/\sigma) \cdot \exp\{-\rho_{\text{H},c}(x)\}$, where $\rho_{\text{H},c}(x)$ is Huber's loss function defined as

$$\rho_{\text{H},c}(x) = \frac{1}{\gamma} \times \begin{cases} |x|^2, & \text{for } |x| \leq c \\ 2c|x| - c^2, & \text{for } |x| > c, \end{cases} \quad x \in \mathbb{F}, \tag{2.54}$$

where c is a user-defined *threshold* that influences the degree of robustness and γ is defined in (2.43). Note that Huber's loss function was also defined (in the real case) in (1.20). Huber's function is convex and differentiable. Huber's score function is of the form

$$\psi_{\text{H},c}(x) = \begin{cases} x, & \text{for } |x| \leq c \\ c \, \text{sign}(x), & \text{for } |x| > c \end{cases}, \quad x \in \mathbb{F}.$$

The threshold c is usually chosen so that the minimizer of (2.52) attains a user-defined ARE w.r.t. the LSE under Gaussian errors (see Chapter 1). To obtain 95 percent (or 85 percent) ARE for the Gaussian noise case, the thresholds are chosen according to $c_{.95} = 1.215$ and $c_{.85} = 0.515$ for the complex-valued case and $c_{.95} = 1.345$ and $c_{.85} = 0.7317$ for the real-valued case.

Tukey's (biweight) loss function: Along with Huber's loss function, one of the most commonly used robust loss functions is the one proposed by Tukey (1977) for the real-valued case. A complex analog can be constructed (Ollila, 2016a). Tukey's loss function is defined as

$$\rho_{\text{T},c}(x) = \frac{1}{\gamma \cdot 3} \min \left\{ 1, 1 - \left(1 - \frac{|x|^2}{c^2} \right)^3 \right\}, \quad x \in \mathbb{F}, \tag{2.55}$$

where γ is as in (2.43) and c is a user-defined *threshold* that influences the degree of robustness of the estimator. Tukey's loss function is also called the biweight loss function, and note that (2.55) coincides with the definition (1.22) for the real-valued case ($\mathbb{F} = \mathbb{R}$). Tukey's loss function is a smooth (possessing continuous second derivatives)

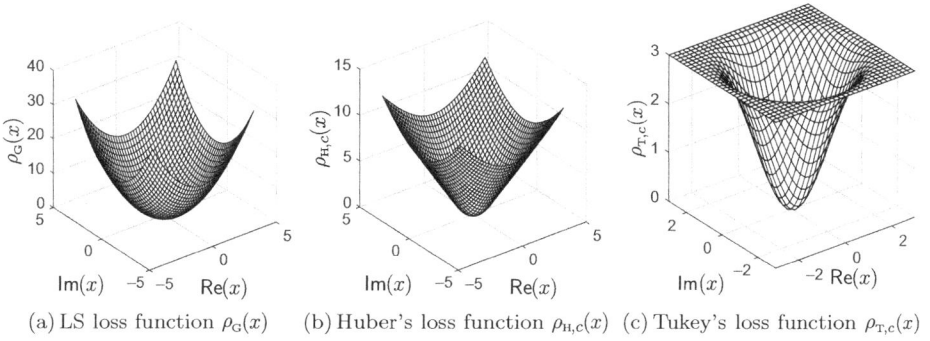

(a) LS loss function $\rho_G(x)$ (b) Huber's loss function $\rho_{H,c}(x)$ (c) Tukey's loss function $\rho_{T,c}(x)$

Figure 2.6 Loss functions for $x \in \mathbb{C}$. LS and Huber's loss functions are convex, whereas Tukey's loss function is bounded and nonconvex.

nonconvex function and bounded, which makes it very robust to large outliers. The respective score function is

$$\psi_{T,c}(x) = \begin{cases} x\left(1 - (|x|/c)^2\right)^2 & \text{for } |x| \leq c \\ 0, & \text{for } |x| > c \end{cases}, \quad x \in \mathbb{F}. \tag{2.56}$$

Thus, large residuals $r_i = y_i - \mathbf{x}_{[i]}^\top \boldsymbol{\beta}$, $i = 1, \ldots, N$, are completely rejected, that is, they are assigned a zero weight by $\psi_{T,c}(r_i)$. Because the loss function is bounded, the respective score function is a *strongly redescending function*, that is, increasing in an interval near zero but leveling off to zero as $|x|$ increases (and reaching 0 at $|x| = c$) as shown in Figure 1.4. Again, the threshold c is usually chosen such that the minimizer of (2.52) attains a user-defined ARE w.r.t. the LSE under Gaussian errors. To obtain 85 percent ARE, the thresholds are chosen according to $c_{.85} = 3.0$ and $c_{.85} = 3.4437$ for the complex-valued and real-valued case, respectively. Finally, note that Tukey's *M*-estimator is not an MLE for any continuous symmetric (or circularly symmetric) error distribution.

The LS, Huber's, and Tukey's loss functions are depicted in Figure 2.6 for the complex-valued case. Note that the LS and Huber's loss functions are convex, whereas Tukey's loss function is bounded and nonconvex. Moreover, note that Huber's loss function is quadratic near zero but linearly increasing as $|x| > c$.

2.5.3   Computation Using the Iteratively Reweighted Least Squares Algorithm

So far we have treated σ as a known parameter. In practice of course, the scale parameter σ is unknown and an *auxiliary estimate*, $\hat{\sigma}$, is needed. LS loss and LAD loss functions do not require an auxiliary scale estimate as the scale parameter factorizes out from the objective function in (2.52). Huber's and Tukey's loss functions obviously require an estimate of scale.

Thus, the first step is to compute an auxiliary scale estimate $\hat{\sigma}$, which in turn requires the computation of an initial robust regression estimate $\hat{\boldsymbol{\beta}}_{\text{init}}$. This estimate allows $\hat{\sigma}$ to be computed as a robust scale statistic of the obtained residuals $\hat{r}_i = y_i - \mathbf{x}_{[i]}^\top \hat{\boldsymbol{\beta}}_{\text{init}}$. The standard choice is to use the *median absolute deviation (MAD)* of the residuals,

$$\hat{\sigma} = b \cdot \text{med}\{|\hat{r}_i| \mid \hat{r}_i \neq 0\}_{i=1}^{N} \tag{2.57}$$

where the scaling constant b is used to obtain a Fisher-consistent scale estimate for the case of Gaussian noise. The constant b is $b = 1.20112$ for the complex-valued case and $b = 1.4826$ for the real-valued case (see (1.25)). The initial robust regression estimate $\hat{\beta}_{\text{init}}$ can further be used as an initial start $\beta^{(0)}$ in the IRWLS algorithm explained in the following text.

A common choice (cf. Maronna et al., 2006) for $\hat{\beta}_{\text{init}}$ is the LAD estimate $\hat{\beta}_{\text{LAD}}$. Because the LAD estimate produces at least p residuals that are zero, they are ignored in (2.57). For nonconvex loss functions, for example, Tukey's loss, the IRWLS algorithm is guaranteed to converge only to a local minimum, that is, it finds a solution to (2.53). To converge to a good (robust) local minimum, a robust initial start is necessary.

The IRWLS method is an old tool for computing M-estimates of regression. The algorithm can be used for both the real- and complex-valued cases. It is based on viewing an M-estimator as an adaptively (re-)weighted LSE. This originates by expressing the score function as $\psi(x) = x \cdot W(x)$, where

$$W(x) = \begin{cases} \psi(x)/x, & \text{if } x \neq 0 \\ 1, & \text{if } x = 0. \end{cases} \tag{2.58}$$

Note that the case of $x = 0$ needs to be handled separately as $\psi(x)/x$ is not defined when $x = 0$. The choice of $W(0) = 1$ is used here because the commonly used robust loss functions, such as Huber's or Tukey's, are either linear or approximately linear near the origin. Note that the same principle was used in (1.24). With the definition

$$w_i \equiv w_i(\hat{\beta}) = W\left(\frac{r_i(\hat{\beta})}{\hat{\sigma}}\right) \quad \text{and} \quad \mathbf{W} = \text{diag}(w_1, \dots, w_N),$$

The M-estimating equation (2.53) can be rewritten as

$$\left(\sum_{i=1}^{N} w_i \mathbf{x}_{[i]}^{*} \mathbf{x}_{[i]}^{\top}\right) \hat{\beta} = \sum_{i=1}^{N} w_i \mathbf{x}_{[i]}^{*} y_i \tag{2.59}$$

or in matrix form as

$$\hat{\beta} = (\mathbf{X}^{H} \mathbf{W} \mathbf{X})^{-1} \mathbf{X}^{H} \mathbf{W} \mathbf{y}. \tag{2.60}$$

It is important to note the difference between (2.60) and a weighted LSE because the weights w_i in (2.60) are not fixed, but depend on the unknown solution $\hat{\beta}$ through the residuals $r_i(\hat{\beta})$. The IRWLS algorithm, detailed in Algorithm 4, solves a weighted LS problem repeatedly with the weights adaptively updated at each iteration.

Note that the location (or location-scale model) is a special case of the linear regression model in (2.8) when the design matrix is $\mathbf{X} = \mathbf{1}$. Therefore Mreg, for this choice of $\mathbf{X}$, can be used to compute a robust M-estimate of location both for the real-valued case and for the complex-valued case. Function Mreg in the RobustSP toolbox uses the pseudo-code given in Algorithm 4 to compute the M-estimate of regression with an auxiliary scale using either Tukey's or Huber's loss function.

Algorithm 4: Mreg computes an *M*-estimate of regression based on an auxiliary scale.

---

input : $\mathbf{y} \in \mathbb{F}^N$, $\mathbf{X} \in \mathbb{F}^{N \times (p+1)}$, $\hat{\boldsymbol{\beta}}_{\text{init}} \in \mathbb{F}^{p+1}$
output : $\hat{\boldsymbol{\beta}} \in \mathbb{F}^{p+1}$ that solves (2.53) based on an auxiliary scale estimate $\sigma = \hat{\sigma}$
initialize: $N_{\text{iter}} \in \mathbb{N}$, $\delta > 0$, $\boldsymbol{\beta}^{(0)} = \hat{\boldsymbol{\beta}}_{\text{init}}$, MAD scale $\hat{\sigma}$ in (2.57) based on
$$\mathbf{r}^{(0)} = \mathbf{y} - \mathbf{X}\boldsymbol{\beta}^{(0)}$$

1 **for** $n = 0, 1, \ldots, N_{\text{iter}}$ **do**
2   update weights
$$w_i^{(n)} \leftarrow W\big(r_i^{(n)}/\hat{\sigma}\big), \quad i = 1, \ldots, N$$

3   Compute $\hat{\boldsymbol{\beta}}$ that solves weighted normal equations,
$$(\mathbf{X}^H \mathbf{W}^{(n)} \mathbf{X})\hat{\boldsymbol{\beta}} = \mathbf{X}^H \mathbf{W}^{(n)} \mathbf{y},$$
    where $\mathbf{W}^{(n)} = \text{diag}(w_1^{(n)}, \ldots, w_N^{(n)})$.

4   update regression coefficient
$$\boldsymbol{\beta}^{(n+1)} \leftarrow \hat{\boldsymbol{\beta}}$$

5   **if** $\|\boldsymbol{\beta}^{(n+1)} - \boldsymbol{\beta}^{(n)}\|_2 / \|\boldsymbol{\beta}^{(n)}\|_2 < \delta$ **then**
6   **return** $\hat{\boldsymbol{\beta}} \leftarrow \boldsymbol{\beta}^{(n+1)}$
7   update residual vector
$$\mathbf{r}^{(n+1)} \leftarrow \mathbf{y} - \mathbf{X}\boldsymbol{\beta}^{(n+1)}$$

2.6 Joint *M*-estimation of Regression and Scale Using Huber's Criterion

M-estimation with an auxiliary scale is somewhat cumbersome as considerable computational effort is needed to first compute the robust initial estimate of regression and the related auxiliary scale statistics. It is, however, possible to perform *M*-estimation by jointly estimating the unknown regression vector $\boldsymbol{\beta}$ and the scale σ. This can be achieved elegantly using Huber's criterion (Huber and Ronchetti, 2009, section 7.7).

First, note that the ML approach for jointly solving the unknown parameters $\boldsymbol{\beta}$ and σ leads to minimizing the negative log-likelihood function (ignoring the scaling constant C in the density):

$$
\begin{aligned}
L_{\text{ML}}(\boldsymbol{\beta}, \sigma) &= -\sum_{i=1}^{N} \ln \left\{ \frac{1}{\sigma^{2/\gamma}} g\left(\frac{|y_i - \mathbf{x}_{[i]}^\top \boldsymbol{\beta}|^2}{\sigma^2} \right) \right\} \\
&= \frac{2N}{\gamma} \ln \sigma + \sum_{i=1}^{N} \rho_{\text{ML}} \left(\frac{y_i - \mathbf{x}_{[i]}^\top \boldsymbol{\beta}}{\sigma} \right),
\end{aligned}
\tag{2.61}
$$

where $\rho_{\text{ML}}(x)$ is the ML loss function given in (2.45). It is then possible to replace the ML loss with a robust loss function, for example, Huber's loss function, to define a more general joint *M*-estimate of regression and scale. The negative log-likelihood function, however, is not convex with respect to $(\boldsymbol{\beta},\sigma)$. This is easy to see by simply noting that $L_{\mathrm{ML}}(\boldsymbol{\beta},\sigma)$ is not convex in σ for a fixed $\boldsymbol{\beta}$. Another problem is that even with a robust ρ function, such as Huber's loss function, the associated scale estimate will not be robust, for example, possessing a bounded IF. This problem was noted in the previous chapter where it was pointed out that the associated scale estimate solves an M-estimating equation (1.19) that is not robust due to the term $\psi(x)x$, which does not remain bounded.

Huber and Ronchetti (2009) proposed an elegant method to circumvent this problem, and this method can be generalized to the complex-valued regression case as detailed in Ollila (2015b). See also (Owen, 2007) and (Ollila, 2015b,a) for further information on Huber's approach for sparse estimation for the linear model case. The goal, then, is to minimize *Huber's criterion*,

$$\underset{\boldsymbol{\beta}\in\mathbb{F}^{p+1},\sigma>0}{\text{minimize}}\left\{L_{\mathrm{HUB}}(\boldsymbol{\beta},\sigma)=\frac{2N}{\gamma}(\alpha\sigma)+\sum_{i=1}^{N}\rho\left(\frac{y_i-\mathbf{x}_{[i]}^{\top}\boldsymbol{\beta}}{\sigma}\right)\sigma\right\},\quad(2.62)$$

where ρ is assumed to be a convex and differentiable loss function and $\alpha>0$ is a fixed *scaling factor* that is used to obtain Fisher-consistency of $\hat\sigma$ when the errors are i.i.d. Gaussian ($v_i\sim\mathcal{N}(0,\sigma^2)$). We discuss later how α is computed. The minimizer $(\hat{\boldsymbol{\beta}},\hat\sigma)$ of L_{HUB} is referred to as *HUB-Reg* estimates (of regression and scale) based on a loss function ρ.

An important feature of Huber's criterion function (2.62) is that it is jointly convex in $(\boldsymbol{\beta},\sigma)$ given that the loss function $\rho(\cdot)$ is convex. In addition, the minimizer $\hat{\boldsymbol{\beta}}$ preserves the same theoretical robustness properties (such as bounded IF) as the M-estimator based on an auxiliary scale $\hat\sigma$.

An important concept is that of the so-called *pseudo-residual*, which is associated with the score function $\psi(t)=\frac{\partial}{\partial x^*}\rho(x)$ (or $\psi(x)=\rho'(x)$ in the real-valued case), and is defined as

$$\mathbf{r}_\psi\equiv\mathbf{r}_\psi(\boldsymbol{\beta},\sigma)=\psi\left(\frac{\mathbf{y}-\mathbf{X}\boldsymbol{\beta}}{\sigma}\right)\sigma,\quad(2.63)$$

where the ψ function in (2.63) acts coordinatewise on the vector $\mathbf{r}/\sigma$, so $[\psi(\mathbf{r}/\sigma)]_i=\psi(r_i/\sigma)$. Note that if $\rho(\cdot)$ is the conventional LS loss, then $\psi(x)=x$, and $\mathbf{r}_\psi$ coincides with the conventional residual vector, so $\mathbf{r}_\psi=\mathbf{y}-\mathbf{X}\boldsymbol{\beta}=\mathbf{r}$. The multiplier σ in (2.63) is needed to map the residuals back to the original scale of the data.

Because the optimization problem is convex, the global minimizer $(\hat{\boldsymbol{\beta}},\hat\sigma)$ is a stationary point of (2.62). Thus, $(\hat{\boldsymbol{\beta}},\hat\sigma)$ can be found by solving the M-estimating equations, obtained by setting the gradient of $L_{\mathrm{HUB}}(\boldsymbol{\beta},\sigma)$, w.r.t. its arguments, to zero, which yields the following estimating equations:

$$\langle\mathbf{x}_j,\hat{\mathbf{r}}_\psi\rangle=0\quad\text{for } j=0,1,\ldots,p\quad(2.64)$$

$$\frac{1}{N(2\alpha)}\sum_{i=1}^{N}\chi\left(\frac{y_i-\mathbf{x}_{[i]}^{\top}\hat{\boldsymbol{\beta}}}{\hat\sigma}\right)=\frac{1}{\gamma}\quad(2.65)$$

where $\chi : \mathbb{F} \to \mathbb{R}_0^+$ is defined as

$$\chi(x) = 2\rho_0'(|x|^2)|x|^2 - \rho_0(|x|^2). \tag{2.66}$$

Here we have used $\rho(x) = \rho_0(|x|^2)$ and (2.51).

Example 7 Consider solving equations (2.64) and (2.65) for the case in which the LS loss $\rho(x) = \rho_0(|x|^2) = \frac{1}{\gamma}|x|^2$. Because $\psi(x) = x$ and $\hat{\mathbf{r}}_\psi = \hat{\mathbf{r}}$, it follows that (2.64) reduces to the normal equations given in (2.13). Hence, the minimizer $\hat{\boldsymbol{\beta}}$ of Huber's criterion is the LSE $\hat{\boldsymbol{\beta}}_{\mathrm{LS}}$. Furthermore, because $\rho_0(t) = \frac{1}{\gamma}t$ and $\rho_0'(t) = \frac{1}{\gamma}$, the χ function in (2.66) is $\chi(x) = \frac{1}{\gamma}|x|^2$, so (2.65) becomes

$$\hat{\sigma}^2 = \frac{1}{N(2\alpha)} \sum_{i=1}^{N} |y_i - \mathbf{x}_{[i]}^\top \hat{\boldsymbol{\beta}}|^2. \tag{2.67}$$

We may now choose the consistency factor α as $\alpha = \frac{1}{2}$ so that the preceding solution $\hat{\sigma}$ coincides with the classical ML estimate $\hat{\sigma}_{\mathrm{SD}}$ in (2.11). Thus, interestingly, the classical estimates $(\hat{\boldsymbol{\beta}}_{\mathrm{LS}}, \hat{\sigma}_{\mathrm{SD}})$ are the global (joint) minimizers of Huber's criterion (2.62) when the LS-loss is used as the loss function. Therefore, Huber's criterion can be seen as a novel device to convexify the nonconvex ML criterion function $L_{\mathrm{ML}}(\boldsymbol{\beta}, \sigma)$ as both problems have the same optimum when $\rho(\cdot)$ is the LS loss function. Naturally, for other ML loss functions, Huber's criterion does not generally yield the corresponding ML estimates.

So far, we have not discussed how to determine the scaling factor α. The scaling factor α is introduced in Huber's criterion solely for the purpose to ensure that the obtained scale estimate $\hat{\sigma}$ is Fisher-consistent for the unknown scale σ when the error-terms $v_i, i = 1, \ldots, N$ are i.i.d. and from a (real or complex) Gaussian distribution $\mathcal{N}(0, \sigma^2)$. Due to (2.65), this goal is achieved if we choose α as the value

$$\alpha = \frac{\gamma}{2} \mathsf{E}[\chi(v)], \quad v \sim \mathcal{N}(0, 1). \tag{2.68}$$

Hence, when using the LS loss, $\chi(v) = \frac{1}{\gamma}|v|^2$ (cf. Example 7), Fisher consistency is obtained if $\alpha = \frac{1}{2}\mathsf{E}[|v|^2] = \frac{1}{2}$, where we have used $\mathsf{E}[|v|^2] = 1$ for a (real- or complex-valued) standard Gaussian random variable $v \sim \mathcal{N}(0, 1)$.

Example 8 Consider finding estimates of Huber's criterion function $L_{\mathrm{HUB}}(\boldsymbol{\beta}, \sigma)$ when using Huber's loss function $\rho_{\mathrm{H},c}(x)$. From Example 7, we know that the classical estimates $(\hat{\boldsymbol{\beta}}_{\mathrm{LS}}, \hat{\sigma}_{\mathrm{SD}})$ are obtained when $c \to \infty$ because Huber's loss function is equivalent to the Gaussian loss function in the limit. First, note that Huber's loss function in (2.54) can be written as $\rho_{\mathrm{H},c}(x) = \rho_0(|x|^2)$ for the case of

$$\rho_0(t) = \frac{1}{\gamma} \times \begin{cases} t, & \text{for } t \leq c^2 \\ 2c\sqrt{t} - c^2, & \text{for } t > c^2, \end{cases} \quad t \in \mathbb{R}_0^+.$$

Hence, the χ function in (2.66) becomes

$$\chi_{\text{H},c}(x) = \frac{1}{\gamma} \times |\psi_{\text{H},c}(x)|^2 = \frac{1}{\gamma} \times \begin{cases} |x|^2, & \text{for } |x| \leq c \\ c^2, & \text{for } |x| > c \end{cases}, \quad x \in \mathbb{F}. \tag{2.69}$$

Substituting (2.69) into (2.65) yields the following estimating equation for the scale term:

$$\frac{1}{N(2\alpha)} \sum_{i=1}^{N} \left| \psi_{\text{H},c}\left(\frac{y_i - \mathbf{x}_{[i]}^{\top}\hat{\boldsymbol{\beta}}}{\hat{\sigma}} \right) \hat{\sigma} \right|^2 = \hat{\sigma}^2.$$

This can be written compactly using vector notation as

$$\hat{\sigma}^2 = \frac{1}{N(2\alpha)} \| \hat{\mathbf{r}}_\psi \|_2^2, \tag{2.70}$$

where $\hat{\mathbf{r}}_\psi = \mathbf{r}_\psi(\hat{\boldsymbol{\beta}}, \hat{\sigma})$ denotes the pseudo-residual vector at the solution. The consistency factor $\alpha = \alpha(c)$ can be computed in closed form. Using (2.68) and (2.69), if follows, from elementary calculus, that

$$2 \cdot \alpha = \gamma \mathsf{E}\left[|\psi_{\text{H},c}(x)|^2 \right]$$
$$= \begin{cases} c^2(1 - F_{\chi_2^2}(2c^2)) + F_{\chi_4^2}(2c^2), & \text{complex-valued case, } \mathbb{F} = \mathbb{C} \\ c^2(1 - F_{\chi_1^2}(c^2)) + F_{\chi_3^2}(c^2), & \text{real-valued case, } \mathbb{F} = \mathbb{R}, \end{cases} \tag{2.71}$$

where $F_{\chi_k^2}$ denotes the cdf of the χ_k^2-distribution. Note that the threshold parameter c determines the value of α. Often, we use $c_{.95} = 1.1214$ and $c_{.95} = 1.345$ to achieve 95 percent efficiency for the case of complex- and real-valued Gaussian errors, respectively. The associated scaling constants are then $\alpha = \alpha(c_{.95}) = 0.7156$ and $\alpha = 0.7102$, respectively.

Note that in (2.62), it also is possible to use $(N - p - 1)$ in the first term as a multiplier of σ instead of N. This adjustment will imply that $(N - p - 1)$ is used for scaling in (2.65) and in (2.70) instead of N. This adjustment will also correct for the bias in the small sample size case. A similar adjustment can be used for $L_{\text{ML}}(\boldsymbol{\beta}, \sigma)$ in (2.61) by replacing $N \log \sigma$ by $(N - p) \log \sigma$. This correction yields the classical *unbiased* estimates $(\hat{\boldsymbol{\beta}}_{\text{LS}}, s^2)$, where $s^2 = [N/(N - p - 1)]\hat{\sigma}_{\text{SD}}^2$, as minimizer of the ML criterion in (2.61), as well as Huber's criterion (2.62) when ρ in both cases is the Gaussian loss function $\rho(x) = \frac{1}{\gamma}|x|^2$.

2.6.1   Minimization-Majorization Algorithm

Here, we present the framework of block-wise Minimization-Majorization (MM) algorithms. For more details about the MM algorithm and its analysis see Razaviyayn et al. (2013), Hunter and Lange (2004), and Sun et al. (2017).

Suppose we want to find arg $\min_{\theta \in \Theta} L(\theta)$. An MM algorithm is an iterative procedure given by

$$\theta^{(n+1)} = T(\theta^{(n)}), \quad T(\theta^{(n)}) = \arg\min_{\theta \in \Theta} g(\theta|\theta^{(n)}), \qquad (2.72)$$

where the majorization surrogate function $g(\cdot|\cdot)$ satisfies

$$g(\theta|\theta') \geq L(\theta) \quad \text{and} \quad g(\theta'|\theta') = L(\theta').$$

Then, the following result holds.:

THEOREM 1   *Suppose L and g are differentiable functions, L is bounded from below, and T is continuous. Then, any accumulation point of the sequence $\theta^{(n)}$ is a stationary point of $L(\theta)$ if it lies in the interior of Θ.*

The MM algorithm has been extended to a block-wise MM algorithm. Let θ be partitioned into $\theta = (\theta_1, \theta_2)$, where $\theta_1 \in \Theta_1$, $\theta_2 \in \Theta_2$ and $\Theta = \Theta_1 \times \Theta_2$. At the $(n+1)$th iteration, the blocks are updated in a cyclic manner as follows:

$$\theta_2^{(n+1)} = T_2(\theta_1^{(n)}, \theta_2^{(n)}), \qquad (2.73)$$
$$T_2(\theta_1^{(n)}, \theta_2^{(n)}) = \arg\min_{\theta_2 \in \Theta_2} g_2(\theta_2|\theta_1^{(n)}, \theta_2^{(n)})$$
$$\theta_1^{(n+1)} = T_1(\theta_1^{(n)}, \theta_2^{(n+1)}), \qquad (2.74)$$
$$T_1(\theta_1^{(n)}, \theta_2^{(n+1)}) = \arg\min_{\theta_1 \in \Theta_1} g_1(\theta_1|\theta_1^{(n)}, \theta_2^{(n+1)})$$

where the majorization surrogate functions $g_i(\theta_i|\theta_1', \theta_2') = g_i(\theta_i|\theta')$, $i \in \{1, 2\}$ satisfy

$$g_i(\theta_i'|\theta_1', \theta_2') = L(\theta_1', \theta_2') \quad \forall (\theta_1', \theta_2') \in \Theta, \qquad (2.75)$$
$$g_1(\theta_1|\theta_1', \theta_2') \geq L(\theta_1, \theta_2') \quad \forall \theta_1 \in \Theta_1, \forall (\theta_1', \theta_2') \in \Theta, \qquad (2.76)$$
$$g_2(\theta_2|\theta_1', \theta_2') \geq L(\theta_1', \theta_2) \quad \forall \theta_2 \in \Theta_2, \forall (\theta_1', \theta_2') \in \Theta. \qquad (2.77)$$

Furthermore, when L and g_i are differentiable functions, it is possible to impose the constraint

$$\nabla_{\theta_i} g_i(\theta_i \mid \theta_1', \theta_2')\big|_{\theta_i = \theta_i'} = \nabla_{\theta_i} L(\theta)\big|_{\theta = \theta'}, \quad i \in \{1, 2\}. \qquad (2.78)$$

It should be noted that (2.78) implies that condition B3 of Razaviyayn et al. (2013) holds but not vice versa (because B3 does not imply that g_i and L are differentiable functions). The challenging part in designing a blockwise MM algorithm is, naturally, in finding appropriate surrogate functions $g_i(\cdot|\theta')$, $i \in \{1, 2\}$. A common choice is a quadratic function of the form $b_0 + b_1\theta + b_2\theta^2$ as it results in a simple update formula in (2.72).

2.6.2   Minimization-Majorization Algorithm for Huber's Criterion

We now construct an MM algorithm for obtaining the stationary solution $(\hat{\beta}, \hat{\sigma})$ in (2.64) and (2.65) of Huber's criterion $L_{\text{HUB}}(\beta, \sigma)$. Recall that we have assumed that $\rho(x) = \rho_0(|x|^2) : \mathbb{F} \to \mathbb{R}_0^+$ is a convex and differentiable function. For the developed MM algorithm to work, some conditions on the used convex loss function $\rho(x) = \rho_0(|x|^2)$ need to be imposed as described in the following text:

CONDITION 1
(a) $\rho(x)/|x| = \rho_0(|x|^2)/|x|$ *is a concave function.*
(b) ρ *is twice differentiable and the second-order (mixed) derivative of the loss function,*

$$\psi'(x) = \frac{\partial}{\partial x}\psi(x) = \frac{\partial^2}{\partial x \partial x^*}\rho(x) : \mathbb{F} \to \mathbb{R}_0^+, \tag{2.79}$$

satisfies $0 \leq \psi'(x) \leq 1$. *Note that* $\psi'(x) = \gamma^2 \rho_0''(|x|^2)|x|^2 + \gamma \rho_0'(|x|^2)$.

Note that Huber's loss function $\rho_{\text{H},c}(x)$, for example, satisfies Condition 1. We use the following shorthand notation:

$$\mathbf{r} = \mathbf{y} - \mathbf{X}\boldsymbol{\beta}, \quad \mathbf{r}' = \mathbf{y} - \mathbf{X}\boldsymbol{\beta}' \quad \text{and} \quad \mathbf{r}'_\psi = \psi\left(\mathbf{r}'/\sigma'\right)\sigma'.$$

Note that here $\boldsymbol{\beta}'$ and σ' denote values of previous iterates, so only $\mathbf{r}$ depends on the unknown parameter $\boldsymbol{\beta}$. For minimizing Huber's criterion $L_{\text{HUB}}(\boldsymbol{\beta}, \sigma) = \sum_i \rho(r_i/\sigma)\sigma + \frac{2N}{\gamma}(\alpha\sigma)$, we utilize the following surrogate function for the scale term:

$$g_2(\sigma|\boldsymbol{\beta}', \sigma') = a' + b'\frac{1}{\sigma} + \frac{2N}{\gamma}\alpha\sigma. \tag{2.80}$$

In (2.80), $a' + b'\sigma^{-1}$ is used to majorize $\sum_i \rho(r_i/\sigma)\sigma$, where a' and b' are constants that depend on the previous iterates $\boldsymbol{\beta}'$ and σ'. These terms can be found by solving the pair of equations (2.75) and (2.78) w.r.t. a' and b' and by substituting the obtained values back into (2.80). This yields (after simplifying) the following surrogate function

$$g_2(\sigma|\boldsymbol{\beta}', \sigma') = L_{\text{HUB}}(\boldsymbol{\beta}', \sigma') + \frac{2N}{\gamma}\alpha(\sigma - \sigma') + \sum_{i=1}^{N}\chi\left(\frac{r_i'}{\sigma}\right)\left[\frac{(\sigma')^2}{\sigma} - \sigma'\right].$$

Due to its construction, the function $g_2(\sigma|\boldsymbol{\beta}', \sigma')$ satisfies (2.75) and (2.78). Furthermore, under Condition 1a, it also satisfies (2.77) and, hence, it is a valid majorization function. To see this, observe that the difference function is given by

$$h_2(\sigma) = g_2(\sigma|\boldsymbol{\beta}', \sigma') - L_{\text{HUB}}(\boldsymbol{\beta}', \sigma) = a_0 + \frac{b_0}{\sigma} - \sum_{i=1}^{N}\rho\left(\frac{r_i'}{\sigma}\right)\sigma$$

for some constant a_0 and b_0. But by construction of the majorization function g_2, the difference function $h_2(\sigma)$ satisfies $\frac{\partial}{\partial\sigma}h_2(\sigma)|_{\sigma=\sigma'} = 0$. Because $h_2(\sigma)$ is a convex function in $1/\sigma$ under Condition 1a and because $h_2(\sigma') = 0$, it follows that $h_2(\sigma) \geq 0$, that is, $g_2(\sigma|\boldsymbol{\beta}', \sigma') \geq L_{\text{HUB}}(\boldsymbol{\beta}', \sigma)$, for all $\sigma > 0$.

The minimizer of $g_2(\sigma|\cdot,\cdot)$ can be found in closed form and is given by

$$\left(\sigma^{(n+1)}\right)^2 = \arg\min_{\sigma>0} g_2(\sigma|\boldsymbol{\beta}^{(n)},\sigma^{(n)})$$

$$= \frac{\left(\sigma^{(n)}\right)^2}{2\alpha} \frac{\gamma}{N} \sum_{i=1}^{N} \chi\left(\frac{r_i^{(n)}}{\sigma^{(n)}}\right),$$

where $r_i^{(n)}$ denotes the *i*th element of $\mathbf{r}^{(n)} = \mathbf{y} - \mathbf{X}\boldsymbol{\beta}^{(n)}$. For example, if $\rho(x)$ is Huber's function $\rho_{\text{H},c}(x)$, then we can use (2.69) to write the preceding update in an intuitive form

$$\left(\sigma^{(n+1)}\right)^2 = \frac{1}{N(2\alpha)} \sum_{i=1}^{N} \left| \psi_{\text{H},c}\left(\frac{y_i - \mathbf{x}_{[i]}^{\top}\boldsymbol{\beta}^{(n)}}{\sigma^{(n)}}\right)\sigma^{(n)} \right|^2 = \frac{1}{N(2\alpha)} \left\| \mathbf{r}_{\psi}^{(n)} \right\|_2^2$$

where $\mathbf{r}_{\psi}^{(n)} = \psi_{\text{H},c}\left(\mathbf{r}^{(n)}/\sigma^{(n)}\right)\sigma^{(n)}$ and α is given in (2.71).

For constructing a majorization surrogate function $g_1(\boldsymbol{\beta}|\boldsymbol{\beta}',\sigma')$, we first find a surrogate function for $\rho(\frac{r_i}{\sigma'})$ using

$$\rho_M(r_i/\sigma) = a_i' + 2b_i'\text{Re}(r_i/\sigma') + |r_i/\sigma'|^2,$$

where the constants a_i' and b_i' depend on the previous iterates $\boldsymbol{\beta}'$ and σ' and are found by solving the pair of equations (2.75) and (2.78) w.r.t. a_i' and b_i'. After finding these solutions, we obtain a surrogate function of the form

$$g_1(\boldsymbol{\beta}|\boldsymbol{\beta}',\sigma') = (2\alpha)N\sigma' + \sum_{i=1}^{N}\left(a_i' + 2b_i'\text{Re}\left(\frac{r_i}{\sigma'}\right) + \left|\frac{r_i}{\sigma'}\right|^2\right)\sigma'$$

$$= \text{const} + \sum_{i=1}^{N}\left(2[r_i' - r_{\psi,i}']\text{Re}(r_i) + |r_i|^2\right)\frac{1}{\sigma'}$$

where the constant term does not depend on $\boldsymbol{\beta}$, $\mathbf{r}' = \mathbf{y} - \mathbf{X}\boldsymbol{\beta}'$ and $\mathbf{r}_{\psi}' = \psi(\mathbf{r}'/\sigma')\sigma'$. Due to the construction of the majorization function g_1, it satisfies (2.75) and (2.78). Furthermore, under Condition 1b, g_1 also satisfies (2.76). To see this, consider the difference function

$$h_1(\boldsymbol{\beta}) = g_1(\boldsymbol{\beta}|\boldsymbol{\beta}',\sigma') - L_{\text{HUB}}(\boldsymbol{\beta},\sigma')$$

$$= \text{const} + \sum_{i=1}^{N}\left(2[r_i' - r_{\psi,i}']\text{Re}(r_i) + |r_i|^2\right)\frac{1}{\sigma'} - \sum_{i=1}^{N}\rho\left(\frac{r_i}{\sigma'}\right)\sigma'.$$

The Hessian matrix of the difference function $h_1 : \mathbb{F}^{p+1} \to \mathbb{R}$ is

$$\mathbf{H}_1 = \frac{\partial}{\partial\boldsymbol{\beta}}\left(\frac{\partial h_1}{\partial\boldsymbol{\beta}}\right)^{\text{H}} = \frac{1}{\sigma'}\sum_{i=1}^{N}\{1 - \psi'(r_i/\sigma')\}\mathbf{x}_i^*\mathbf{x}_i^{\top},$$

where the function $\psi'(\cdot)$ is defined in (2.79). Due to Condition 1b, the matrix $\mathbf{H}_1$ is a positive semidefinite matrix and thus the difference function h_1 is a convex function.

These results and the fact that $h_1(\boldsymbol{\beta}') = \mathbf{0}$ imply that $g_1(\boldsymbol{\beta}|\boldsymbol{\beta}',\sigma') \geq L_{\text{HUB}}(\boldsymbol{\beta},\sigma'), \forall \boldsymbol{\beta} \in \mathbb{F}^{p+1}$. Furthermore, the minimizer of $g_1(\boldsymbol{\beta}|\cdot,\cdot)$ can be found in closed form:

$$\boldsymbol{\beta}^{(n+1)} = \arg\min_{\boldsymbol{\beta} \in \mathbb{F}^{p+1}} g_1(\boldsymbol{\beta}|\boldsymbol{\beta}^{(n)}, \sigma^{(n+1)})$$

$$= \boldsymbol{\beta}^{(n)} + (\mathbf{X}^{\mathsf{H}}\mathbf{X})^{-1}\mathbf{X}^{\mathsf{H}}\mathbf{r}_\psi^{(n+1)}$$

where $\mathbf{r}_\psi^{(n+1)} = \psi(\mathbf{r}^{(n)}/\sigma^{(n+1)})\sigma^{(n+1)}$ denotes the pseudo-residual vector calculated at $(\boldsymbol{\beta}^{(n)}, \sigma^{(n+1)})$.

Thus, we have devised an MM algorithm that can be used to compute the global minimum $(\hat{\boldsymbol{\beta}}, \hat{\sigma})$ of Huber's criterion $L_{\text{HUB}}(\boldsymbol{\beta}, \sigma)$. The pseudo-code of the MM algorithm described previously, named `hubreg`, is given in Algorithm 5 for the case that the loss function ρ is Huber's function (2.54). A MATLAB$^{\copyright}$ function of the same name is available in the RobustSP toolbox.

Algorithm 5: hubreg, solves (2.62) when $\rho = \rho_{\text{H},c}$ using the MM algorithm.

---

input   : $\mathbf{y} \in \mathbb{F}^N$, $\mathbf{X} \in \mathbb{F}^{N \times (p+1)}$, c, $\boldsymbol{\beta}^{(0)}$, $\sigma^{(0)}$

output : $(\hat{\boldsymbol{\beta}}, \hat{\sigma})$, the minimizer of Huber's criterion $L_{\text{HUB}}(\boldsymbol{\beta}, \sigma)$ when $\rho = \rho_{\text{H},c}$.

initialize: $N_{\text{iter}} \in \mathbb{N}$, $\delta > 0$, $\alpha = \alpha(c)$ in (2.71), $\mathbf{X}^+ = (\mathbf{X}^{\mathsf{H}}\mathbf{X})^{-1}\mathbf{X}^{\mathsf{H}}$.

1 **for** $n = 0, 1, \ldots, N_{\text{iter}}$ **do**

2 | update residual

$$\mathbf{r}^{(n)} = \mathbf{Y} - \mathbf{X}\boldsymbol{\beta}^{(n)}$$

update pseudo-residual

$$\mathbf{r}_\psi^{(n)} = \psi_{\text{H},c}\left(\frac{\mathbf{r}^{(n)}}{\sigma^{(n)}}\right)\sigma^{(n)}$$

update scale

$$\sigma^{(n+1)} = \frac{1}{\sqrt{N(2\alpha)}}\left\|\mathbf{r}_\psi^{(n)}\right\|_2$$

3 | (re)update the pseudo-residual

$$\mathbf{r}_\psi^{(n+1)} = \psi_{\text{H},c}\left(\frac{\mathbf{r}^{(n)}}{\sigma^{(n+1)}}\right)\sigma^{n+1}$$

4 | regress $\mathbf{X}$ on $\mathbf{r}_\psi^{(n+1)}$

$$\boldsymbol{\delta}^{(n)} = \mathbf{X}^+\mathbf{r}_\psi^{(n+1)}$$

5 | update regression vector

$$\boldsymbol{\beta}^{(n+1)} = \boldsymbol{\beta}^{(n)} + \boldsymbol{\delta}^{(n)}$$

6 | **if** $\|\boldsymbol{\delta}^{(n)}\|/\|\boldsymbol{\beta}^{(n)}\| < \delta$ **then**

7 | $\lfloor$ **return** $(\hat{\boldsymbol{\beta}}, \hat{\sigma}) \leftarrow (\boldsymbol{\beta}^{(n+1)}, \sigma^{(n+1)})$

Recalling that the location (or location-scale model) is a special case of the linear regression model in (2.8) when the design matrix is $\mathbf{X} = \mathbf{1}$, hubreg can be used to compute robust (real or complex) joint M-estimates of location and scale.

2.7 Measures of Robustness

There has been an ongoing discussion for a few decades on how to best address robust linear regression problems. No single estimator is preferred over all others and, in practice, choosing the best approach results in prioritizing properties, such as the BP, relative efficiency, and practical computability. Sometimes, computing and comparing multiple estimates may serve as a cross-check (Huber and Ronchetti, 2009). This section introduces different types of outliers, and the most important measures of robustness, that is, the IF and the BP for linear regression.

2.7.1 Outliers in the Linear Regression Model

In the regression model that is defined in (2.7), a data point is defined by $\mathbf{z}_i = (y_i, \mathbf{x}_{[i]}), i \in \{1, \ldots, N\}$. A data point may be outlying in terms of the y_i and if it deviates from the majority of points in the y-direction, it is called *vertical outlier*. If a data point deviates in the x-direction, it is called a *leverage point* by analogy to the concept of leverage in mechanics. Deviation in the x-direction means that it is far away from the bulk of $\mathbf{x}_{[i]}, i \in \{1, \ldots, N\}$, in the p-dimensional space (factor space). There are "good" leverage points that are on the true regression line and "bad" leverage points that are not. The LSE is not robust against any type of outliers. M-estimators are robust against vertical outliers, which are outliers in the residuals, but they become nonrobust even for a single outlier in $\mathbf{x}_{[i]}$.

Figures 2.7 and 2.8 illustrate the two different kinds of outlying observations (x_i, y_i), which do not follow the linear pattern of the majority of the data for a simple regression with an intercept and where the residuals are

$$y_i - \beta_0 - x_i\beta_1, \quad i = 1, \ldots, N.$$

2.7.2 $(p+1)$-dimensional Influence Function

The IF, as a measure of robustness, was introduced in Chapter 1 for the single-channel location and scale model. It shows the approximate behavior of an estimator when the sample contains a small fraction ε of identical outliers. The IF of an estimator $\hat{\boldsymbol{\beta}} \in \mathbb{F}^{p+1}$ of a $(p+1)$-dimensional parameter vector $\boldsymbol{\beta}$ is defined exactly as in the one-dimensional case.

DEFINITION 8 *The $(p+1)$-dimensional IF is defined according to*

$$\mathsf{IF}(\mathbf{z}; \mathbf{T}(F), F) = \lim_{\varepsilon \downarrow 0} \frac{\mathbf{T}(F_\varepsilon) - \mathbf{T}(F)}{\varepsilon} = \left[\frac{\partial \mathbf{T}(F_\varepsilon)}{\partial \varepsilon} \right]_{\varepsilon=0} \tag{2.81}$$

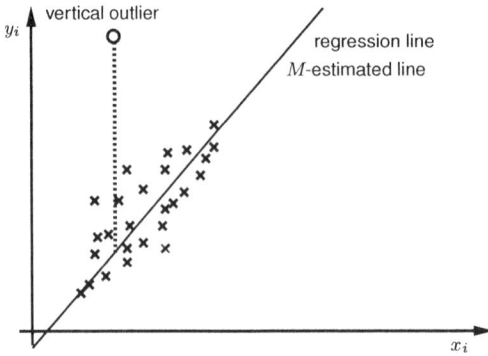

Figure 2.7 Illustration of a vertical outlier, which is an outlier in the residuals, for the case of a simple regression. M-estimators are robust against vertical outliers and for this case the estimated regression line is very close to the true line determined by $\beta_0 + x\beta_1$.

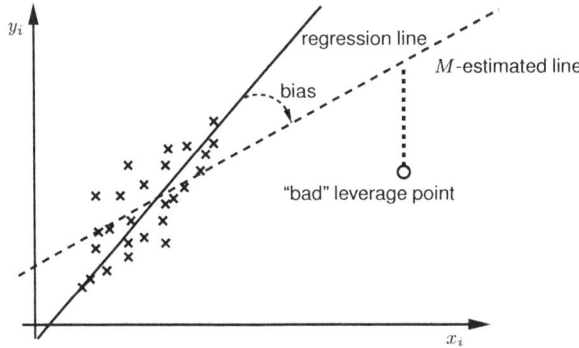

Figure 2.8 Illustration of a leverage point for the case of a simple regression. M-estimators are nonrobust with a BP equal to zero in the presence of a leverage point, which causes a large tilt in the estimated regression line.

In (2.81), it is assumed that $\mathbf{X}$ is random and that the distribution of $v_i, i \in \{1, \ldots, N\}$ does not depend on $\mathbf{x}_{[i]}^{\top}$. Moreover, $\mathbf{T}(F)$ and $\mathbf{T}(F_\varepsilon)$ are the functional representations of the estimator $\hat{\boldsymbol{\beta}}$ when the data $\mathcal{Z} = \{\mathbf{z}_i = (y_i, \mathbf{x}_{[i]}); i = 1, \ldots, N\}$ is distributed, respectively, following F and the contaminated distribution $F_\varepsilon = (1 - \varepsilon)F + \varepsilon\delta_{\mathbf{z}}$ with $\mathbf{z} = \delta_{(\mathbf{x}_0, y_0)}$ being the point-mass probability on $(\mathbf{x}_0, y_0) \in \mathbb{F}^{p+2}$ and ε the fraction of contamination.

For example, the IF of an M-estimator with known σ, as defined in (2.53), for the real-valued case, is

$$\mathsf{IF}(\mathbf{z}; \mathbf{T}(F), F) = \frac{\sigma}{b}\psi\left(\frac{y_0 - \mathbf{x}_0^{\top}\mathbf{T}(F)}{\sigma}\right) \cdot \boldsymbol{\Sigma}_{\mathbf{x}}^{-1}\mathbf{x}_0 \qquad (2.82)$$

where

$$b = \mathsf{E}\left[\frac{v}{\sigma}\right]$$

and

$$\Sigma_{\mathbf{x}} = \mathsf{E}[\mathbf{x}\mathbf{x}^{\top}].$$

When σ has been estimated, it has been shown in Maronna et al. (2006), for the case of v having a symmetrical distribution, that (2.82) holds with $\hat{\sigma}$ replacing σ.

The IF of an M-estimator is unbounded in $\mathbf{z} = (\mathbf{x}_0, y_0)$. To understand this, consider the following factorization:

$$\mathsf{IF}(\mathbf{z}; \mathbf{T}(F), F) = \mathsf{IR}((\mathbf{x}_0, y_0); \mathbf{T}(F), F) \cdot \mathsf{IP}(\mathbf{x}_0; \mathbf{T}(H), H)$$

where

$$\mathsf{IR}((\mathbf{x}_0, y_0); \mathbf{T}(F), F) = \frac{\sigma}{b} \psi \left(\frac{y_0 - \mathbf{x}_0^{\top} \mathbf{T}(F)}{\sigma} \right),$$

$$\mathsf{IP}(\mathbf{x}_0; \mathbf{T}(H), H) = \Sigma_{\mathbf{x}}^{-1} \mathbf{x}_0,$$

and H denotes the distribution of $\mathbf{x}_0$. The influence of an outlier $(\mathbf{x}_0, y_0)$ on the residual $y_0 - \mathbf{x}_0^{\top} \mathbf{T}(F)$ is measured by $\mathsf{IR}((\mathbf{x}_0, y_0); \mathbf{T}(F), F)$, which is bounded for bounded ψ functions. Therefore, unlike for the LSE, the bias of an M-estimator remains bounded when $v_i, i = 1, \ldots, N$, is generated from a heavy-tailed distribution. However, $\mathsf{IP}(\mathbf{x}_0; \mathbf{T}(H), H)$, which measures the $(p + 1)$-valued influence of an outlier $\mathbf{x}_0$ in factor space, is unbounded. Because of this, a single $\mathbf{x}_0$ can completely dominate the fit. For this reason, outliers in factor space are called leverage points because they tilt the regression line in their direction (see Figure 2.8).

2.7.3   Breakdown Point

Achieving a high BP in regression models is much more challenging than in single-channel location and scale estimation. As shown in Section 2.7.2, it is not sufficient to apply a bounded ψ function to the residuals in the presence of leverage points. As discussed in Rousseeuw and Leroy (2005), a leverage point may not even have a large residual. A practitioner may define the breakdown as the smallest fraction of contamination that causes the estimator to take on values that are arbitrarily far from the true value. This idea is formalized in the finite-sample BP as follows.

DEFINITION 9   *Finite-Sample BP*
Let $\mathcal{Z} = \{\mathbf{z}_i = (y_i, \mathbf{x}_{[i]}); i = 1, \ldots, N\}$ denote a data set that consists of N samples and let $\hat{\boldsymbol{\beta}}(\mathcal{Z})$ stand for an estimator of $\boldsymbol{\beta}$ based on $\mathcal{Z}$. Furthermore, let $\mathcal{Z}_m$ be the data set that is obtained by replacing m of the original data points $\mathbf{z}_i = (y_i, \mathbf{x}_{[i]}), i = 1, \ldots, N$ by arbitrary values.

The maximum possible finite-sample bias that is introduced by m outliers is

$$\text{bias}\left(m; \hat{\boldsymbol{\beta}}, \mathcal{Z}\right) = \sup_{\mathcal{Z}_m \in \check{\mathcal{Z}}_m} \left\| \hat{\boldsymbol{\beta}}(\mathcal{Z}_m) - \hat{\boldsymbol{\beta}}(\mathcal{Z}) \right\|, \tag{2.83}$$

where $\check{\mathcal{Z}}_m$ is the set of all possible $\mathcal{Z}_m$.

Then, according to Donoho and Huber (1983), the finite-sample BP is defined as

$$\varepsilon_N^*(\hat{\boldsymbol{\beta}}, \mathcal{Z}) = \min \left\{ \frac{m}{N} : \text{bias}\left(m; \hat{\boldsymbol{\beta}}, \mathcal{Z}\right) < \infty \right\}. \tag{2.84}$$

The limit version of (2.84) is the asymptotic BP, which has been defined with respect to a metric by Davies (1993). This article also contains an excellent discussion of the breakdown behavior of most existing robust regression estimators.

2.8   Positive Breakdown Point Regression Estimators

A number of estimators of the regression parameters with a positive breakdown were proposed in the 1980s, such as the least-median of squares (LMS) (Rousseeuw, 1984), the least trimmed squares (LTS) (Rousseeuw and Van Driessen, 2006), S-estimators (Rousseeuw and Yohai, 1984), τ-estimators (Yohai and Zamar, 1988), and *MM*-estimators (Yohai, 1987). In this section, we only describe estimators that are regression, affine, and scale equivariant:

DEFINITION 10   *Regression Equivariance*
*Let **u** be any $p \times 1$ vector. Then, $\hat{\boldsymbol{\beta}}$ is regression equivariant if*

$$\hat{\boldsymbol{\beta}}(\mathbf{X}, \mathbf{y} + \mathbf{X}\mathbf{u}) = \hat{\boldsymbol{\beta}}(\mathbf{X}, \mathbf{y}) + \mathbf{u}.$$

DEFINITION 11   *Scale Equivariance*
Let c be a scalar. Then, $\hat{\boldsymbol{\beta}}$ is scale equivariant if

$$\hat{\boldsymbol{\beta}}(\mathbf{X}, c\mathbf{y}) = c\hat{\boldsymbol{\beta}}(\mathbf{X}, \mathbf{y}).$$

This property is useful because it implies that rescaling the data $\mathbf{y}$ results in rescaling the predicted values $\hat{\mathbf{y}}$ and the residuals $\mathbf{y} - \hat{\mathbf{y}}$ by the same factor.

DEFINITION 12   *Affine Equivariance*
*Let **A** be any $p \times p$ non-singular matrix. Then, $\hat{\boldsymbol{\beta}}$ is affine equivariant if*

$$\hat{\boldsymbol{\beta}}(\mathbf{X}\mathbf{A}, \mathbf{y}) = \mathbf{A}^{-1}\hat{\boldsymbol{\beta}}(\mathbf{X}, \mathbf{y}).$$

Affine equivariance is very useful because it implies that both the predicted values $\hat{\mathbf{y}}$ and the residuals $\mathbf{y} - \hat{\mathbf{y}}$ are invariant under an affine transformation of the predictor matrix $\mathbf{X}$.

All of the positive breakdown estimators that are described in this section are consistent and asymptotically normal. The maximum possible finite-sample BP that is achievable by regression equivariant estimators is (Rousseeuw and Leroy, 2005, p. 125)

$$\varepsilon_N^*(\hat{\boldsymbol{\beta}}, \mathcal{Z}) = \frac{\lfloor (N - (p+1))/2 \rfloor + 1}{N}, \tag{2.85}$$

where $\lfloor (N - (p+1))/2 \rfloor$ is the integer part of $(N - (p+1))/2$. For $N \to \infty$ and fixed values $p < \infty$, it is the case that $\varepsilon_N^*(\hat{\boldsymbol{\beta}}, \mathcal{Z}) = 50$ percent. Equation (2.85) implies that the estimator $\hat{\boldsymbol{\beta}}$ remains bounded whenever strictly more than $1/2(N+p)$ observations are contaminated.

2.8.1   Least-Median of Squares and Least Trimmed Squares Estimator

Amongst the first estimators with a positive BP is the LMS, which replaces the sum of the LSE by the median

$$\underset{\boldsymbol{\beta} \in \mathbb{R}^{p+1}}{\text{minimize}} \left\{ \text{med}\{r_i^2\} \right\}, \quad i \in \{1, \dots, N\}, \tag{2.86}$$

where $r_i^2 = (y_i - \mathbf{x}_{[i]}^\top \boldsymbol{\beta})^2$. Its finite-sample BP approaches 50 percent for large values of N. However, the LMS has a low convergence rate of $N^{-1/3}$ and its efficiency for the Gaussian distribution case is low. To overcome its low convergence rate and to arrive at a higher asymptotic efficiency, Rousseeuw proposed the LTS estimator, which is defined as

$$\hat{\boldsymbol{\beta}} = \arg \min_{\boldsymbol{\beta} \in \mathbb{R}^{p+1}} \sum_{i=1}^{h} r_{(i)}^2 \quad h \in \mathbb{N}, \, h < N, \tag{2.87}$$

where $r_{(i)}^2$ denotes ordered squared residuals, $r_{(1)}^2 \leq r_{(2)}^2 \leq \dots \leq r_{(N)}^2$. The LTS estimators minimize an α-trimmed squares scale, where $\alpha = N - h$ of the largest squared residuals are trimmed. Unfortunately, (2.87) is a nonconvex optimization problem and computing such estimates is computationally very expensive due to the large number of possible h-subsets. In the simplest case, an h-subset is a random subsample of the data points $(y_i, \mathbf{x}_{[i]})$, $i \in \{1, \dots, N\}$ of size h. Alternative ways to create h-subsets are discussed in Rousseeuw and Van Driessen (2006). To speed up the computation, the authors proposed an approximation algorithm that does not consider all possible h-subsets of the original data set. The algorithm does not solve the original problem stated in (2.87) but provides an approximation for the optimum provided that a sufficient number of subsets is drawn. Starting from an arbitrarily chosen h-subset, another h-subset is constructed that yields a lower value of the objective function (2.87). This algorithm is applied to many initial h-subsets until convergence, and the subset, which gives the minimum value of the objective function, is chosen as an approximate value for the LTS estimate (see Rousseeuw and Van Driessen [2006] for more details).

For $h = \lfloor N/2 \rfloor + \lfloor (p+2)/2 \rfloor$ the finite-sample BP for LTS estimators, for an observation in a general position, is

$$\varepsilon_N^*(\hat{\boldsymbol{\beta}}, \mathcal{Z}) = \frac{\lfloor (N - (p+1))/2 \rfloor + 1}{N},$$

which is the maximum attainable value for any regression equivariant estimator (see (2.85)). The observations y_i, $i = 1, \ldots, N$ are in a *general position*, if any $p + 1$ of them give a unique determination of $\boldsymbol{\beta} \in \mathbb{R}^{p+1}$. In the case of simple regression, for example, this means that no two points may coincide or determine a vertical line. When the observations come from a continuous distribution, the event that the observations are in a general position has probability one.

While this subsection details the real-valued case, naturally, the LMS and LTS can be defined for the complex-valued case as well. For the LMS, one would simply replace r_i^2 by $|r_i|^2$, that is, the modulus squared of the residuals, whereas for the LTS, analogously, one would replace $r_{(i)}^2$ by $|r_{(i)}|^2$.

2.8.2   *S*-Estimators and τ-Estimators

This section briefly describes *S*-estimators (Rousseeuw and Yohai, 1984) and τ-estimators (Yohai and Zamar, 1988) for the real-valued case. The key idea of these estimators is to minimize a robust scale estimate of the residuals. These high-BP regression estimators share the flexibility and good asymptotic properties of *M*-estimators and have a convergence rate of $N^{-1/2}$.

S-estimators are defined as minimizers of an *M*-scale estimate of the residuals:

$$\hat{\boldsymbol{\beta}} = \arg \min_{\boldsymbol{\beta} \in \mathbb{R}^{p+1}} \hat{\sigma}(\mathbf{r}) \qquad (2.88)$$

with $\hat{\sigma}(\mathbf{r})$ being the *M*-scale estimate (1.19) and $\mathbf{r} = (r_1, \ldots, r_N)^\top$, where $r_i \equiv r_i(\boldsymbol{\beta}) = y_i - \mathbf{x}_{[i]}^\top \boldsymbol{\beta}$. For real-valued data, an approximate solution for the optimization problem stated in (2.88) can be obtained using an IRWLS algorithm (Salibián-Barrera and Yohai, 2012).

τ-estimators are defined as minimizers of a τ-scale estimate of the residuals:

$$\hat{\boldsymbol{\beta}} = \arg \min_{\boldsymbol{\beta} \in \mathbb{R}^{p+1}} \hat{\sigma}_\tau(\mathbf{r}), \qquad (2.89)$$

where the τ-scale estimate is given by

$$\hat{\sigma}_\tau(\mathbf{r}) = \sqrt{\hat{\sigma}^2(\mathbf{r}) \cdot \frac{1}{N} \sum_{i=1}^{N} \rho_2\left(\frac{r_i}{\hat{\sigma}(\mathbf{r})}\right)}. \qquad (2.90)$$

Here, $\hat{\sigma}$ is an *M*-scale estimate (1.19). A common choice is Tukey's, that is, $\psi = \psi_1$ from (1.23). A small value is chosen for the tuning constant c_1 to obtain a highly robust (but not very efficient) scale estimate. By contrast, ρ_2, as defined in (1.22), is tuned for high efficiency under the Gaussian model by letting c_2, the tuning constant of ρ_2, be larger than c_1.

τ-estimators have an asymptotic BP of 50 percent, but in contrast to *S*-estimators, they are able to *simultaneously* achieve a high ARE and a high BP by using two differently tuned redescending functions.

Asymptotically, the τ-estimator is an M-estimator, as defined in (2.52), whose ψ function is a weighted sum of two ψ functions (Yohai and Zamar, 1988):

$$\psi_\tau(r_i) = W(\boldsymbol{\beta}) \cdot \psi_1\left(\frac{r_i}{\hat{\sigma}(\mathbf{r})}\right) + \psi_2\left(\frac{r_i}{\hat{\sigma}(\mathbf{r})}\right), \tag{2.91}$$

with data-dependent weights

$$W(\boldsymbol{\beta}) = \frac{\sum_{i=1}^N \left(2\rho_2\left(\frac{r_i}{\hat{\sigma}(\mathbf{r})}\right) - \psi_2\left(\frac{r_i}{\hat{\sigma}(\mathbf{r})}\right) \cdot \frac{r_i}{\hat{\sigma}(\mathbf{r})}\right)}{\psi_1\left(\frac{r_i}{\hat{\sigma}(\mathbf{r})}\right) \cdot \frac{r_i}{\hat{\sigma}(\mathbf{r})}}, \tag{2.92}$$

where $\psi_2(x) = \frac{d\rho_2(x)}{dx}$. If $2\rho_2(x) - \psi_2(x)x \geq 0$, then $W(\boldsymbol{\beta}) \geq 0$. For real-valued data, an approximation for the optimal solution to (2.89) can be computed using an IRWLS algorithm (Salibián-Barrera et al., 2008b).

2.8.3   *MM*-Estimators

We briefly describe *MM*-estimators (Yohai, 1987) for the real-valued case. Like τ-estimators, they are able to *simultaneously* achieve a high ARE and a high BP by using two differently tuned redescending functions.

   MM-estimators are two step-estimators:

1.   Compute an *S*-estimate that solves (2.88) to obtain an initial estimate $\hat{\boldsymbol{\beta}}_1$, as well as the corresponding residual scale $\hat{\sigma}_1(\mathbf{r})$. Choose a small value for the parameter of the ψ_1 function of the M-scale.
2.   Improve upon the solution $\hat{\boldsymbol{\beta}}_1$ by evaluating

$$\underset{\boldsymbol{\beta}\in\mathbb{R}^{p+1}}{\text{minimize}}\left\{\sum_{i=1}^N \rho_2\left(\frac{y_i - \mathbf{x}_{[i]}^\top\boldsymbol{\beta}}{\hat{\sigma}_1}\right)\right\} \tag{2.93}$$

   using a ρ_2 function that is tuned for high efficiency. In particular, $\rho_2(x) \leq \rho_1(x)$ and $\sup \rho(x) = \sup \rho_1(x)$ (Yohai, 1987). Here, $\psi_1(x) = \frac{d\rho_1(x)}{dx}$.

MM-estimators possess a high asymptotic efficiency when the residuals are normally distributed. They also have an asymptotic BP of 50 percent.

2.9   Simulation Studies

2.9.1   Study 1: Randomly Flipped Measurements

To compare the performance of the classical estimates of regression and scale $(\hat{\boldsymbol{\beta}}_{\text{LS}}, \hat{\sigma}_{\text{SD}})$ with estimates from robust regression methods, the LAD estimate $\hat{\boldsymbol{\beta}}_{\text{LAD}}$ and the HUB-Reg estimates of regression and scale $(\hat{\boldsymbol{\beta}}, \hat{\sigma})$ that minimize $L_{\text{HUB}}(\boldsymbol{\beta}, \sigma)$, the following simulation study is proposed: Using a Monte Carlo (MC) trial, generate a data set $(\mathbf{y}, \mathbf{X})$ based on $N = 500$ measurements and with $p = 250$ predictors.

1. Randomly generate the true regression vector $\boldsymbol{\beta} \in \mathbb{C}^p$ where $\beta_j = r_j e^{J\theta_j}$ are independent, $r_j \overset{i.i.d.}{\sim} \mathcal{U}(0, 10)$ and $\theta_j \overset{i.i.d.}{\sim} \mathcal{U}(0, 2\pi)$ where r_j is independent of θ_j $(j = 1, \ldots, p)$.

2. Generate the standardized (unit scale) complex error vector as $\mathbf{v}_0 \sim \mathcal{N}_N(\mathbf{0}, \mathbf{I})$, and the design matrix $\mathbf{X}$ with standard complex Gaussian entries $x_{ij} \overset{i.i.d.}{\sim} \mathcal{N}(0, 1)$, $i = 1, \ldots, N, j = 1, \ldots, p$.

3. Generate the (noncorrupted) response vector as $\tilde{\mathbf{y}} = \mathbf{X}\boldsymbol{\beta} + \sigma\mathbf{v}_0$, where the error scale σ is chosen such that the signal-to-noise power ratio is SNR $= 10 \log_{10}(\|\mathbf{X}\boldsymbol{\beta}\|_2^2/(\sigma^2 \|\mathbf{v}_0\|_2^2)) = 20$ dB.

4. With probability ε, flip the orientation of the ith response y_i to $y_i \leftarrow e^{J\pi}\tilde{y}_i$, and with probability $1 - \varepsilon$ set $y_i \leftarrow \tilde{y}_i$ which yields the (corrupted) observed response vector $\mathbf{y}$.

The design matrix $\mathbf{X}$ and the response vector $\mathbf{y}$ can then be determined as observed measurements of the regression model. Note that $\mathbf{y}$, on average, has a fraction εN of (nonobvious) wrong measurements. The reason for this is that the good response $\tilde{\mathbf{y}}$ and the corrupted response $\mathbf{y}$ have the same distribution because $e^{J\theta}y_i \overset{d}{=} y_i$ for all $\theta \in \mathbb{R}$. This means that the corrupted measurements are difficult to clean from the data. The aim is to recover the true $\boldsymbol{\beta}$ and the error scale σ as accurately as possible. The probability of flips, ε, is allowed to vary from 0.01 to 0.10 and the empirical (averaged over MC trials) relative mean-squared error (RMSE) for the regression coefficients

$$\|\hat{\boldsymbol{\beta}} - \boldsymbol{\beta}\|_2^2/\|\boldsymbol{\beta}\|_2^2$$

is plotted in Figure 2.9 for the LAD estimate $\hat{\boldsymbol{\beta}}_{\text{LAD}}$ and the HUB-Reg estimates of regression and scale with $c_{.95}$ and $c_{.85}$, respectively. A similar plot is depicted for the scale estimates, but now the empirical RMSE is defined as $\log_{10}^2(\hat{\sigma}/\sigma)$, and again averaged over all MC trials. As can be seen in Figure 2.9a and 2.9c, when the average fraction of corrupted measurements ε increases, the LSE $\hat{\boldsymbol{\beta}}_{\text{LS}}$ and the corresponding scale estimate $\hat{\sigma}_{\text{SD}}$ deviate significantly, whereas HUB-Reg estimates of regression and scale provide highly accurate estimates even in the presence of a large fraction of corrupted observations. Figure 2.9b is an expanded version of Figure 2.9a, which facilitates the comparison of the considered robust estimators. It is important to note, for $\varepsilon = 0$, that the performance difference between the HUB-Reg and LSE estimates for $\boldsymbol{\beta}$ is negligible. The LAD estimator $\hat{\boldsymbol{\beta}}_{\text{LAD}}$, however, is not very accurate for very small values of ε, but starts to show reliable performance as the fraction of corrupted measurements increase. For $\varepsilon = 0.04$ (or $\varepsilon = 0.08$) the performance of LAD and HUB-Reg using $c_{.95}$ (or $c_{.85}$) are practically the same.

Finally, to illustrate the computation times, we plot in Figure 2.9d the average computation times in seconds for each ε. Note, on average, that the LSE estimate $\hat{\boldsymbol{\beta}}_{\text{LS}}$ is computed in 0.012 seconds, whereas the HUB-Reg estimates of regression and scale $(\hat{\boldsymbol{\beta}}, \hat{\sigma})$ are computed in 0.049 seconds. Because HUB-Reg starts by first computing the LSE and the sample standard deviation as initial start, the extra computation time is thus only 0.037 (= 0.049 − 0.012) seconds on average. Both the LSE and HUB-Reg estimates are obtained very quickly, but this is not the case for the LAD estimate, which

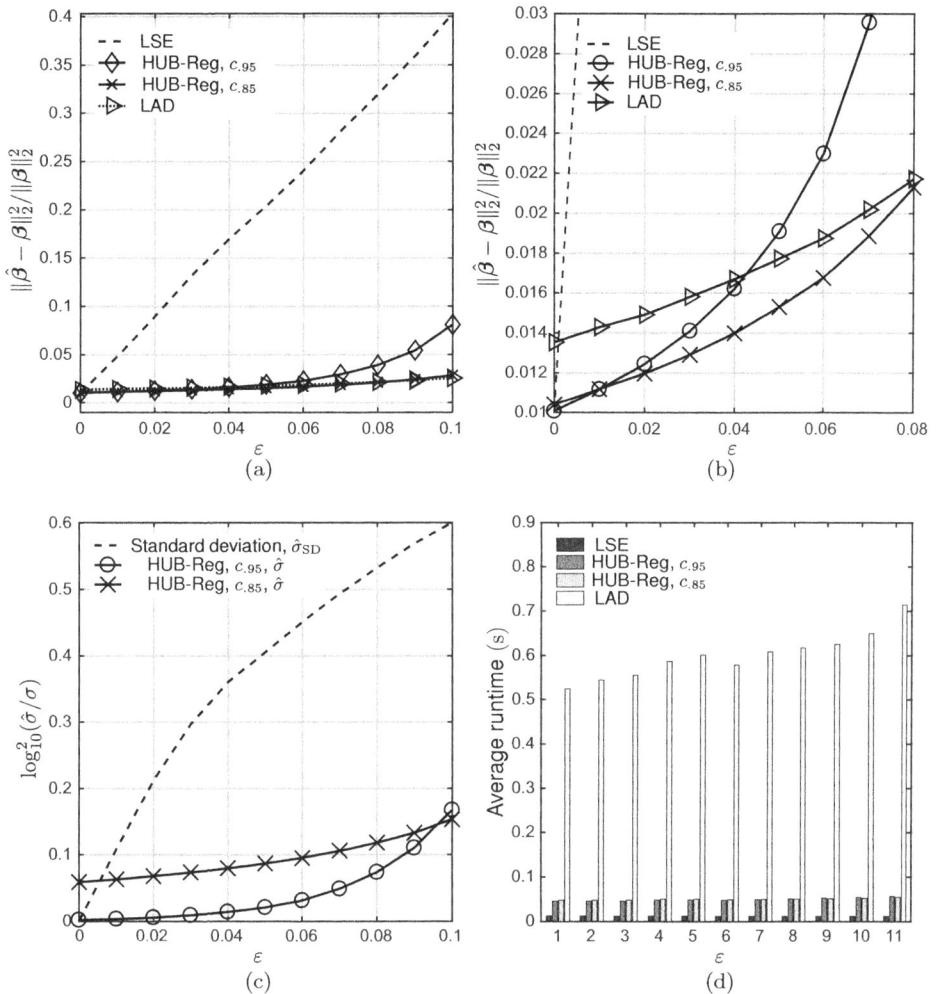

Figure 2.9 (a) RMSE based on LSE, LAD, and HUB-Reg joint M-estimates of regression and scale; (b) an expanded view of the region of interest in (a); (c) RMSE of $\hat{\sigma}_{SD}$ and HUB-Reg estimate of scale $\hat{\sigma}$ and (d) computation times.

takes on average 0.604 seconds. Thus, if one is interested in a regression estimate only, then the HUB-Reg estimate using $c_{.85}$ is a good option due to its robustness and fast computation times.

2.9.2   Study 2: Localization of Mobile User Equipment

To illustrate the applicability of the methods that have been discussed in this chapter, consider the following practical example of the localization of user equipment (UE), that is, an object or human being associated with a wireless transmitter device, using different base stations (BS).

Table 2.1 Positions of the $M = 10$ BSs for Study 2.

BS	1	2	3	4	5	6	7	8	9	10
x-position (m)	2500	1000	4500	1500	3000	1750	4000	4000	1000	2000
y-position (m)	5000	3500	1750	4000	4500	1000	750	3500	2000	250

In many applications, such as, for example, emergency services, intelligent transport systems, or any service that relies on geographic or location information, the precise location of every device is highly desirable. If line-of-sight (LOS) channels between the UE and the BS exist, high-positioning accuracy can be obtained, for example, by using triangulation techniques.

However, the LOS assumption often does not hold in practice because the direct path from the UE to the BS may be blocked. For example, in urban areas and hilly terrains, reflections due to obstacles such as buildings and trees result in an indirect signal path. This is referred to as NLOS propagation and this can lead to erroneous signal parameter estimates. For this case, the NLOS measurements are modeled as outliers. As an example, consider the localization, based on time-of-arrival (TOA) measurements, where the ith nonlinear range measurement at the mth BS is modeled by

$$r_{m,i} = h_m(\boldsymbol{\beta}) + v_{m,i}, \quad m = 1, \ldots, M, \quad i = 1, \ldots, N. \quad (2.94)$$

Here, $\boldsymbol{\beta} = (x_{\mathrm{UE}}, y_{\mathrm{UE}})^\top$ is the unknown position of the UE that is assumed to remain constant for N subsequent measurements,

$$h_m(\boldsymbol{\beta}) = \sqrt{(x_{\mathrm{UE}} - x_{\mathrm{BS},m})^2 + (y_{\mathrm{UE}} - y_{\mathrm{BS},m})^2} \quad (2.95)$$

is the distance from the UE to the mth BS and $v_{m,i}$ is the range error at the mth BS for measurement i. As described in Cheung et al. (2004), by introducing a range variable $R = \sqrt{x_{\mathrm{UE}}^2 + y_{\mathrm{UE}}^2}$, (2.94) can be expressed as a set of linear equations in $x_{\mathrm{UE}}, y_{\mathrm{UE}}$ and R^2. This allows for the application of the linear regression estimators that have been discussed in this chapter.

The positions $(x_{\mathrm{BS},m}, y_{\mathrm{BS},m})^\top$ of the $m \in \{1, \ldots, M\}$ BS are assumed to be known, and, for this example, are given in Table 2.1. Because $(x_{\mathrm{BS},m}, y_{\mathrm{BS},m})^\top$ does not contain outliers, robustness against leverage points is not required.

The NLOS effects are modeled as i.i.d. random variables with pdf

$$f(v_i|\mu_v, \sigma_v) = (1 - \varepsilon)f_{\mathrm{LOS}}(v_i|0, \sigma_{\mathrm{LOS}}) + \varepsilon f_{\mathrm{NLOS}}(v_i|\mu_{\mathrm{NLOS}}, \sigma_{\mathrm{NLOS}}). \quad (2.96)$$

Here, the LOS condition is well modeled by a zero mean Gaussian $f_{\mathrm{LOS}}(v_i|0, \sigma_{\mathrm{LOS}})$, where in this study, $\sigma_{\mathrm{LOS}} = 10\,\mathrm{m}$. The NLOS condition induces a positive mean and a larger variance of the TOA measurements, which can be modeled, for example, by an exponential or a shifted Gaussian distribution. In this study, as in Hammes and Zoubir (2011), $f_{\mathrm{NLOS}}(v_i|\mu_{\mathrm{NLOS}}, \sigma_{\mathrm{NLOS}})$ is assumed to be an exponential distribution. ε is the probability that a channel from the UE to the BS is in NLOS condition, and we assume that each BS has collected $N = 5$ measurements, and that there are $M = 10$ BS. The position of the UE is varied in each MC experiment where, for the kth trial and assuming a uniform distribution, $2000\,\text{m} \leq x(k) \leq 3000\,\text{m}$ and $2000\,\text{m} \leq y(k) \leq 3000\,\text{m}$.

As a performance metric, the (empirical) mean circular positioning error (MCPE) using 1000 MC runs

$$\text{MCPE} = \frac{1}{1000} \sum_{k=1}^{1000} \sqrt{(\hat{x}_{\text{UE}}(k) - x_{\text{UE}}(k))^2 + (\hat{y}_{\text{UE}}(k) - y_{\text{UE}}(k))^2}$$

is evaluated for the following estimators:

1. LSE, as defined in (2.14);
2. M-estimator with auxiliary scale estimate, as defined in (2.53) with Huber's monotone ψ function (2.54), with $c_{.95}$;
3. M-estimator with auxiliary scale estimate, as defined in (2.53) with Tukey's redescending ψ function (2.54), with $c_{.95}$;
4. S-estimator, as defined in (2.88) using Tukey's M-scale and the algorithm specified by
    Salibián-Barrera and Yohai (2012) using their default parameters;
5. LAD estimator, as defined in (2.15);
6. Rank-LAD estimator, as defined in (2.17) and
7. Huber's joint M-estimator of regression and scale (HUB-Reg) as defined in (2.62), with $c_{.95}$.

Figure 2.10 shows the MCPE of the preceding estimators for a fixed NLOS probability of $\varepsilon = 0.4$ and varying $0 \leq \sigma_{\text{NLOS}} \leq 500\,\text{m}$. For the Gaussian case, which corresponds to $\sigma_{\text{NLOS}} = \sigma_{\text{LOS}}$, all estimators, except for the S-estimator, which is tuned to a high BP, have a similar performance. Increasing the value of σ_{NLOS} corresponds to larger variability and bias of the TOA measurements that form the underlying data for the localization algorithms. With such a setup, the localization error for the Huber's estimator and for the LAD estimator increases strongly. The rank-LAD and Tukey's redescending estimators show good resistance against NLOS outliers, but all estimators are outperformed by the S-estimator. As shown in Table 2.2, this gain in performance for impulsive noise comes with a considerable increase in computational complexity.

Figure 2.11 shows the MCPE of the preceding estimators for $\sigma_{\text{NLOS}} = 250\,\text{m}$ and $0 \leq \varepsilon \leq 0.4$. Again, for the Gaussian case, which corresponds to $\varepsilon = 0$, all estimators except for the S-estimator perform similarly. Given its low computational complexity and guaranteed convergence due to the convex loss function, the rank-LAD provides excellent results for this setup. Again, Tukey's redescending M-estimator outperforms Huber's monotone M-estimator, as well as the LAD estimator. Best results for $\varepsilon > 0$ are delivered by the S-estimator.

Table 2.2 lists the average runtimes of the MATLAB$^{©}$ implementations of the algorithms for the localization task on an Intel$^®$ Core$^{™}$ i5 CPU 760, 2.80 GHz computer. The robustness of the estimators against impulsive noise when compared to the LSE comes at the expense of an increase in computational cost. The runtime of the S-estimator, for example, is more than 500 times longer than that of the LSE. The M-estimators and and

Table 2.2 Average runtimes of the MATLAB© implementations of the localization algorithms on an Intel® Core™ i5 CPU 760, 2.80 GHz computer. Here, *M*-Huber refers to Huber's *M*-estimate with auxiliary scale, whereas Joint *M*-Huber denotes the joint regression and scale estimate.

	LSE	*M*-Huber	*M*-Tukey	*S*	LAD	rank-LAD	Joint *M*-Huber
Average runtime (s)	$1.8697 \cdot 10^{-4}$	0.0039	0.0034	0.1030	0.0028	0.0095	$4.3333 \cdot 10^{-4}$
Runtime/LSE runtime	1	20.96	18.28	553.76	15.05	51.07	2.31

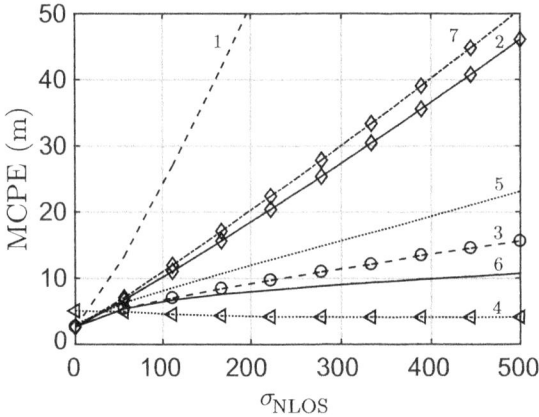

Figure 2.10 MCPE of different estimators as a function of the NLOS standard deviation. 1 = LSE, 2 = Huber's *M*-estimate with auxiliary scale, 3 = Tukey's *M*-estimate, 4 = *S*-estimate, 5 = LAD estimate, 6 = rank-LAD estimate, and 7 = Huber's joint *M*-estimate of regression and scale (HUB-Reg).

the estimators based on absolute deviations (LAD and rank-LAD) increase the runtime by a factor between 15 and 50. These estimators are further outperformed by Huber's joint *M*-estimator of regression and scale, for which the runtime, on average, is only increased by a factor of 2.3. As shown in Figures 2.11 and 2.10, this computational efficiency comes at the cost of a small performance loss compared to Huber's *M*-estimator with an auxiliary scale estimate.

2.10   Concluding Remarks

In this chapter, robust estimation methods in the linear model were considered. First we reviewed the LS method that minimizes the RSS criterion function. We then illustrated how robustness can be imposed by replacing the nonrobust squared (ℓ_2-) loss function with a more robust robust absolute value (ℓ_1-) loss function, which yields the LAD regression estimator. The closely related rank LAD estimator was obtained by minimizing the LAD criterion for the pairwise differences of the residuals. We then

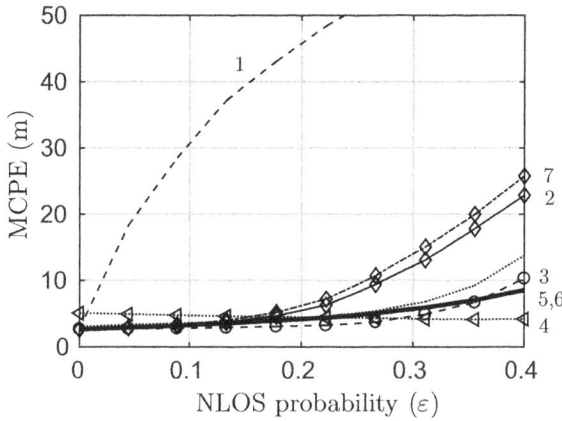

Figure 2.11 MCPE of different estimators as a function of the NLOS probability. 1 = LSE, 2 = Huber's M-estimate with auxiliary scale, 3 = Tukey's M-estimate, 4 = S-estimate, 5 = LAD estimate, 6 = rank-LAD estimate, and 7 = Huber's joint M-estimate of regression and scale.

reviewed ML-estimation and M-estimation of regression. The most commonly used robust loss functions in M-estimation, such as Tukey's and Huber's loss function, were discussed in detail. We then considered joint M-estimation of the regression and scale using Huber's criterion. In all cases, practical algorithms to compute the estimators were derived. We illustrated how the influence function and the breakdown point can be used to measure the robustness of the regression parameter estimators and reviewed the positive breakdown point estimators. Finally, simulation studies were used to illustrate the impact of outliers on different regression estimators.

3 Robust Penalized Regression in the Linear Model

In this chapter, we review and detail sparse and robust regression methods. Sparse regression methods have become increasingly important in modern data analysis due to both the measurement and the feature space being high-dimensional. In high-dimensional regression problems, the regression model is often *ill-posed*, that is, the number of inputs (covariates, predictors) p often exceeds the number of outputs (responses, measurements) N ($p > N$). In such cases, LS estimation no longer provides a unique solution, and infinitely many solutions are possible.

The popularity of sparse regression methods took off after the introduction of the Lasso (Least Absolute Shrinkage and Selection Operator) regression method proposed by Tibshirani (1996). Its influence has been significant in the natural sciences, engineering, and related fields. Sparse regression approaches are closely related to sparse signal reconstruction (SSR) and compressed sensing (CS) (Donoho, 2006; Candes and Wakin, 2008), where $\boldsymbol{\beta}$ (often referred to as signal vector) is assumed to be k-sparse (i.e., containing only k non-zero coefficients, $k \ll p$) or compressible (i.e., having only few large coefficients). The importance of SSR and CS is based on the fact that in many applications, the measured signal (e.g., an image) can have a *sparse* representation with respect to a suitable basis set (e.g., wavelet basis set). Making use of the underlying hidden sparsity can provide several advantages for both data compression and denoising.

3.1 Sparse Regression and Outline of the Chapter

As in Chapter 2, we consider the linear regression model,

$$y_i = \beta_0 + \mathbf{x}_{[i]}^{\top}\boldsymbol{\beta} + v_i, \quad i = 1, \ldots, N,$$

where $\beta_0 \in \mathbb{F}$ is the unknown intercept, $\boldsymbol{\beta} \in \mathbb{F}^p$ is the unknown regression coefficient vector, $y_i \in \mathbb{F}$ are the outputs (responses), and $\mathbf{x}_{[i]} \in \mathbb{R}^p$ are the inputs (predictors), $i = 1, \ldots, N$. Using matrix-vector notation, we can write the linear model as $\mathbf{y} = \beta_0 \mathbf{1} + \mathbf{X}\boldsymbol{\beta} + \mathbf{v}$, where $\mathbf{X} \in \mathbb{F}^{N \times p}$ is the input matrix with $\mathbf{x}_{[i]}$ as its row vectors and $\mathbf{y} = (y_1, \ldots, y_N)^{\top}$.

We then seek a solution $(\hat{\beta}_0, \hat{\boldsymbol{\beta}})$ to a penalized optimization problem,

$$\underset{(\beta_0, \boldsymbol{\beta}) \in \mathbb{F} \times \mathbb{F}^p}{\text{minimize}} L(\beta_0, \boldsymbol{\beta}) + \lambda P(\boldsymbol{\beta}) \tag{3.1}$$

where $L(\beta_0, \boldsymbol{\beta})$ is the *criterion function* (e.g., the RSS criterion) that measures the distance between $\mathbf{y}$ and the fit $\hat{\mathbf{y}} = \hat{\beta}_0\mathbf{1} + \mathbf{X}\hat{\boldsymbol{\beta}}$. The second term in (3.1), $P(\boldsymbol{\beta})$, is the *regularizer term* that incorporates the constraints, or prior structure, into the model and $\lambda \geq 0$ is the positive tuning or *penalty parameter*. This parameter is chosen by the user and balances the interplay between unrestricted regression estimation and the constraint. The regularizer term is commonly of the following form

$$P(\boldsymbol{\beta}) = \sum_{i=1}^{p} w_j P_j(\beta_j). \tag{3.2}$$

It consists of fixed positive *weights* w_j, chosen by the user, and the *shrinkage penalty function* $P_j(\beta_j)$ that measures the distance of β_j from 0. The penalty function $P_j(\cdot)$ is used to penalize large values of β_j and, with a proper choice of the penalty function and the penalty parameter, the solution $\hat{\boldsymbol{\beta}}$ can have coefficients that are shrunk to zero. The weights w_j, on the other hand, can be used to penalize some coefficients more heavily and/or to leave some coefficients unpenalized (when $w_j = 0$ for some j). Thus, the weights w_j allow the incorporation of prior knowledge into the optimization problem. Unless otherwise specified, however, we take $w_j \equiv 1$. Finally, note that shrinkage penalization is not applied to the intercept β_0. This is the main reason why the intercept β_0 and the regression coefficient vector $\boldsymbol{\beta} \in \mathbb{F}^p$ are explicitly separated in the criterion function $L(\cdot)$.

There are alternative formulations of the penalized optimization problem (3.2). The constrained form of (3.1) is

$$\underset{(\beta_0, \boldsymbol{\beta}) \in \mathbb{F} \times \mathbb{F}^p}{\text{minimize}} L(\beta_0, \boldsymbol{\beta}) \text{ subject to } P(\boldsymbol{\beta}) \leq t \tag{3.3}$$

where $L(\beta_0, \boldsymbol{\beta})$ and $P(\boldsymbol{\beta})$ are as defined earlier. Here, the *regularization threshold* $t \geq 0$ bounds the magnitude of the regression coefficients. For convex $L(\beta_0, \boldsymbol{\beta})$ and $P(\boldsymbol{\beta})$, these problems are equivalent; that is, for each value of t there is a corresponding λ that yields the same solution for (3.1) (Hastie et al., 2015). The choice of the optimization problem depends mainly on the application. Here we mainly work with the penalized form (3.1) but use the equivalent constrained form (3.3) in some instants.

The solution $(\hat{\beta}_0(\lambda), \hat{\boldsymbol{\beta}}(\lambda))$ to (3.1) is indexed by the penalty parameter $\lambda \geq 0$ (or, equivalently due to a one-to-one relation with (3.3), by the regularization threshold t). The penalty (or shrinkage) parameter plays a crucial role and its judicious selection is fundamental for obtaining good prediction or model selection. It controls the trade-off between the constraint (penalty) $P(\boldsymbol{\beta})$ and minimization of the data fit term $L(\beta_0, \boldsymbol{\beta})$; the larger λ is, the greater is the shrinkage of the coefficients toward zero. The case of $\lambda = 0$ is consistent with no shrinkage/regularization and (3.1) reduces to the conventional unrestricted estimation problem studied in Chapter 2. For each λ (or t), we have a solution and a set of λs trace out a *path of solutions*. In many applications/problems, one may wish to depict the whole solution path, that is, the graph of $[\hat{\boldsymbol{\beta}}(\lambda)]_j$ as a function of λ for each $j = 1, \ldots, p$.

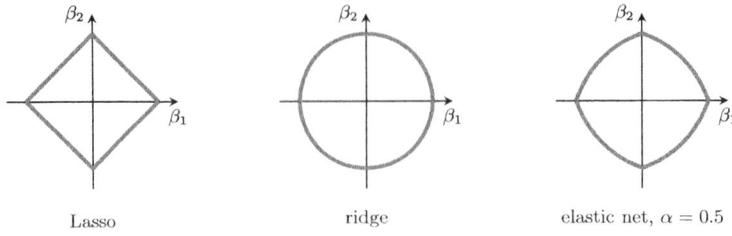

Figure 3.1 The constraint regions of the Lasso, ridge, and EN estimators.

The most commonly used shrinkage regularizer is the ℓ_1-norm of $\boldsymbol{\beta}$,

$$P_1(\boldsymbol{\beta}) = \|\boldsymbol{\beta}\|_1 = \sum_{j=1}^{p} |\beta_j|. \tag{3.4}$$

This corresponds to the case $w_j \equiv 1$ and $P_j(\beta_j) = |\beta_j|$, $\forall j = 1,\ldots,p$ in (3.2). In his seminal paper, Tibshirani (1996) proposed and studied the optimization problem (3.3) for $L(\beta_0, \boldsymbol{\beta})$ being specified by a RSS criterion and $P(\boldsymbol{\beta})$ having an ℓ_1-penalty. The Lasso, has become a benchmark method for sparse regression. There are many generalizations and extensions of the Lasso method such as the elastic net (EN) (Zou and Hastie, 2005), the fused Lasso (FL) (Tibshirani et al., 2005), the adaptive Lasso (Zou, 2006), the bridge estimator (Fu, 1998), and many others. A good review can be found in Hastie et al. (2015). In this chapter, robust alternatives to the Lasso estimator, and its extensions, are reviewed.

3.2   Extensions of the Lasso Penalty

An older idea of regularization of the regression model (and an inspiration for the Lasso) is *ridge regression (RR)* (Hoerl and Kennard, 1970), which uses a RSS criterion with an ℓ_2-norm and with $P_2(\boldsymbol{\beta}) = \|\boldsymbol{\beta}\|_2^2$, as the penalty function. It solves the problem,

$$\underset{(\beta_0, \boldsymbol{\beta}) \in \mathbb{F} \times \mathbb{F}^p}{\text{minimize}} \left\{ \frac{1}{2} \|\mathbf{y} - \beta_0 \mathbf{1} - \mathbf{X}\boldsymbol{\beta}\|_2^2 + \lambda \|\boldsymbol{\beta}\|_2^2 \right\}. \tag{3.5}$$

More generally, one could utilize an ℓ_q-norm penalty function,

$$P_q(\boldsymbol{\beta}) = \|\boldsymbol{\beta}\|_q^q = \sum_{j=1}^{p} |\beta_j|^q, \quad q \in (0, \infty).$$

In fact, prior to Lasso, Frank and Friedman (1993) proposed solving the RSS critertion with an ℓ_q-penalty, $q \geq 0$. They called the estimator the *Bridge estimator*. The ℓ_q-norm $\|\boldsymbol{\beta}\|_q$ is convex for $q \geq 1$ and nonconvex for $0 \leq q < 1$.

The constraint regions of unit $(\ell_q)^q$ balls, $\sum_{j=1}^{p} |\beta_j|^q \leq 1$, for $q = 1$ (Lasso penalty) and $q = 2$ (RR penalty) are illustrated in Figure 3.1. Also shown is the EN penalty function defined by (3.6) when $\alpha = 0.5$.

There are cases when the Lasso ($q = 1$) is not optimal and an ℓ_q penalty with $q > 1$ may provide a better solution. The well-known shortcomings of the Lasso are (Hastie et al., 2015):

1.   It does not cope well with *multicollinearity*. For this case, the coefficient paths tend to be erratic and can show wild behavior.
2.   It *does not select groups*. It may be useful to select the whole group if one variable amongst them is selected.
3.   If $p > N$, the Lasso selects at most N variables

For example, in microarray studies, it holds that $p \gg N$, and groups of genes in the same biological pathway tend to be expressed together, and, hence, measures of their expression tend to be strongly correlated (Hastie et al., 2015). One can think of these genes as forming a group. RR, can handle well-correlated predictors and has a tendency to select the whole group. It does not, however, give a sparse solution. As a remedy for the shortcomings of the Lasso, Zou and Hastie (2005) proposed the popular *EN* regression that uses a convex combination of ridge and Lasso penalties:

$$P_{\mathrm{EN}}(\boldsymbol{\beta}) = \frac{1}{2}(1 - \alpha)\|\boldsymbol{\beta}\|_2^2 + \alpha\|\boldsymbol{\beta}\|_1 = \sum_{i=1}^{p}\left\{\frac{1}{2}(1 - \alpha)|\beta_i|^2 + \alpha|\beta_i|\right\} \qquad (3.6)$$

where $\alpha \in [0, 1]$ is a user-chosen EN tuning parameter that can be varied. For $\alpha = 1$, one obtains the Lasso and for $\alpha = 0$ one obtains the RR estimator. The EN parameter α can be set on subjective grounds or one can include a (coarse) grid of values of α in a cross-validation scheme.

It is also possible to use weights in the Lasso penalty according to

$$P_1(\boldsymbol{\beta}; \mathbf{w}) = \sum_{j=1}^{p} w_j|\beta_j| \qquad (3.7)$$

where $w_1, \ldots, w_p$ are user-specified nonnegative weights. The standard Lasso is obtained when $w_1 = \ldots = w_p = 1$. The weighted Lasso penalty can naturally be incorporated into the EN as well. Such an approach was used, for example, by Tabassum and Ollila (2016). When the weights w_j also depend on the data, the approach is referred to as the *adaptive Lasso* (Zou, 2006) or the adaptive EN (Tabassum and Ollila, 2016).

3.3   The Lasso and the Elastic Net

Lasso solves the optimization problem

$$\underset{(\beta_0, \boldsymbol{\beta}) \in \mathbb{F} \times \mathbb{F}^p}{\text{minimize}} \; \frac{1}{2}\|\mathbf{y} - \beta_0\mathbf{1} - \mathbf{X}\boldsymbol{\beta}\|_2^2 + \lambda\|\boldsymbol{\beta}\|_1. \qquad (3.8)$$

The solution will often be denoted by $(\hat{\beta}_{0,\mathrm{LS}}(\lambda), \hat{\boldsymbol{\beta}}_{\mathrm{LS}}(\lambda))$ when there is a possibility for confusion. For $\lambda = 0$ (no penalization), the Lasso estimator reduces to the LSE,

$\hat{\boldsymbol{\beta}}_{LS}(0) = \hat{\boldsymbol{\beta}}_{LS}$ and $\hat{\beta}_{LS,0}(0) = \hat{\beta}_{0,LS}$. Lasso is essentially an ℓ_1-penalized LSE and it is often referred to as LS-Lasso. The EN solves the optimization problem,

$$\underset{(\beta_0,\boldsymbol{\beta})\in\mathbb{F}\times\mathbb{F}^p}{\text{minimize}} \frac{1}{2}\|\mathbf{y} - \beta_0\mathbf{1} - \mathbf{X}\boldsymbol{\beta}\|_2^2 + \lambda\left\{\frac{1}{2}(1-\alpha)\|\boldsymbol{\beta}\|_2^2 + \alpha\|\boldsymbol{\beta}\|_1\right\}. \tag{3.9}$$

Lasso (3.8) and RR (3.5) are special cases corresponding, respectively, to $\alpha = 1$ and $\alpha = 0$.

The reason why Lasso enforces sparse solutions can be visualized by using the constrained form (3.3) of the Lasso problem. For a moment, assume that $N > p$ and that $\mathbf{X}$ is of full rank. After centering the response $\mathbf{y}$ and the feature vectors so they have zero mean, the Lasso solution $\hat{\boldsymbol{\beta}}_{LS}(\lambda)$ can be equivalently found by solving

$$\underset{\boldsymbol{\beta}\in\mathbb{F}^p}{\text{minimize}} \frac{1}{2}(\hat{\boldsymbol{\beta}}_{LS} - \boldsymbol{\beta})^{\mathsf{H}}(\mathbf{X}^{\mathsf{H}}\mathbf{X})(\hat{\boldsymbol{\beta}}_{LS} - \boldsymbol{\beta}) \text{ subject to } \|\boldsymbol{\beta}\|_1 \le t$$

for some t that is a one-to-one mapping of the given fixed λ. Note that the equicontours of the RSS criterion are ellipsoids that are centered (minimized) at the point $\boldsymbol{\beta} = \hat{\boldsymbol{\beta}}_{LS}$ as is well known. Now the Lasso solution can be found as the point where the equicontours first touch the edge of the constraint region (the ℓ_1-ball). This is illustrated in Figure 3.2, where the optimization landscape is pictured for the $p = 2$ case. The ridge and the Lasso constraints (ℓ_1 and ℓ_2-balls) are illustrated with the diamond and the circle centered at the origin. Due to the sharp corners of the diamond-shaped ℓ_1-constraint, the elliptical equicontours of the RSS criterion are likely to touch the constraint region at an axis point, thus making the other coordinate equal to zero. In higher dimensions, this phenomenon becomes increasingly more likely as the ℓ_1-constraint resembles an inflated pin cushion. This feature of the Lasso makes it a very powerful sparsity-inducing variable selection method. On the contrary, the ridge constraint (ℓ_2-ball) does not have sharp edges. As a consequence, the RR estimates are only shrunk and the coefficients do not obtain a value that is exactly zero.

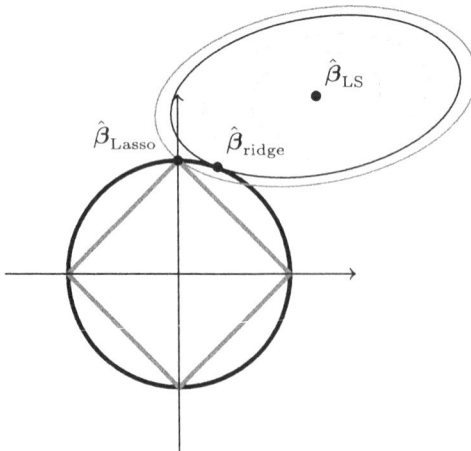

Figure 3.2 Penalized LS solutions for the Lasso and RR.

When solving the Lasso/EN optimization problem, it is common practice to standardize the feature vectors $\mathbf{x}_j \in \mathbb{F}^N$, which are the column vectors of $\mathbf{X} \in \mathbb{F}^{N \times p}$ ($j = 1, \ldots, p$). This is often advisable, as different input variables can have different scales, and standardizing brings them to the same scale. However, standardizing is not necessary if the input variables are measured with the same units. If the intercept is included in the linear model, then prior to standardizing the feature vectors, both the feature vectors and the response variable are centered. Centering is convenient as it allows the omission of the intercept in the Lasso optimization problem (3.8). The found estimate $\hat{\boldsymbol{\beta}}_{LS}(\lambda)$ based on the (centered and) standardized data set is then transformed to the original scale and the intercept is estimated at the last stage.[1]

The steps for solving the Lasso/EN optimization problem are detailed in the following text:

Step 1 (optional) If the intercept is included in the linear model, then the predictors and responses are centered, that is,

$$\mathbf{x}_j \leftarrow \mathbf{x}_j - \bar{x}_j \mathbf{1} \quad \text{and} \quad \mathbf{y} \leftarrow \mathbf{y} - \bar{y}\mathbf{1},$$

($j = 1, \ldots, p$) , where $\bar{x}_j = \frac{1}{N} \sum_{i=1}^N x_{ij}$ and $\bar{y} = \frac{1}{N} \sum_{i=1}^N y_i$ are the sample means of the response variables and the predictors, respectively.

Step 2 Standardize the feature vectors: $\mathbf{x}_j \leftarrow \mathbf{x}_j/\delta_j$ for $j = 1, \ldots, p$, where $\delta_j = \|\mathbf{x}_j\|_2$.

Note: After Steps 1 and 2, it follows that

$$\bar{\mathbf{x}} = (\bar{x}_1, \ldots, \bar{x}_p) = \frac{1}{n} \sum_{i=1}^N \mathbf{x}_{[i]} = \mathbf{0}_{p \times 1}, \qquad \mathbf{x}_j^H \mathbf{x}_j = 1, \quad \text{for all } j = 1, \ldots, p,$$

and $\bar{y} = 0$.

Step 3 Based on the (centered and) standardized data set $\mathcal{Z} = (\mathbf{y}, \mathbf{X})$, solve

$$\hat{\boldsymbol{\beta}}_{LS}(\lambda, \alpha) = \arg \min_{\boldsymbol{\beta} \in \mathbb{F}^p} \frac{1}{2} \|\mathbf{y} - \mathbf{X}\boldsymbol{\beta}\|_2^2 + \lambda \left\{ \frac{1}{2}(1 - \alpha)\|\boldsymbol{\beta}\|_2^2 + \alpha \|\boldsymbol{\beta}\|_1 \right\}.$$

Step 4 Transform the regression coefficient back to the original scale:

$$\hat{\beta}_{LS,j}(\lambda, \alpha) \leftarrow \frac{\hat{\beta}_{LS,j}(\lambda, \alpha)}{\delta_j} \quad \text{for } j = 1, \ldots, p.$$

If the intercept is in the model, it is computed as $\hat{\beta}_{0,LS}(\lambda, \alpha) = \bar{y} - \bar{\mathbf{x}}^\top \hat{\boldsymbol{\beta}}_{LS}(\lambda, \alpha)$.

The solution to Step 3 is discussed in Section 3.3.3.

3.3.1 Simple Linear Regression and Soft-Thresholding

Consider the simple linear regression model ($p = 1$), so the data set is $\mathcal{Z} = \{(y_i, x_i)\}_{i=1}^N$. We can assume, without loss of generality, that there is no intercept (otherwise assume

---

[1] This holds true for the RSS criterion, but not necessarily for other criteria.

that the responses and predictors are centered). We standardize the predictor so that $\langle \mathbf{x}, \mathbf{x} \rangle = \sum_{i=1}^{N} |x_i|^2 = 1$ holds. The EN optimization problem is

$$\underset{\beta \in \mathbb{F}}{\text{minimize}} \; \frac{1}{2} \sum_{i=1}^{N} |y_i - \beta x_i|^2 + \lambda \left\{ \alpha |\beta| + \frac{(1-\alpha)}{2} |\beta|^2 \right\} \qquad (3.10)$$

After reparametrizing, $\lambda_1 = \lambda \alpha \in \mathbb{R}_0^+$ and $\lambda_2 = \lambda(1 - \alpha) \in \mathbb{R}_0^+$, and noting that $\frac{1}{2} \sum_{i=1}^{N} |y_i - \beta x_i|^2 = C + \frac{1}{2} |\beta - \langle \mathbf{x}, \mathbf{y} \rangle|^2$ for a constant C not dependent on β, we may state the optimization problem (3.10) in an equivalent form

$$\underset{\beta \in \mathbb{F}}{\text{minimize}} \; \frac{1}{2} |\beta - \langle \mathbf{x}, \mathbf{y} \rangle|^2 + \lambda_1 |\beta| + \frac{\lambda_2}{2} |\beta|^2. \qquad (3.11)$$

When $\lambda = 0$ (no penalization), the solution to (3.11) is the LSE, $\hat{\beta}_{\text{LS}} = \langle \mathbf{x}, \mathbf{y} \rangle = \sum_{i=1}^{N} x_i^* y_i$. To describe the solution for the preceding problem, we need to define the *soft-thresholding (ST) operator*,

$$\text{soft}_\lambda(x) = \text{sign}(x)(|x| - \lambda)_+, \quad x \in \mathbb{F} \; (= \mathbb{C}, \text{or } \mathbb{R}) \qquad (3.12)$$

where $(t)_+ = \max(t, 0)$ denotes the positive part of $t \in \mathbb{R}$. As we shall see later, the ST operator is a basic building block that is used in many sparse linear regression algorithms. Its extension, the *shrinkage-soft-thresholding (SST) operator*,

$$\mathcal{S}(x; \lambda_1, \lambda_2) = \frac{\text{soft}_{\lambda_1}(x)}{1 + \lambda_2}, \quad x \in \mathbb{F}$$

expands the ST operator using ridge-style shrinking. The importance of these operators are revealed in the Lemma in the following text.

LEMMA 1   *The solution to the EN optimization problem (3.11) exists, is unique, and is given by*

$$\hat{\beta}_{\text{LS}}(\lambda, \alpha) = \mathcal{S}(\hat{\beta}_{\text{LS}} ; \lambda_1, \lambda_2) = \mathcal{S}(\langle \mathbf{x}, \mathbf{y} \rangle; \lambda_1, \lambda_2)$$

where $\lambda_1 = \lambda \alpha \in \mathbb{R}_0^+$ and $\lambda_2 = \lambda(1 - \alpha) \in \mathbb{R}_0^+$.

Proof   Let us denote the EN criterion function in (3.10) by $L_{\text{EN}}(\beta; \lambda, \alpha)$ and write $\hat{\beta}_{\text{LS}} = \langle \mathbf{x}, \mathbf{y} \rangle$ for the LSE. If $\beta \neq 0$, the target is differentiable and its complex conjugate derivative can be set equal to zero to obtain

$$0 = \frac{\partial}{\partial \beta^*} L_{\text{EN}}(\beta; \lambda, \alpha) \; \Leftrightarrow \; 0 = \frac{1}{2}(\beta - \hat{\beta}_{\text{LS}}) + \frac{\lambda_2}{2}\beta + \frac{\lambda_1}{2}\frac{\beta}{|\beta|}$$

$$\Leftrightarrow 0 = \beta\left(1 + \frac{\lambda_1}{|\beta|} + \lambda_2\right) - \hat{\beta}_{\text{LS}}$$

which shows that the solution must satisfy

$$\beta = \frac{\hat{\beta}_{\text{LS}}}{1 + (\lambda_1/|\beta|) + \lambda_2} = \frac{\hat{\beta}_{\text{LS}}|\beta|}{|\beta|(1 + \lambda_2) + \lambda_1}. \qquad (3.13)$$

Taking the modulus of both sides, yields the identity

$$|\beta|(1 + \lambda_2) + \lambda_1 = |\hat{\beta}_{\mathrm{LS}}| \quad \Leftrightarrow \quad |\beta| = \frac{|\hat{\beta}_{\mathrm{LS}}| - \lambda_1}{1 + \lambda_2} \tag{3.14}$$

as long as $|\hat{\beta}_{\mathrm{LS}}| > \lambda_1$. Substitution of (3.14) into (3.13) yields a closed-form solution to the EN optimization problem (3.10) as

$$\beta = \frac{\hat{\beta}_{\mathrm{LS}}(|\hat{\beta}_{\mathrm{LS}}| - \lambda_1)}{|\hat{\beta}_{\mathrm{LS}}|(1 + \lambda_2)} = \frac{\hat{\beta}_{\mathrm{LS}}}{|\hat{\beta}_{\mathrm{LS}}|} \cdot \frac{|\hat{\beta}_{\mathrm{LS}}| - \lambda_1}{1 + \lambda_2} = \mathrm{sign}(\hat{\beta}_{\mathrm{LS}}) \frac{|\hat{\beta}_{\mathrm{LS}}| - \lambda_1}{1 + \lambda_2}$$

given that $|\hat{\beta}_{\mathrm{LS}}| > \lambda_1$. When $|\hat{\beta}_{\mathrm{LS}}| \leq \lambda_1$, the solution is zero. Thus, for all values of $\hat{\beta}_{\mathrm{LS}}$

$$\beta = \mathrm{sign}(\hat{\beta}_{\mathrm{LS}}) \frac{(|\hat{\beta}_{\mathrm{LS}}| - \lambda_1)_+}{1 + \lambda_2} = \frac{\mathrm{soft}_{\lambda\alpha}(\hat{\beta}_{\mathrm{LS}})}{1 + \lambda(1 - \alpha)}$$

which concludes the proof.   □

Several remarks are in order. First, note that increasing the penalty λ will shrink the EN estimator $\hat{\beta}_{\mathrm{LS}}(\lambda, \alpha)$ linearly toward zero and, when $\lambda_1 = \lambda\alpha$ exceeds the magnitude of the LSE $|\hat{\beta}_{\mathrm{LS}}|$, the EN estimator becomes zero. Second, notice that the sign of the EN estimator is the same as the sign of the LSE, so the EN and the LSE share the same phase information, that is, $\mathrm{Arg}(\hat{\beta}_{\mathrm{LS}}) = \mathrm{Arg}(\hat{\beta}_{\mathrm{LS}}(\lambda, \alpha))$, only the magnitude of the EN is shrunk. Third, for fixed λ, decreasing α from its maximum of one toward its minimum of zero has the effect of only shrinking $\hat{\beta}_{\mathrm{LS}}(\lambda, \alpha)$ due to the scaling of the coefficient, but the estimator does not become zero, unless it is zero already from the start.

REMARK 3   *In the real-valued case ($\mathbb{F} = \mathbb{R}$), $\mathrm{sign}(x)$ reduces to the conventional sign function, as defined in (1.11). The ST operator ($\alpha = 1$) can be written in the form*

$$\mathrm{soft}_\lambda(x) = \begin{cases} x - \lambda & \text{if } x > \lambda \\ 0 & \text{if } |x| \leq \lambda \\ x + \lambda & \text{if } x < -\lambda, \end{cases}$$

which is shown in Figure 3.3 by the dotted line. The effect of extra shrinkage, due to the SST operator $\mathcal{S}(x; \lambda_1, \lambda_2)$, is illustrated in this figure for the case of $\alpha = 0.5$, but with the λ parameter being the same.

3.3.2   Subgradient Equations for the Lasso/Elastic Net

Suppose for a moment that the model does not have an intercept as defined by $\beta_0 = 0$. The RSS criterion function $L_{\mathrm{RSS}}(\boldsymbol{\beta}) = \frac{1}{2}\|\mathbf{y} - \mathbf{X}\boldsymbol{\beta}\|_2^2$ is convex (in fact strictly convex if $N > p$) and $\mathbb{R}$-differentiable, but the ℓ_1-penalty function $\|\boldsymbol{\beta}\|_1$ is not $\mathbb{R}$-differentiable at a point where at least one coordinate β_j is zero. Recall from Section 2.1 that the subdifferential of the modulus $|\beta_j|$ is

$$\partial|\beta_j| = \begin{cases} \frac{1}{2}\mathrm{sign}(\beta_j), & \text{for } \beta_j \neq 0 \\ \frac{1}{2}s & \text{for } \beta_j = 0 \end{cases}$$

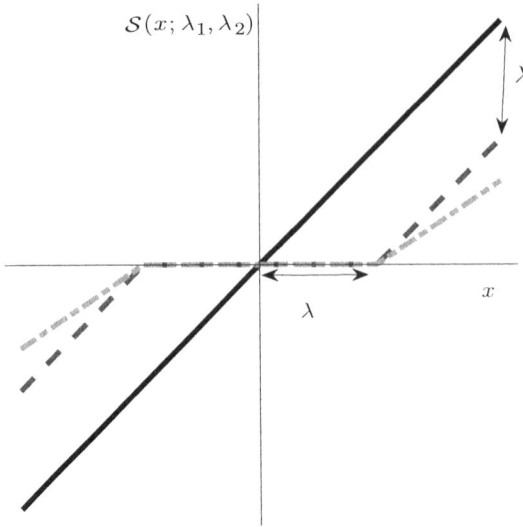

Figure 3.3 The SST operator $\mathcal{S}(x; \lambda_1, \lambda_2)$ for $\lambda_1 = \lambda\alpha$ and $\lambda_2 = \lambda(1 - \alpha)$ when $x \in \mathbb{R}$. The black solid line shows the case when $\lambda = 0$ (no penalization). The dashed line depicts the special case of $\alpha = 1$, which corresponds to the ST-operator $\text{soft}_\lambda(x)$. The dash-dotted line corresponds to the case of $\alpha = 0.5$ for the EN.

where s is some complex number satisfying $|s| \leq 1$. Thus, the subdifferential of $|\beta_j|$ is the usual complex conjugate derivative when $\beta_j \neq 0$, that is, $\partial|\beta_j| = \frac{\partial}{\partial\beta_j^*}|\beta_j|$ for $\beta_j \neq 0$.

Thus, $\hat{\boldsymbol{\beta}}$ is a solution to (3.8) if it verifies the zero subgradient equations

$$-\mathbf{x}_j^{\mathsf{H}}(\mathbf{y} - \mathbf{X}\hat{\boldsymbol{\beta}}) + \lambda\hat{s}_j = 0 \quad \text{for } j = 1, \ldots, p \tag{3.15}$$

where $\hat{s}_j \in \text{sign}(\hat{\beta}_j)$, meaning that it is equal to $\text{sign}(\hat{\beta}_j)$ if $\hat{\beta}_j \neq 0$ and some number inside the unit circle, $\hat{s}_j \in S = \{x \in \mathbb{F} : |x| \leq 1\}$, otherwise. Thus, $\hat{s}_j$ is two times an element of the subdifferential of $|\beta_j|$ evaluated at $\hat{\beta}_j$.

In the single predictor ($p = 1$) case, the Lasso estimate $\hat{\beta}$ solves the subgradient equation

$$-\langle \mathbf{x}, \mathbf{y} - \hat{\beta}\mathbf{x} \rangle + \lambda\hat{s} = 0 \Leftrightarrow -\langle \mathbf{x}, \mathbf{y} \rangle + \hat{\beta}\langle \mathbf{x}, \mathbf{x} \rangle = -\lambda\hat{s}$$
$$\Leftrightarrow \hat{\beta} = \langle \mathbf{x}, \mathbf{y} \rangle - \lambda\hat{s} \tag{3.16}$$

where it is assumed that $\mathbf{x}$ is standardized so that $\langle \mathbf{x}, \mathbf{x} \rangle = 1$. Thus, any pair $(\hat{\beta}, \hat{s})$ that solves equation (3.16) is a solution of the original minimization problem. Solving this equation then, gives the solution, $\hat{\beta}_{\text{LS}}(\lambda) = \text{soft}_\lambda(\langle \mathbf{x}, \mathbf{y} \rangle)$, derived earlier in Lemma 1.

3.3.3   Computation of the Lasso/Elastic Net

Next we describe the cyclic coordinatewise descent (CCD) algorithm (Friedman et al., 2007), which can be used to compute the Lasso solution path (at a grid of penalty parameter values) efficiently. Coordinate descent (CD) is an old optimization technique that uses the simple idea of optimizing one parameter (coordinate) at a time. It is particularly useful in the cases in which the single parameter problem is easy to solve. The CCD algorithm can be used to solve the general sparse linear regression objective function

$$f(\beta_1, \ldots, \beta_p) = L(\beta_1, \ldots, \beta_p) + \lambda \sum_{j=1}^{p} P_j(\beta_j)$$

when the criterion function $L(\boldsymbol{\beta})$ is convex and differentiable and the penalty term $P(\boldsymbol{\beta}) = \lambda \sum_{j=1}^{p} P_j(\beta_j)$ is convex (but not necessarily differentiable). The CCD algorithm solves coordinatewise minimization problems,

$$\hat{\beta}_j \leftarrow \underset{\beta_j \in \mathbb{F}}{\arg \min} f(\hat{\beta}_1, \hat{\beta}_2, \ldots, \hat{\beta}_{j-1}, \beta_j, \hat{\beta}_{j+1}, \ldots, \hat{\beta}_p), \qquad (3.17)$$

for each coordinate at a time (e.g., $j = 1, \ldots, p$) and cyclically repeats the procedure until convergence. It was shown in Tseng (2001) that the CCD algorithm converges to a global minimizer of the optimization problem.

The CCD algorithm is useful for penalized regression problems for at least two reasons. First and foremost, the coefficient update in (3.17) can be often computed in closed form by using simple ST as in Lemma 1 for Lasso/EN. Second, previous estimates in the decreasing sequence of penalty values provide good start values for subsequent iterations. Such start values result in the computation of a solution path to be much faster than computing a single solution for some arbitrary λ. Also, most coordinates at each update that are zero never become nonzero. This allows various enhancements and computational tricks to speed up the computations. See Hastie et al. (2015) for details.

Cyclic Coordinate Descent Algorithm

We describe here how Step 3 of the Lasso/EN can be solved using the CCD algorithm. The features are standardized (Step 2), that is, $\sum_{i=1}^{N} |x_{ij}|^2 = 1$ holds. Furthermore, if the intercept is in the model, then the data is assumed to be centered (Step 1), that is, $\sum_{i=1}^{N} y_i = 0$ and $\sum_{i=1}^{N} x_{ij} = 0$ for $j = 1, \ldots, p$.

The CCD algorithm for Lasso/EN was developed by Friedman et al. (2007) and Wu and Lange (2008) but can be extended to the complex-valued case in a straightforward manner. Define the *partial residuals* as

$$r_i^{(j)} = y_i - \sum_{k \neq j} x_{ik} \hat{\beta}_k$$

for $j = 1, \ldots, p$. The EN criterion function can be written in a partial residual form according to

$$L(\boldsymbol{\beta}; \lambda, \alpha) = \frac{1}{2} \sum_{i=1}^{N} \left| y_i - \underbrace{\sum_{k \neq j} x_{ik}\beta_k - x_{ij}\beta_j}_{\text{partial residual}} \right|^2 + \lambda \left\{ \alpha|\beta_j| + \frac{(1-\alpha)}{2}|\beta_j|^2 \right\}$$

$$+ \lambda \sum_{k \neq j} \left\{ \alpha|\beta_k| + \frac{(1-\alpha)}{2}|\beta_k|^2 \right\}.$$

When optimizing for β_j (holding other coefficients β_k fixed at their current iterates $\hat{\beta}_k$, $k \neq j$), the last term inside the parentheses can be treated as a constant. Consequently, by Lemma 1, we get a closed-form solution for the CCD coefficient update

$$\hat{\beta}_j \leftarrow \arg\min_{\beta_j} \frac{1}{2} \sum_{i=1}^{N} |r_i^{(j)} - x_{ij}\beta_j|^2 + \lambda \left\{ \alpha|\beta_j| + \frac{(1-\alpha)}{2}|\beta_j|^2 \right\} = \mathcal{S}(\langle \mathbf{x}_j, \mathbf{r}^{(j)} \rangle; \lambda_1, \lambda_2),$$

where $\lambda_1 = \lambda\alpha$ and $\lambda_2 = \lambda(1-\alpha)$. The preceding update for $\hat{\beta}_j$ can be written in a computationally more convenient, but equivalent, form:

$$\hat{\beta}_j \leftarrow \mathcal{S}(\hat{\beta}_j + \langle \mathbf{x}_j, \mathbf{r} \rangle; \lambda_1, \lambda_2) \tag{3.18}$$

where $\mathbf{r} = \mathbf{y} - \mathbf{X}\hat{\boldsymbol{\beta}}$ denotes the current full residual vector.

To apply the CCD method, one needs an initial start $\hat{\boldsymbol{\beta}}_{\text{init}} \in \mathbb{F}^p$ and then the CCD algorithm updates each coordinate at a time (e.g., $j = 1, \ldots, p$) using (3.18) and cyclically repeats the procedure until convergence. A pseudo-code for the CCD based EN algorithm is given in Algorithm 6. It assumes that the predictors are (centered and) standardized ($\mathbf{x}_j^{\mathsf{H}}\mathbf{x}_j = 1$). The MATLAB© implementation in RobustSP toolbox is enet.

Algorithm 6: enet: computes the EN solution using the CCD algorithm for (centered and) standardized data.

---

input   : $\mathbf{y} \in \mathbb{F}^N, \mathbf{X} \in \mathbb{F}^{N \times p}, \hat{\boldsymbol{\beta}}_{\text{init}}, \lambda > 0, \alpha \in (0, 1]$
output   : $\hat{\boldsymbol{\beta}} = \hat{\boldsymbol{\beta}}(\lambda, \alpha) \in \mathbb{F}^p$, a solution to EN optimization problem
initialize : $\hat{\boldsymbol{\beta}} \leftarrow \hat{\boldsymbol{\beta}}_{\text{init}}, \lambda_1 = \lambda\alpha$ and $\lambda_2 = \lambda(1-\alpha)$

1 **while** *stopping criteria not met* **do**

2   **for** $j = 1$ **to** p **do**

3          $\mathbf{r} \leftarrow \mathbf{y} - \mathbf{X}\hat{\boldsymbol{\beta}}$

4          $\hat{\beta}_j \leftarrow \mathcal{S}(\hat{\beta}_j + \langle \mathbf{x}_j, \mathbf{r} \rangle; \lambda_1, \lambda_2)$

Pathwise Coordinate Descent

In practice, the goal is to solve for the complete Lasso/EN path, that is, to solve $\hat{\boldsymbol{\beta}}_{\mathrm{LS}}(\lambda, \alpha)$ for a large range of λ values for a given fixed α. To achieve this, the Lasso/EN solutions are computed at $L + 1$ grid points of penalty values,

$$[\lambda] = \{\lambda_0, \ldots, \lambda_L\}, \quad \lambda_0 > \lambda_1 > \ldots > \lambda_L, \tag{3.19}$$

where the sequence $\{\lambda_i\}$ is monotonically decreasing from λ_0 to $\lambda_L \approx 0$ on a log scale. By default, we use $\lambda_L = \epsilon\lambda_0$, so $\lambda_j = \epsilon^{j/L}\lambda_0 = \epsilon^{1/L}\lambda_{j-1}$. We set $\lambda_0 = \lambda_{\max}(\alpha)$, where $\lambda_{\max}(\alpha)$ denotes the smallest penalty parameter λ for which the zero solution is obtained, that is, $\hat{\boldsymbol{\beta}}_{\mathrm{LS}}(\lambda_0, \alpha) = \mathbf{0}$, but $\hat{\boldsymbol{\beta}}_{\mathrm{LS}}(\lambda, \alpha) \neq \mathbf{0}$ for $\lambda < \lambda_0$. It is easy to show that

$$\lambda_{\max}(\alpha) = \max_j \frac{|\langle \mathbf{x}_j, \mathbf{y}\rangle|}{\alpha}.$$

The path algorithm starts from λ_0 and then decrease to λ_L using previous estimates as warm starts for the next λ on the grid for fixed $\alpha \in [0, 1]$.

The pseudo-code for the algorithm, referred to as pathwise CD EN algorithm, is detailed in Algorithm 7. The MATLAB© implementation in RobustSP toolbox is enet-path.

Algorithm 7: enetpath: computes the pathwise CD Lasso/EN algorithm.

> **input**   : $\mathbf{y} \in \mathbb{F}^N, \mathbf{X} \in \mathbb{F}^{N \times p}, \alpha \in (0, 1]$
> **output** : $\hat{\boldsymbol{\beta}}_{\mathrm{LS}}(\lambda, \alpha) \in \mathbb{F}^p$ computed on L grid points of λ from $\lambda_{\max}$ to $\epsilon\lambda_{\max}$
>
> **initialize:** $L = 120$ (grid size), $\epsilon = 10^{-3}$
> 1 $\lambda_0 \leftarrow \max_j \frac{|\langle \mathbf{x}_j, \mathbf{y}\rangle|}{\alpha}$.
> 2 $\hat{\boldsymbol{\beta}}_{\mathrm{init}} \leftarrow \mathbf{0}$
> 3 **for** $\lambda \in [\lambda] \equiv (\epsilon^{1/L}\lambda_0, \epsilon^{2/L}\lambda_0, \ldots, \epsilon\lambda_0)$ **do**
> 4 $\quad$ $\hat{\boldsymbol{\beta}}_{\mathrm{LS}}(\lambda, \alpha) \leftarrow$ enet$(\mathbf{y}, \mathbf{X}, \hat{\boldsymbol{\beta}}_{\mathrm{init}}, \lambda, \alpha)$ with enet given in Algorithm 6
> 5 $\quad$ $\hat{\boldsymbol{\beta}}_{\mathrm{init}} \leftarrow \hat{\boldsymbol{\beta}}_{\mathrm{LS}}(\lambda, \alpha)$

3.4   The Least Absolute Deviation-Lasso and the Rank-Lasso

We now consider the Lasso penalized regression of LAD and Rank-LAD criteria. The LAD-Lasso estimate is the solution of the convex ℓ_1-penalized LAD criterion,

$$(\hat{\beta}_{0,\mathrm{LAD}}(\lambda), \hat{\boldsymbol{\beta}}_{\mathrm{LAD}}(\lambda)) = \underset{\beta_0 \in \mathbb{F}, \boldsymbol{\beta} \in \mathbb{F}^p}{\arg\min} \sum_{i=1}^{N} \left| y_i - \beta_0 - \mathbf{x}_{[i]}^{\top}\boldsymbol{\beta} \right| + \lambda\|\boldsymbol{\beta}\|_1. \tag{3.20}$$

LAD-Lasso was proposed by Wang et al. (2007) for the real-valued case, and it has been studied and used in many subsequent papers, for example, Gao and Huang (2010). The

Rank-Lasso estimate is similarly defined as the solution of the convex ℓ_1-penalized rank regression criterion (Kim et al., 2015):

$$\hat{\boldsymbol{\beta}}_R(\lambda) = \arg \min_{\boldsymbol{\beta} \in \mathbb{F}^p} \sum_{i<j} |r_i - r_j| + \lambda \|\boldsymbol{\beta}\|_1,$$

where $r_i = y_i - \mathbf{x}_{[i]}^\top \boldsymbol{\beta}$ denotes the (noncentered) residuals.

3.4.1   Simple Linear Regression ($p = 1$)

To illustrate a simple linear regression, we first consider solving the LAD-Lasso criterion for the single predictor ($p = 1$) case without intercept ($\beta_0 = 0$). The starting point is to add an artificial $(N + 1)$th data point, with output $y_{N+1} = 0$ and input $x_{N+1} = \lambda$, to the data set $\mathcal{Z}$. We refer to this data set of $N+1$ observations as *augmented data* and the additional data point $z_{N+1} = (y_{N+1}, x_{N+1})$ as the *pseudo-observation*. As $\lambda|\beta| = |0 - \beta\lambda|$, the penalized LAD criterion can be rewritten according to

$$J_\lambda(\beta) = L_{\text{LAD}}(\beta) + \lambda|\beta| = \sum_{i=1}^{N+1} |b_i - \beta| \, w_i,$$

where $b_i = y_i/x_i$ and $w_i = |x_i|$, $i = 1, \ldots, N + 1$, are the slopes and the respective weights. This implies, for a given penalty $\lambda > 0$, that the LAD-Lasso solution is a weighted median of slopes $b_i \in \mathbb{F}$ with weighs $w_i > 0$, $i = 1, \ldots, N + 1$. The pseudo-observation corresponds to a slope $b_{N+1} = 0$ and a weight $w_{N+1} = \lambda$. For the real-valued case, (2.21) can be used to determine the solution and in the complex-valued case, (2.29) is iterated but with the augmented data set being used. Adding the Lasso penalty to the LAD or Rank-LAD criteria does not imply extra computational effort.

In the real-valued case, it is possible to compute the smallest penalty value, $\lambda_{\max}$, that shrinks the LAD-Lasso estimate to zero, in closed form. Given the ordered observations $b_{(i)}$ of original slopes, and their respective weights $w_{(i)}$, the rank of the slope $b_{N+1} = 0$ can be determined according to $k = \#\{b_i \leq 0; i = 1, \ldots, N\}$, where $\#\{\cdot\}$ denotes the cardinality of a set. The penalty $\lambda_{\max}$, can then be computed as

$$\lambda_{\max} = \left| 2 \sum_{i=1}^{k-1} w_{(i)} - \sum_{i=1}^{N} w_{(i)} \right|.$$

In the complex-valued case, a closed-form expression for $\lambda_{\max}$ is not possible.

When an intercept is included in the model, the LAD criterion can be written as

$$L_{\text{LAD}}(\boldsymbol{\beta}) = \sum_{i=1}^{N} \left| y_i - \mathbf{x}_{[i]}^\top \boldsymbol{\beta} \right|,$$

where $\boldsymbol{\beta} = (\beta_0, \beta_1)^\top$ and $\mathbf{x}_{[i]}^\top = (1, x_i)$. We then add a pseudo-observation $z_{N+1} = (y_{N+1}, \mathbf{x}_{N+1})$ to the data set, where $y_{N+1} = 0$ and $\mathbf{x}_{N+1} = (0, \lambda)$. The LAD-Lasso objective function can then be written as

$$L_\lambda(\boldsymbol{\beta}) = L_{\mathrm{LAD}}(\beta_0, \beta_1) + \lambda|\beta_1| = \sum_{i=1}^{N+1} \left| y_i - \mathbf{x}_{[i]}^\top \boldsymbol{\beta} \right|$$

and the computation of LAD-Lasso can be done using the alternating median algorithm described in Section 2.4.2.

Example 9 We continue our explorations with the data of Example 6. Here, we compute the LAD-Lasso solution when the penalty parameter is $\lambda = \lambda_{\max}$. Figure 3.4 depicts the original data along with the pseudo-observation $z_{N+1} = (y_{N+1}, x_{N+1}) = (0, \lambda_{\max})$. Also the LAD regression line (solid line) and the obtained LAD-Lasso line (dotted line) are shown. Recall that the LAD-Lasso estimate $\hat{\beta}_{\mathrm{LAD}}(\lambda)$ is computed as the weighted median of slopes $\{b_i\}_{i=1}^{N+1}$. A comparison of Figure 3.4 with Figure 2.5 shows a dramatic change. The pseudo-observation along the x-axis pulls the LAD line toward it. As a consequence $\hat{\beta}_{\mathrm{LAD}}(\lambda) = 0$.

3.4.2 The Computation of Least Absolute Deviation-Lasso and Rank-Lasso Estimates: $p > 1$ Case

In the $p > 1$ case, an augmented data set

$$\mathbf{y}_a = \begin{pmatrix} \mathbf{y} \\ \mathbf{0} \end{pmatrix} \in \mathbb{F}^{N+p}, \quad \mathbf{X}_a(\lambda) = \begin{pmatrix} \mathbf{1}_N & \mathbf{X} \\ \mathbf{0} & \lambda\mathbf{I}_p \end{pmatrix} \in \mathbb{F}^{(N+p)\times(p+1)}$$

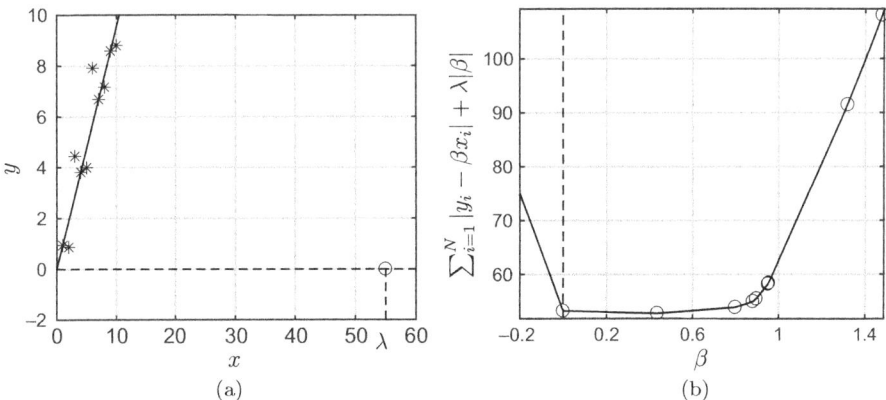

(a)

(b)

Figure 3.4 (a) Augmented data, the LAD fit (solid line) and the LAD-Lasso (dashed line) fit using $\lambda = \lambda_{\max}$. The pseudo-obervation $z_{N+1} = (0, \lambda_{\max})$ is marked by a circle. (b) Graph of $L_\lambda(\beta)$ for $\lambda = \lambda_{\max}$. The vertical line indicates the LAD-Lasso solution $\hat{\beta}_{\mathrm{LAD}}(\lambda) = 0$. The points of discontinuity $\beta = b_i = y_i/x_i$ are plotted with circles ($\circ$) on the graph. At $\lambda = \lambda_{\max}$, the LAD-Lasso solution is $\hat{\beta}_{\mathrm{LAD}}(\lambda) = 0$, which is the slope of the line passing through the pseudo-observation and the origin.

is utilized. Then, the optimization problem (3.20) of LAD-Lasso can be equivalently written as

$$(\hat{\beta}_{0,\text{LAD}}(\lambda), \hat{\boldsymbol{\beta}}_{\text{LAD}}(\lambda)) = \underset{\beta_0 \in \mathbb{F}, \boldsymbol{\beta} \in \mathbb{F}^p}{\arg\min} \left\| \mathbf{y}_a - \mathbf{X}_a(\lambda) \begin{pmatrix} \beta_0 \\ \boldsymbol{\beta} \end{pmatrix} \right\|_1$$

that is, LAD-Lasso solves the unpenalized LAD criterion for the augmented data set $\mathcal{Z} = (\mathbf{y}_a, \mathbf{X}_a(\lambda))$. This means that the IRWLS algorithm discussed in Section 2.4.3 can be used to find the LAD-Lasso estimate. Finally, note that the maximum penalty value, $\lambda_{\max}$, that shrinks all the coefficients to zero can be approximated according to

$$\lambda_{\max} \approx \|\mathbf{X}^{\mathsf{H}}\text{sign}(\mathbf{y})\|_\infty,$$

where $\|\mathbf{a}\|_\infty = \max_i |a_i|$ denotes the ℓ_∞-norm and $\big[\text{sign}(\mathbf{y})\big]_i = \text{sign}(y_i)$. This approximation is based on zero-subgradient equations.

Similarly, the Rank-Lasso estimate can be found by solving

$$\hat{\boldsymbol{\beta}}_{\text{R}}(\lambda) = \underset{\boldsymbol{\beta} \in \mathbb{F}^p}{\arg\min} \|\tilde{\mathbf{y}}_a - \tilde{\mathbf{X}}_a(\lambda)\boldsymbol{\beta}\|_1$$

based on *augmented pairwise measurements*

$$\tilde{\mathbf{y}}_a = \begin{pmatrix} \tilde{\mathbf{y}} \\ \mathbf{0} \end{pmatrix}, \quad \tilde{\mathbf{X}}_a(\lambda) = \begin{pmatrix} \tilde{\mathbf{X}} \\ \lambda \mathbf{I}_p \end{pmatrix}, \tag{3.21}$$

where $\tilde{\mathbf{y}}$ and $\tilde{\mathbf{X}}$ are defined in (2.40). The maximum penalty value is found similarly as

$$\lambda_{\max} \approx \|\tilde{\mathbf{X}}^{\mathsf{H}}\text{sign}(\tilde{\mathbf{y}})\|_\infty.$$

The algorithms for LAD-Lasso and Rank-Lasso are implemented as functions ladlasso and ranklasso in the MATLAB© RobustSP toolbox.

For LAD-Lasso and Rank-Lasso, the solution path can be computed in a manner similar to that specified in the pathwise EN algorithm, described in Algorithm 7. The procedure is described in Algorithm 8 for LAD-Lasso. The function ladlassopath implements this procedure in the robustSP toolbox and a similar algorithm, named ranklassopath, is implemented for Rank-Lasso.

Algorithm 8: ladlassopath: computes the pathwise LAD-Lasso algorithm.

---

    input : $\mathbf{y} \in \mathbb{F}^N$, $\mathbf{X} \in \mathbb{F}^{N \times p}$, grid size $L \in \mathbb{N}^+$, grid parameter $\epsilon \in [0, 1]$

    output: $(\hat{\beta}_{0,\text{LAD}}(\lambda), \hat{\boldsymbol{\beta}}_{\text{LAD}}(\lambda)) \in \mathbb{F}^{p+1}$ computed over L grid points of λ

1   $\lambda_0 \leftarrow \max_j |\langle \mathbf{x}_j, \text{sign}(\mathbf{y}) \rangle|$.

2   $\hat{\boldsymbol{\beta}}_{\text{init}} \leftarrow \mathbf{0}_{p \times 1}$, $\hat{\beta}_{0,\text{init}} \leftarrow \text{med}(\mathbf{y})$.

3   **for** $\lambda \in [\lambda] \equiv (\epsilon^{1/L}\lambda_0, \epsilon^{2/L}\lambda_0, \dots, \epsilon\lambda_0)$ **do**

4       $(\hat{\beta}_{0,\text{LAD}}(\lambda), \hat{\boldsymbol{\beta}}_{\text{LAD}}(\lambda)) \leftarrow \text{ladlasso}(\mathbf{y}, \mathbf{X}, \lambda, \hat{\boldsymbol{\beta}}_{\text{init}}, \hat{\beta}_{0,\text{init}})$

5       $\hat{\boldsymbol{\beta}}_{\text{init}} \leftarrow \hat{\boldsymbol{\beta}}_{\text{LAD}}(\lambda)$, $\hat{\beta}_{0,\text{init}} \leftarrow \hat{\beta}_{0,\text{LAD}}(\lambda)$

3.4.3   The Fused Rank-Lasso

To enforce block-sparsity and smoothness, Kim et al. (2015) considered the FL penalty with the rank-LAD criterion, and solved

$$\underset{\boldsymbol{\beta} \in \mathbb{F}^p}{\text{minimize}} \sum_{i<j} |r_i - r_j| + \lambda \|\boldsymbol{\beta}\|_1 + \lambda_2 \|\mathbf{D}\boldsymbol{\beta}\|_1, \tag{3.22}$$

where $\lambda_1, \lambda_2 \geq 0$ form a pair of fixed regularization parameters, chosen by the user, and

$$\|\mathbf{D}\boldsymbol{\beta}\|_1 = \sum_{j=2}^{p} |\beta_j - \beta_{j-1}|,$$

is the FL penalty where $\mathbf{D}$ is the $(p-1) \times p$ matrix

$$\mathbf{D} = \begin{pmatrix} (\mathbf{e}_2 - \mathbf{e}_1)^\top \\ (\mathbf{e}_3 - \mathbf{e}_2)^\top \\ \vdots \\ (\mathbf{e}_p - \mathbf{e}_{p-1})^\top \end{pmatrix} = \begin{pmatrix} -1 & 1 & 0 & \cdots & 0 & 0 \\ 0 & -1 & 1 & \cdots & 0 & 0 \\ \vdots & \vdots & \vdots & \ddots & \vdots & \vdots \\ 0 & 0 & 0 & \cdots & -1 & 1 \end{pmatrix}.$$

Here, $\mathbf{e}_1, \ldots, \mathbf{e}_p$ denote the basis vectors of $\mathbb{R}^p$, having a 1 at its ith element and 0's elsewhere. The FL penalty is also called total variation denoising in image processing and engineering (e.g., Rudin et al., 1992; Tibshirani et al., 2005). Note, again, that the optimization problem is convex and, hence, a global solution $\hat{\boldsymbol{\beta}}_{\text{R}}(\lambda_1, \lambda_2)$, referred to as the fused Rank-Lasso, can be computed efficiently. If $\lambda_2 = 0$, then we obtain the Rank-Lasso solution presented earlier. FL penalty encourages flatness of the magnitudes as a function of j and local constancy of the coefficient profile. Important practical applications of FL penalization can be found in areas, such as, protein mass spectroscopy, microarray gene expression, or in image denoising (Tibshirani et al., 2005).

The FL penalty can be incorporated into the Rank-LAD criterion $L_{\text{R}}(\boldsymbol{\beta})$ with minimal programming effort. This is in contrast to the RSS-criterion $L_{\text{RSS}}(\boldsymbol{\beta})$ as the FL penalty is not separable in terms of β_j-s and therefore the CCD algorithm cannot be utilized. Instead, specialized computationally involved algorithms need to be used; see Tibshirani et al. (2005) for details. The fused Rank-Lasso, on the contrary, can be computed with the same IRWLS algorithm as the Rank-LAD estimator. Define the *augmented fused measurements* as

$$\tilde{\mathbf{y}}_{\text{fa}} = \begin{pmatrix} \tilde{\mathbf{y}}_a \\ \mathbf{0} \end{pmatrix}, \quad \tilde{\mathbf{X}}_{\text{fa}}(\lambda_1, \lambda_2) = \begin{pmatrix} \tilde{\mathbf{X}}_a(\lambda_1) \\ \lambda_2 \mathbf{D} \end{pmatrix},$$

where $\tilde{\mathbf{X}}_a(\lambda_1)$ and $\tilde{\mathbf{y}}_{\text{fa}}$ are defined in (3.21). It is then easy to verify that the optimization problem (3.22) can be written in an equivalent form

$$\hat{\boldsymbol{\beta}}_{\text{R}}(\lambda_1, \lambda_2) = \underset{\boldsymbol{\beta} \in \mathbb{F}^p}{\text{minimize}} \left\| \tilde{\mathbf{y}}_{\text{fa}} - \tilde{\mathbf{X}}_{\text{fa}}(\lambda_1, \lambda_2)\boldsymbol{\beta} \right\|_1,$$

that is, the fused Rank-Lasso solves a standard LAD criterion based on a data set $(\tilde{\mathbf{y}}_{\text{fa}}, \tilde{\mathbf{X}}_{\text{fa}}(\lambda_1, \lambda_2))$. Hence, the IRWLS algorithm can be used to compute the fused

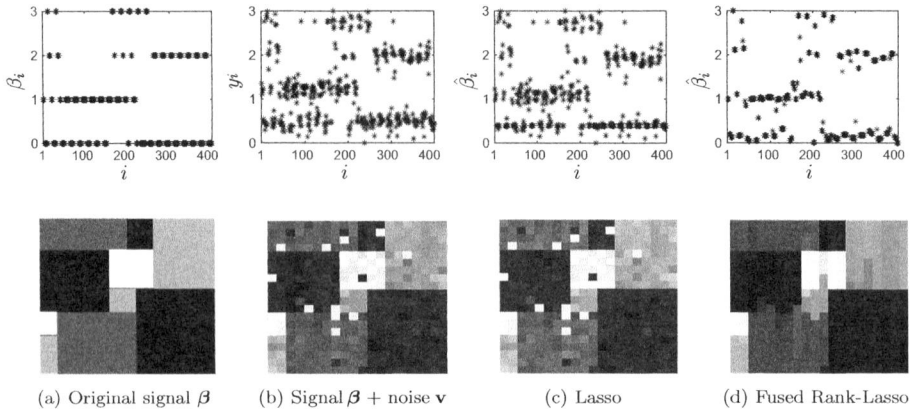

(a) Original signal $\boldsymbol{\beta}$ (b) Signal $\boldsymbol{\beta}$ + noise $\mathbf{v}$ (c) Lasso   (d) Fused Rank-Lasso

Figure 3.5 (a) The signal $\boldsymbol{\beta}$ represents a vectorized gray-scale image of squares shown in the bottom panel. (b) The observed measurement $\mathbf{y}$ is the signal $\boldsymbol{\beta}$ distorted by additive noise $\mathbf{v}$. (c, d) The observed $\mathbf{y}$ is given as an input to Lasso, and fused Rank-Lasso with $\mathbf{X} = \mathbf{I}_N$, and the best penalty parameter λ and (λ_1, λ_2), respectively, are chosen on a grid that provides the minimum MSE $\|\hat{\boldsymbol{\beta}} - \boldsymbol{\beta}\|_2^2$.

Rank-Lasso estimate. In the RobustSP toolbox, the function rankflasso implements this approach.

Image Denoising Example

Figure 3.5a shows a signal $\boldsymbol{\beta}$ comprising of $N = 400$ measurements where each measurement comprises of a block, of varying length, of numbers from the set $\{0, 1, 2, 3\}$. The observed noisy signal $\mathbf{y} = \boldsymbol{\beta} + \mathbf{v}$ is shown in Figure 3.5b. Note that the noise-free signal $\boldsymbol{\beta}$ is sparse, 43.75 percent of the measurements are equal to zero. The data sets are visualized as 20×20 gray-value images shown in the bottom panel. Given the knowledge of the noisy signal/image alone, the objective is to find a good approximation to the original noise-free signal/image. Due to the block-sparse nature of the signal, the FL penalty can offer efficient signal/image denoising. We computed the Lasso and fused Rank-Lasso signal approximations given knowledge of the noisy signal $\mathbf{y}$ and $\mathbf{X} = \mathbf{I}_N$ as the input matrix. The best signal approximation obtained by the Lasso $\hat{\boldsymbol{\beta}}_{1s}(\lambda)$ and the fused Rank-Lasso $\hat{\boldsymbol{\beta}}_{R}(\lambda_1, \lambda_2)$ are shown at Figure 3.5c and 3.5d, respectively. For both of the methods, we performed a grid search of penalty parameters λ and (λ_1, λ_2), and the denoised signals/images are the solutions that have the smallest MSE between the solution and the true (unobserved, noise-free) signal $\boldsymbol{\beta}$. The minimum MSE $\|\hat{\boldsymbol{\beta}} - \boldsymbol{\beta}\|^2$ was 15.8 for the fused Rank-Lasso and 93.2 for the Lasso. This difference in denoising is due to the fact that the fused Rank-Lasso has successfully exploited the spatial smoothness (block sparsity) and the constant coefficient profile of the data. As a result, it has provided significantly better signal approximation than the Lasso, as is illustrated in Figure 3.5d.

3.5   Joint Penalized *M*-estimation of Regression and Scale

Previous approaches have been based on determining an objective function for *M*-estimation in which an ℓ_1-penalty is added to a robust objective function (2.52) such as the LAD or Rank-LAD criterion. *M*-Lasso estimates of regression and scale (Ollila, 2016b,a) follow the estimating equation approach and define the estimators as solutions to the generalized zero subgradient equations of the Lasso problem. These equations depend on the convex and differentiable loss function ρ. To achieve robust estimators, Huber's loss function is utilized. However, when the loss function is the LS-loss, the equations reduce to zero subgradient equations of the Lasso problem (3.8). It is assumed that the regression model does not have an intercept. When the intercept is in the model, then the response and the predictors need to be centered. Estimation of the intercept is addressed in Section 3.5.1.

We assume that the loss function (cf. Definition 7 and discussion in Section 2.5) $\rho(x) : \mathbb{F} \to \mathbb{R}_0^+$ defined by $\rho(x) = \rho_0(|x|^2)$ is a differentiable and convex function, that is, that ρ_0 is differentiable. Recall that the pseudo-residual (2.63) is associated with the score function,

$$\psi(x) = \frac{\partial}{\partial x^*}\rho(x), \quad x \in \mathbb{C} \qquad (\text{resp. } \psi(x) = \rho'(x) \text{ for } x \in \mathbb{R}),$$

and is defined as

$$\mathbf{r}_\psi \equiv \mathbf{r}_\psi(\boldsymbol{\beta}, \sigma) = \psi\left(\frac{\mathbf{y} - \mathbf{X}\boldsymbol{\beta}}{\sigma}\right)\sigma,$$

where the notation is such that the score function ψ acts coordinatewise on the vector $\mathbf{r}/\sigma$, that is, $[\psi(\mathbf{r}/\sigma)]_i = \psi(r_i/\sigma)$. For the LS-loss function $\rho(x) = (1/\gamma)|x|^2$, where γ is defined in (2.43), $\psi(x) = x$, so $\mathbf{r}_\psi$ coincides with the usual definition of the residual, that is, $\mathbf{r}_\psi = \mathbf{y} - \mathbf{X}\boldsymbol{\beta} = \mathbf{r}$.

A natural estimate $\hat{\sigma} = \hat{\sigma}(\lambda)$ of the error scale σ associated with the Lasso estimate $\hat{\boldsymbol{\beta}} = \hat{\boldsymbol{\beta}}(\lambda)$ is

$$\hat{\sigma}^2 = \frac{1}{N}\sum_{i=1}^{N}\left|y_i - \mathbf{x}_{[i]}^\top\hat{\boldsymbol{\beta}}\right|^2 = \frac{1}{N}\|\hat{\mathbf{r}}\|_2^2, \tag{3.23}$$

where $\hat{\mathbf{r}} = \mathbf{y} - \mathbf{X}\hat{\boldsymbol{\beta}}$ denotes the residual vector at the Lasso solution. Thus, the Lasso estimates $(\hat{\boldsymbol{\beta}}, \hat{\sigma})$ of regression and scale are obtained as solutions to

$$\langle \mathbf{x}_j, \hat{\mathbf{r}} \rangle = \lambda \hat{s}_j \quad \text{for } j = 1, \ldots, p \tag{3.24}$$

$$\hat{\sigma}^2 = \frac{1}{N}\|\hat{\mathbf{r}}\|_2^2, \tag{3.25}$$

where $\hat{s}_i \in \text{sign}(\hat{\beta}_i)$, meaning that it is equal to $\text{sign}(\hat{\beta}_j)$ if $\hat{\beta}_j \neq 0$ and some number s inside the unit circle, $s \in S = \{x \in \mathbb{F} : |x| \leq 1\}$, otherwise. The *M*-Lasso estimates are defined as generalizations of the estimating equations (3.24) and (3.25).

DEFINITION 13   *The M-Lasso estimates $(\hat{\boldsymbol{\beta}}, \hat{\sigma}) \in \mathbb{F}^p \times \mathbb{R}^+$ of regression and scale are defined as solutions to the generalized zero subgradient equations,*

$$\langle \mathbf{x}_j, \hat{\mathbf{r}}_\psi \rangle = \lambda \hat{s}_j \quad for \, j = 1, \ldots, p \tag{3.26}$$

$$\frac{1}{N(2\alpha)} \sum_{i=1}^{N} \chi \left(\frac{y_i - \mathbf{x}_{[i]}^\top \hat{\boldsymbol{\beta}}}{\hat{\sigma}} \right) = \frac{1}{\gamma} \tag{3.27}$$

where $\alpha > 0$ is a fixed scaling factor, $\lambda > 0$ is the penalty parameter, γ is a constant defined in (2.43), $\hat{s}_j \in \mathsf{sign}(\hat{\beta}_j)$ and the function $\chi : \mathbb{F} \rightarrow \mathbb{R}_0^+$, is defined as

$$\chi(x) = 2\rho_0'(|x|^2)|x|^2 - \rho_0(|x|^2). \tag{3.28}$$

Here $\rho(x) : \mathbb{F} \rightarrow \mathbb{R}_0^+$ defined by $\rho(x) = \rho_0(|x|^2)$ is a convex and differentiable loss function. Equations (3.26) and (3.27) are referred to as the M-Lasso estimating equations.

Some remarks are in order. First, observe the similarity of (2.64)–(2.65) with (3.26)–(3.27). Indeed, for $\lambda = 0$ (no penalization), these estimating equations are equivalent. Hence, for $N > p$ and $\lambda = 0$, $(\hat{\boldsymbol{\beta}}, \hat{\sigma})$ is the global minimizer of Huber's criterion $L(\boldsymbol{\beta}, \sigma)$ of regression and scale. Second, for LS-loss $\rho(x) = (1/\gamma)|x|^2$ and $\alpha = 1/2$, the M-Lasso estimate is the Lasso estimate $\hat{\boldsymbol{\beta}}_{\mathrm{LS}}(\lambda)$ with the scale statistic in (3.23); see also Example 7. This can be verified by noting that (3.26)–(3.27) reduce to (3.24)–(3.25) for this case. Third, taking the modulus of both sides of (3.26) yields

$$|\langle \mathbf{x}_j, \hat{\mathbf{r}}_\psi \rangle| = \lambda, \quad \text{if } \hat{\beta}_j \neq 0 \tag{3.29}$$

$$|\langle \mathbf{x}_j, \hat{\mathbf{r}}_\psi \rangle| \leq \lambda, \quad \text{if } \hat{\beta}_j = 0. \tag{3.30}$$

Hence, whenever a component, say $\hat{\beta}_j$, of $\hat{\boldsymbol{\beta}}$ becomes nonzero, the corresponding absolute correlation between the pseudo-residual $\hat{\mathbf{r}}_\psi$ and the column $\mathbf{x}_j \in \mathbb{F}^N$ of $\mathbf{X}$, $|\langle \mathbf{x}_j, \hat{\mathbf{r}}_\psi \rangle|$, increases to λ. This is a well-known property of Lasso; see for example, Hastie et al. (2015) or Gerstoft et al. (2015) for the complex-valued case. This property is fulfilled by M-Lasso estimates by definition. Fourth, for Huber's loss function $\rho_{\mathrm{H},c}(x)$ defined in (2.54), the M-lasso estimating equations reduce to the form

$$\langle \mathbf{x}_j, \hat{\mathbf{r}}_\psi \rangle = \lambda \hat{s}_j \quad \text{for } j = 1, \ldots, p \tag{3.31}$$

$$\hat{\sigma}^2 = \frac{1}{N(2\alpha)} \| \hat{\mathbf{r}}_\psi \|_2^2 \tag{3.32}$$

For the scale (3.32), this result was shown in Example 8. One can verify the similarity with the Lasso solutions that verify (3.24)–(3.25).

The scaling factor α can be computed, as detailed in Section 2.6, to be

$$\alpha = (\gamma/2)\mathsf{E}[\chi(v)], \quad v \sim \mathcal{N}(0, 1). \tag{3.33}$$

When one uses Huber's function (2.54) to compute the M-Lasso estimates, the scaling factor $\alpha = \alpha(c)$ has the closed-form solution given in (2.71).

3.5.1 Algorithm

M-Lasso estimates can be computed using the CCD algorithm defined by Ollila (2016a). However, theoretical guarantees of convergence of the algorithm to the correct solution have not been established. The algorithm assumes (as CCD does for Lasso) that the predictors are standardized, so $\mathbf{x}_j^H \mathbf{x}_j = 1$ holds.

Recall that the CCD algorithm repeatedly cycles through the predictors updating one coordinate β_j at a time ($j = 1, \ldots, p$) while keeping others fixed at their current iterate values. At the jth step, the update for $\hat{\beta}_j$ is obtained by ST a conventional CD update $\hat{\beta}_j + \langle \mathbf{x}_j, \hat{\mathbf{r}} \rangle$, where $\hat{\mathbf{r}}$ denotes the residual vector $\hat{\mathbf{r}} = \mathbf{r}(\hat{\boldsymbol{\beta}})$ calculated at the current estimate $\hat{\boldsymbol{\beta}}$. For *M*-Lasso, similar updates are performed, but $\hat{\mathbf{r}}$ is replaced with the pseudo-residual vector $\hat{\mathbf{r}}_\psi$ and the update for scale is calculated prior to cycling through the coefficients. The pseudo-code for computing the *M*-Lasso estimates of regression and scale is given in Algorithm 9 using Huber's loss function. In the RobustSP toolbox, the function hublasso implements the method.

Algorithm 9: hublasso: computes the *M*-Lasso solution using $\rho_{H,c}(x)$.

input : $\mathbf{y} \in \mathbb{F}^N$, $\mathbf{X} \in \mathbb{F}^{N \times p}$, $\lambda > 0$, c, $\hat{\boldsymbol{\beta}}_{\text{init}}, \hat{\sigma}_{\text{init}}$
output : $(\hat{\boldsymbol{\beta}}, \hat{\sigma}) \in \mathbb{F}^p \times \mathbb{R}^+$, solution to equations in (3.31)–(3.32).
1   Compute consistency factor $\alpha = \alpha(c)$ in (2.71)

2   $\hat{\mathbf{r}}_\psi \leftarrow \psi_{H,c}\left(\dfrac{\mathbf{y} - \mathbf{X}\hat{\boldsymbol{\beta}}_{\text{init}}}{\hat{\sigma}_{\text{init}}} \right) \hat{\sigma}_{\text{init}}$

3   **while** *stopping criteria not met* **do**

4   update the scale

$$\hat{\sigma} \leftarrow \frac{1}{\sqrt{N(2\alpha)}} \| \hat{\mathbf{r}}_\psi \|_2$$

5   **for** $j = 1$ **to** p **do**

6   (re)update the pseudo-residual

$$\hat{\mathbf{r}}_\psi \leftarrow \psi_{H,c}\left(\frac{\mathbf{y} - \mathbf{X}\hat{\boldsymbol{\beta}}}{\hat{\sigma}} \right) \hat{\sigma}$$

7         $\hat{\beta}_j \leftarrow \text{soft}_\lambda\left(\hat{\beta}_j + \langle \mathbf{x}_j, \hat{\mathbf{r}}_\psi \rangle \right)$

It is also important to find the smallest penalty parameter $\lambda_{\max}$ that yields the all zeros solution, that is, $\hat{\boldsymbol{\beta}}(\lambda_{\max}) = \mathbf{0}$ and $\hat{\boldsymbol{\beta}}(\lambda) \neq \mathbf{0}$ for $\lambda < \lambda_{\max}$. It is easy to see that $\lambda_{\max}$ is

$$\lambda_{\max} = \max_j \left| \left\langle \mathbf{x}_j, \; \psi\left(\frac{\mathbf{y}}{\hat{\sigma}} \right) \hat{\sigma} \right\rangle \right|,$$

where $\hat{\sigma}$ is a solution to Huber's (scale only) criterion function,

$$\hat{\sigma} = \arg\min_{\sigma > 0} \frac{2N}{\gamma} (\alpha\sigma) + \sum_{i=1}^{N} \rho\left(\frac{y_i}{\sigma} \right) \sigma.$$

Thus, $\hat{\sigma}$ is a solution to

$$\frac{1}{N(2\alpha)} \sum_{i=1}^{N} \chi\left(\frac{y_i}{\hat{\sigma}}\right) = \frac{1}{\gamma}.$$

Finally as mentioned earlier, it is often desirable to solve for the whole solution path. This can be done as in Algorithm 7 for the Lasso/EN. In the RobustSP toolbox, the hublassopath function can be used to compute the whole M-Lasso solution path based on Huber's function. Modest modifications and extensions to M-Lasso are possible. For example, the adaptive M-Lasso approach that utilizes adaptive weights is easily incorporated as described by Ollila (2016a).

When the intercept is in the model, the output $\mathbf{y}$ and the feature vectors $\mathbf{x}_j$, $j = 1$, $\ldots, p$, are centered and standardized. The found estimates are transformed to the original scale at the end. The steps for solving the M-Lasso estimates are illustrated as follows:

> **Step 1 (optional)**   If the intercept is included in the linear model, the predictors and responses are centered;
>
> $$\mathbf{x}_j \leftarrow \mathbf{x}_j - \bar{x}_j \mathbf{1} \quad \text{and} \quad \mathbf{y} \leftarrow \mathbf{y} - \hat{\mu}_y \mathbf{1},$$
>
> $(j = 1, \ldots, p)$, where $\bar{x}_j = \frac{1}{N} \sum_{i=1}^{N} x_{ij}$ are the sample means of the predictors and $\hat{\mu}_y$ is a solution to Huber's criterion in the location-only case (i.e., $\mathbf{X} = \mathbf{1}_N$):
>
> $$(\hat{\mu}_y, \hat{\sigma}) = \underset{\mu \in \mathbb{F}, \sigma > 0}{\arg\min} \frac{2N}{\gamma} (\alpha\sigma) + \sum_{i=1}^{N} \rho\left(\frac{y_i - \mu}{\sigma}\right) \sigma.$$
>
> The solution can be found using the RobustSP toolbox function hubreg detailed in Algorithm 5.
>
> **Step 2**   Standardize the feature vectors: $\mathbf{x}_j \leftarrow \mathbf{x}_j / \delta_j$ for $j = 1, \ldots, p$, where $\delta_j = \|\mathbf{x}_j\|_2$.
> Note: After Steps 1 and 2, it holds that
>
> $$\bar{\mathbf{x}} = (\bar{x}_1, \ldots, \bar{x}_p) = \frac{1}{n} \sum_{i=1}^{N} \mathbf{x}_{[i]} = \mathbf{0}_{p \times 1}, \qquad \mathbf{x}_j^{\mathsf{H}} \mathbf{x}_j = 1, \quad \text{for all } j = 1, \ldots, p.$$
>
> and $\sum_{i=1}^{N} \psi((y_i - \hat{\mu}_y)/\hat{\sigma})\hat{\sigma} = 0$.
>
> **Step 3**   Based on the (centered and standardized) data set $\mathcal{Z} = (\mathbf{y}, \mathbf{X})$, solve the M-Lasso estimates of regression $\hat{\boldsymbol{\beta}}(\lambda)$ and scale $\hat{\sigma}(\lambda)$ using Algorithm 9.
> **Step 4**   Transform the regression coefficient back to the original scale:
>
> $$\hat{\beta}_j(\lambda) \leftarrow \frac{\hat{\beta}_j(\lambda)}{\delta_j} \quad \text{for } j = 1, \ldots, p.$$

If the intercept is in the model, it is computed as $\hat{\beta}_0(\lambda) = \hat{\mu}_y - \bar{\mathbf{x}}^{\top} \hat{\boldsymbol{\beta}}(\lambda)$.

3.6 Penalty Parameter Selection

Information criteria are commonly used for choosing among models while balancing the competing goals of goodness-of-fit and parsimony (simpler model). See Koivunen and Ollila (2014) for a review on model selection approaches. One of the most commonly used criteria is the Bayesian information criterion (BIC) proposed by Schwarz (1978). The generalized Bayesian Information Criterion (gBIC) for penalized regression is commonly defined as

$$\lambda^\star = \arg\min_{\lambda \in [\lambda]} \left\{ 2N \ln \hat{\sigma}(\lambda) + \mathrm{df}(\lambda) \cdot \ln N \right\}$$

where $\mathrm{df}(\lambda)$ is the number of nonzero elements in the penalized estimate $\hat{\boldsymbol{\beta}}(\lambda)$ (where λ denotes the penalty parameter) $\hat{\sigma}(\lambda)$ denotes the estimate of scale of the error distribution, and $[\lambda]$ denotes the grid of penalty values on the interval $(0, \lambda_{\max})$.

Different estimates of scale are naturally used for different penalized estimators. Let $r_i(\hat{\boldsymbol{\beta}}(\lambda)) = y_i - \mathbf{x}_{[i]}^\top \hat{\boldsymbol{\beta}}(\lambda)$, $i = 1, \ldots, N$, denote the residuals corresponding to a penalized regression estimate $\hat{\boldsymbol{\beta}}(\lambda)$. For the Lasso, the conventional estimate is the sample standard deviation

$$\hat{\sigma}(\lambda) = b_\lambda \cdot \sqrt{\frac{1}{N} \sum_{i=1}^{N} |\hat{r}_i(\lambda)|^2}. \tag{3.34}$$

and for the LAD-Lasso, the mean absolute deviation,

$$\hat{\sigma}(\lambda) = b_\lambda \cdot \frac{1}{N} \sum_{i=1}^{N} |\hat{r}_i(\lambda)| \tag{3.35}$$

is used. These are rather natural choices, as they are also the ML estimates of scale when the error distribution is Gaussian and Laplacian, respectively. For Rank-Lasso (3.34), the mean difference of residuals, defined by Gini (1921) as

$$\hat{\sigma}(\lambda) = b_\lambda \cdot \frac{2}{N(N-1)} \sum_{i<j} |\hat{r}_i(\lambda) - \hat{r}_j(\lambda)| \tag{3.36}$$

is used. In the equations (3.34)–(3.36), $b_\lambda > 0$ is a bias correction factor that may also depend on λ. It is used to mitigate the effect of bias in the estimate $\hat{\sigma}$ of σ. The bias in $\hat{\sigma}$ naturally increases as a function of the effective sample size $N/\mathrm{df}(\lambda)$. For example, it is well known that the sample standard deviation is biased, but applying the scale factor

$$b_\lambda = \sqrt{N/(N - \mathrm{df}(\lambda))} \tag{3.37}$$

in (3.34) gives an unbiased estimator of the scale σ in the conventional $(N > p)$ regression model, that is, in the case in which all selected regressors are significant. The bias correction factor is not known in closed form for the LAD and Rank-LAD criteria. Nevertheless, the same bias correction factor, b_λ, is used in (3.37) and also in (3.35) and (3.36) for the LAD-Lasso and Rank-Lasso, respectively.

3.7   Application Example: Prostate Cancer

We consider the benchmark prostate cancer data set ($N = 97, p = 8$) used in many textbooks, for example, Hastie et al. (2001). The goal is to predict the logarithm of the prostate-sepcific antigen (psa) measurement, log (`lpsa`), for men who are about to receive a radical prostatectomy, as a function of the number of clinical measures used. The study had a total of $N = 97$ observations of male patients aged from 41 to 79 years. The predictor variables are the logarithm of the cancer volume (`lcavol`), the logarithm of the prostate weight (`lweight`), the age of the patient (`age`), the logarithm of the benign prostatic hyperplasia amount (`lbph`), the presence or absence of seminal vesicle invasion (`svi`), the logarithm of the capsular penetration (`lbph`), the Gleason grade (`gleason`), and the percent Gleason grade 4 or 5 (`pgg45`).

Figure 3.6 depicts the coefficient paths of the Lasso estimates $\hat{\boldsymbol{\beta}}(\lambda)$ as λ ranges from $(0, \lambda_{\max})$ for the Lasso, LAD-Lasso, Rank-Lasso, and M-Lasso (using Huber's

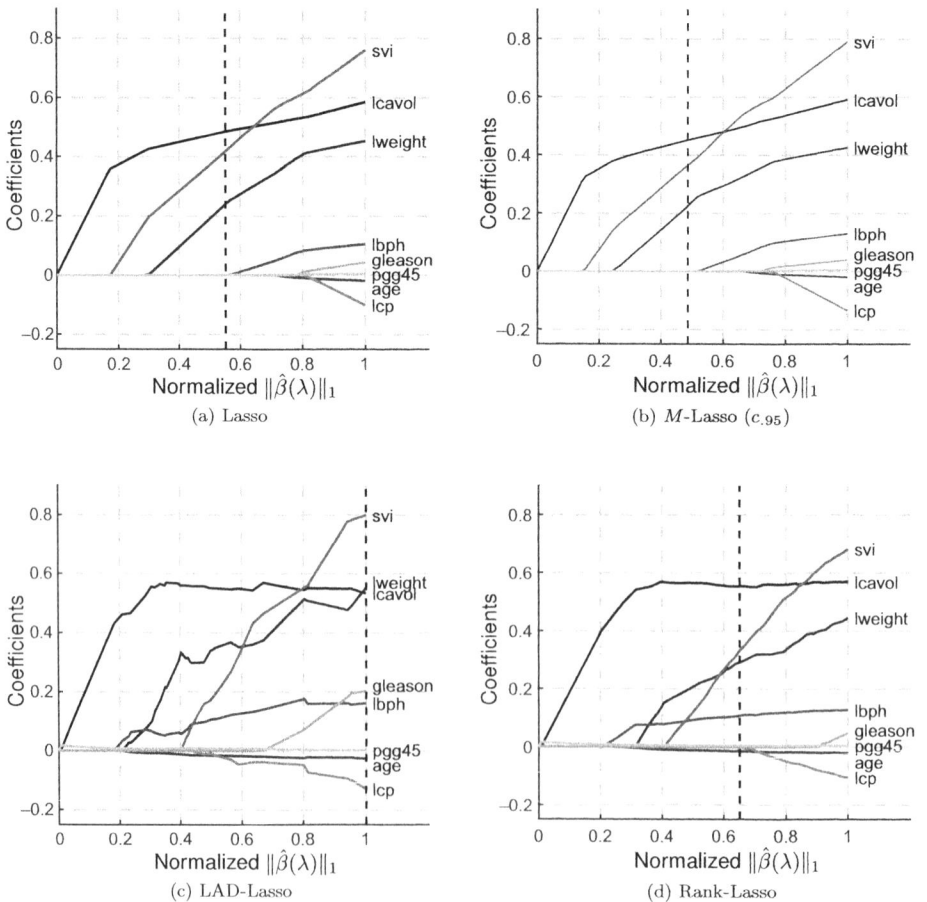

Figure 3.6 Coefficient paths of Lasso estimates for the prostate data set. The vertical line identifies the (generalized) BIC solution. Observe that the Rank-Lasso paths are *smoothed and monotone* versions of the LAD-Lasso paths whereas the Lasso and M-Lasso paths are similar.

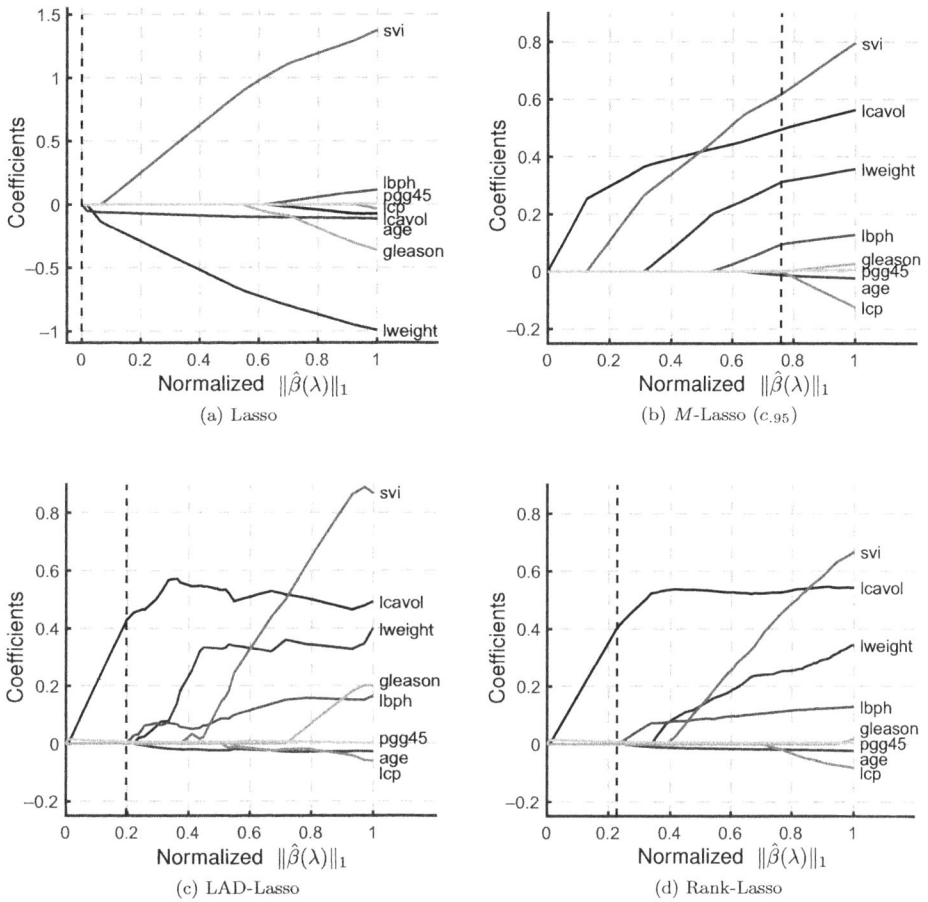

(a) Lasso

(b) M-Lasso ($c_{.95}$)

(c) LAD-Lasso

(d) Rank-Lasso

Figure 3.7 Coefficient paths of Lasso estimates for the prostate data set with an outlier. The vertical line identifies the (generalized) BIC solution. Observe that only the Lasso coefficient paths become completely corrupted.

loss function with threshold $c_{.95}$) estimates. It is instructive to compare the LAD-Lasso and Rank-Lasso coefficient paths because both are based on the ℓ_1 criterion. As can be seen, the LAD approach yields coefficient paths that are nonmonotone and highly nonsmooth with a visible zigzag feature. One can observe, in the outlier-free case that the Rank-Lasso coefficient paths can be described as smoother and monotone versions of the LAD-Lasso coefficient paths. Similarly, we can compare the Lasso and M-Lasso coefficient paths. Because the threshold used was rather high, Huber's loss function is similar to the LS loss function, except at the tails, where it resembles the ℓ_1-loss. Indeed, the results detailed in Figure 3.6 show that coefficient paths of Lasso and M-Lasso are rather similar and somewhat different in shape from LAD-Lasso and Rank-Lasso paths.

We also tested the effect of an outlier by changing y_1 to $y_1^* = 10\max(|y_i|_{i=1}^N)$ and recomputing the estimates. Figure 3.7 depicts the coefficient paths for the data with this outlier. Only the Lasso coefficient paths are completely changed, which is evident from a comparison of Figure 3.6a and 3.7a. Note that the LAD coefficient paths have

Table 3.1 LSE and Lasso solutions obtained by using the BIC.

	LSE	Lasso	Rank-Lasso	LAD-Lasso	M-Lasso
intercept	0.669	0.856	1.416	−0.295	0.985
lcavol	0.587	0.485	0.554	0.533	0.451
lweight	0.454	0.240	0.290	0.552	0.224
age	−0.020		−0.015	−0.027	
lbph	0.107		0.108	0.161	
svi	0.766	0.417	0.331	0.799	0.365
lcp	−0.105			−0.129	
gleason	0.045			0.203	
pgg45	0.005		0.005	0.004.	

changed more visibly than the Rank-Lasso coefficient paths. Although both methods are robust, Rank-Lasso appears *more stable locally*, that is, a small change in λ does not imply a large effect on the solution. This example illustrates the increased stability and robustness of Rank-Lasso compared to LAD-Lasso.

The (generalized) BIC solution for each method was computed, and the results are shown in Figures 3.6 and 3.7 as dotted vertical lines. The LSE and the Lasso solutions arising from use of the BIC are given in Table 3.1 for the original (noncorrupted) prostate cancer data. As can be seen, the solutions vary with the estimator used and such variation can be due to: firstly, the strong correlations that are present between the predictors. For example, the correlation between gleason and pgg45 is 0.752, but high correlations are found between other covariates as well. The correlation between svi and lcp is 0.673 and 0.675 between lcavol and lcp. The condition number is 16.9, which indicates a modest form of collinearity in the covariates. This indicates that other penalties, such as the EN or bridge penalty, might perform better for the data and provide a more stable solution. The EN penalty was utilized for prostate cancer data by Zou and Hastie (2005) and the bridge penalty was used by Fu (1998). This instability of the Lasso solutions is also observed when computing the optimal penalty parameter using cross-validation (CV). We found that the selected penalty parameter value $\lambda^\star$ chosen by five-fold CV was highly dependent on the folds selected. Due to this instability, the BIC was used in this example for penalty parameter selection as the results would be difficult to reproduce using CV.

3.8   Concluding Remarks

In this chapter, penalized (regularized) robust estimation methods in the linear model were considered. Penalized regression estimators are classically obtained by adding a penalty function to the criterion (objective) function. The penalty term shrinks the coefficients toward zero and the associated penalty parameter controls the amount of shrinkage. The most popular penalty function is the ℓ_1-norm that together with the RSS criterion gives the popular Lasso method. We also discussed extensions of the

ℓ_1-norm penalty such as the EN or the weighted ℓ_1-norm penalty function. We discussed how penalized robust regression methods, such as LAD-Lasso and Rank-Lasso, can be constructed in a similar manner by introducing the ℓ_1-norm penalty to the associated criterion function. The M-Lasso estimators are defined as a generalization of zero subgradient equations of the Lasso. In all cases, practical algorithms to compute the penalized regression estimators were derived. Finally, we illustrated the performance of different estimators with a real prostate cancer data set.

4 Robust Estimation of Location and Scatter (Covariance) Matrix

Many data mining, signal processing, and classic multivariate analysis techniques require an estimate of the covariance matrix or some nonlinear function of it (e.g., the inverse covariance matrix or its eigenvalues/eigenvectors). Given an i.i.d. sample $\mathbf{x}_1, \ldots, \mathbf{x}_N$ of real or complex p-variate observations from an (unspecified) p-variate distribution $\mathbf{x} \sim F$, the *sample mean* and *sample covariance matrix* (SCM) defined according to

$$\bar{\mathbf{x}} = \frac{1}{N} \sum_{i=1}^{N} \mathbf{x}_i \quad \text{and} \quad \mathbf{S} = \frac{1}{N} \sum_{i=1}^{N} (\mathbf{x}_i - \bar{\mathbf{x}})(\mathbf{x}_i - \bar{\mathbf{x}})^{\mathsf{H}} \tag{4.1}$$

are the most commonly used estimators for the unknown population mean $\boldsymbol{\mu} = \mathsf{E}[\mathbf{x}]$ and the covariance matrix $\boldsymbol{\Sigma} = \mathrm{cov}(\mathbf{x}) = \mathsf{E}[(\mathbf{x} - \boldsymbol{\mu})(\mathbf{x} - \boldsymbol{\mu})^{\mathsf{H}}]$. When the samples are randomly chosen from a multivariate (real or complex) normal distribution $\mathcal{N}_p(\boldsymbol{\mu}, \boldsymbol{\Sigma})$, these classical descriptive statistics provide a sufficient summary of the data set and correspond to the maximum likelihood estimates of the unknown parameters $\boldsymbol{\mu}$ and $\boldsymbol{\Sigma}$. We assume that the covariance matrix is nonsingular and, hence, positive definite, denoted $\boldsymbol{\Sigma} \succ 0$ or $\boldsymbol{\Sigma} \in \mathbb{S}_{++}^p$, where $\mathbb{S}_{++}^p$ is used to denote the set of all positive definite real (or complex) symmetric (or Hermitian) matrices.

It is well known, though, that the statistics $(\bar{\mathbf{x}}, \mathbf{S})$ are not robust against nonnormality, especially against longer tailed distributions, and that they are highly influenced by outliers. A variety of robust covariance matrix estimators have been proposed in the literature since the fundamental contribution of Maronna (1976) on M-estimators (short for ML-type estimators) for location and the covariance matrix. Of importance are the S-estimators (Davies, 1987), MM-estimators (Tatsuoka and Tyler, 2000), minimum volume ellipsoid (MVE) and minimum covariance determinant (MCD) estimators (Rousseeuw, 1985), and multivariate sign and rank-based covariance matrix estimators (Visuri et al., 2000a; Ollila et al., 2003a, 2004), among others.

This chapter focuses only on M-estimators for the covariance (scatter) matrix as these estimators have been by far the most popular estimators in the signal processing literature. This is partially due to the fact that they can be generalized to the complex-valued data case in a fairly straightforward manner, and the theory of ML- and M-estimation for the (complex or real) covariance matrix can be developed in parallel. In addition, the global minimizer of the M-estimation objective function can be found by a rather simple fixed-point algorithm whereas other estimators lack an algorithm that finds a global solution. The usefulness of robust covariance estimators is illustrated for the case of signal detection using a normalized matched filter. As detailed in Chapter 5, robust covariance estimators are extensively used in array signal processing applications.

4.1   Complex Vector Space Isomorphism and Complex Distributions

Complex distributions, and the related statistics that characterize them, have been a topic of active research in signal processing; see for example, Picinbono (1994), van den Bos (1995), Eriksson et al. (2010), Ollila (2010), Schreier and Scharf (2010), Adali et al. (2011), and Ollila et al. (2011a). In this section, we provide a short summary of definitions and results that are relevant for this chapter.

A complex vector is denoted by $\mathbf{x} = \mathbf{x}_r + \jmath \mathbf{x}_i$, where $\jmath = \sqrt{-1}$ denotes the imaginary unit and $\mathbf{x}_r = \mathsf{Re}(\mathbf{x})$ and $\mathbf{x}_i = \mathsf{Im}(\mathbf{x})$ in $\mathbb{R}^p$ denote the real and imaginary part of $\mathbf{x}$. The complex conjugate and the ℓ_2-norm of $\mathbf{x}$ are defined as $\mathbf{x}^* = \mathbf{x}_r - \jmath \mathbf{x}_i$ and $\|\mathbf{x}\|_2 = (\mathbf{x}^H \mathbf{x})^{1/2}$. The ℓ_2-norm for a single variable is known as the *modulus*, $|x| = (x^* x)^{1/2} = (x_r^2 + x_i^2)^{1/2}, \forall x = x_r + \jmath x_i \in \mathbb{C}$.

First, the isomorphism $[\cdot]_\mathbb{R}$ between $\mathbb{C}^p$ and $\mathbb{R}^{2p}$ (Andersen et al., 1995) is given by

$$[\mathbf{x}]_\mathbb{R} = \begin{pmatrix} \mathbf{x}_r \\ \mathbf{y}_i \end{pmatrix}.$$

Next, let $\mathbf{C} = \mathbf{A} + \jmath \mathbf{B} \in \mathbb{C}^{n \times p}$, where $\mathbf{A}, \mathbf{B} \in \mathbb{R}^{n \times p}$. Also define the mapping $\{\cdot\}_\mathbb{R} : \mathbb{C}^{n \times p} \mapsto \mathbb{R}^{2n \times 2p}$ as

$$\{\mathbf{C}\}_\mathbb{R} = \begin{pmatrix} \mathbf{A} & -\mathbf{B} \\ \mathbf{B} & \mathbf{A} \end{pmatrix}. \tag{4.2}$$

In terms of notation, the real $2n \times 2p$ matrix on the right hand side of (4.2) is said to have a *complex form*. Next, observe the obvious relation

$$[\mathbf{C}\mathbf{x}]_\mathbb{R} = \{\mathbf{C}\}_\mathbb{R}[\mathbf{x}]_\mathbb{R}, \tag{4.3}$$

that is, a ($\mathbb{C}$-)linear transform $\mathbf{C}\mathbf{x}$ of a vector $\mathbf{x}$ in $\mathbb{C}^p$ corresponds to a restricted linear transform of a vector $[\mathbf{x}]_\mathbb{R}$ in $\mathbb{R}^{2p}$, namely, multiplication of $[\mathbf{x}]_\mathbb{R} \in \mathbb{R}^{2p}$ by the real matrix $\{\mathbf{C}\}_\mathbb{R} \in \mathbb{R}^{2n \times 2p}$ of complex form.

The following theorem (mostly adapted from Ollila et al., 2012) gives some basic results relating to the mappings defined previously.

THEOREM 2   *Let $\mathbf{C} \in \mathbb{C}^{p \times p}$ and $\mathbf{x} \in \mathbb{C}^p$. The following properties hold:*

(a) $\{\mathbf{DE}\}_\mathbb{R} = \{\mathbf{D}\}_\mathbb{R}\{\mathbf{E}\}_\mathbb{R}$ *for* $\mathbf{D} \in \mathbb{C}^{n \times p}$, $\mathbf{E} \in \mathbb{C}^{p \times q}$.
(b) $\{a\mathbf{C} + b\mathbf{D}\}_\mathbb{R} = a\{\mathbf{C}\}_\mathbb{R} + b\{\mathbf{D}\}_\mathbb{R}$ *for* $\mathbf{D} \in \mathbb{C}^{p \times p}, a, b \in \mathbb{R}$.
(d) $\mathbf{x}^H \mathbf{C}\mathbf{x} = [\mathbf{x}]_\mathbb{R}^\top \{\mathbf{C}\}_\mathbb{R}[\mathbf{x}]_\mathbb{R}$
(e) $\mathbf{C}$ *is Hermitian if and only if* $\{\mathbf{C}\}_\mathbb{R}$ *is symmetric.*
(f) $\mathbf{C}$ *is unitary if and only if* $\{\mathbf{C}\}_\mathbb{R}$ *is orthogonal.*
(g) $\mathbf{C}$ *is nonsingular if and only if* $\{\mathbf{C}\}_\mathbb{R}$ *is nonsingular. Furthermore* $\{\mathbf{C}^{-1}\}_\mathbb{R} = \{\mathbf{C}\}_\mathbb{R}^{-1}$.

(h) $\mathbf{C}$ *is positive definite Hermitian if and only if* $\{\mathbf{C}\}_{\mathbb{R}}$ *is positive definite symmetric.*
       The eigenvalues of $\{\mathbf{C}\}_{\mathbb{R}}$ *are the eigenvalues of* $\mathbf{C}$ *with multiplicity 2 and* $|\{\mathbf{C}\}_{\mathbb{R}}| = |\mathbf{C}|^2$.

(i) $\mathbf{C} = \mathbf{A}\mathbf{A}^H$ *if and only if* $\{\mathbf{C}\}_{\mathbb{R}} = \{\mathbf{A}\}_{\mathbb{R}}\{\mathbf{A}\}_{\mathbb{R}}^{\top}$ *for* $\mathbf{C} \in \mathbb{S}_{++}^p$.

A complex random vector (r.v.) $\mathbf{x} = (x_1, \ldots, x_p)^{\top} = \mathbf{x}_r + J\mathbf{y}_i \in \mathbb{C}^p$ comprises of a pair of real r.v.'s $\mathbf{x}_r$ and $\mathbf{x}_i$ in $\mathbb{R}^p$. The distribution of $\mathbf{x}$ on $\mathbb{C}^p$ determines the joint real $2p$-variate distribution of $\mathbf{x}_r$ and $\mathbf{y}_i$ on $\mathbb{R}^{2p}$ and conversely due to the isomorphism between $\mathbb{C}^p$ and $\mathbb{R}^{2p}$. Hence, its distribution is identified with the joint (real $2p$-variate) distribution of the composite real r.v. $[\mathbf{x}]_{\mathbb{R}}$:

$$F_{\mathbf{x}}(\mathbf{x}') = \mathsf{Prob}([\mathbf{x}]_{\mathbb{R}} \leq [\mathbf{x}']_{\mathbb{R}}),$$

In a similar manner, the pdf of $\mathbf{x} = \mathbf{x}_r + J\mathbf{x}_i$ is identified with the joint pdf $f(\mathbf{x}) = f([\mathbf{x}]_{\mathbb{R}}) \equiv f(\mathbf{x}_r, \mathbf{x}_i)$ of the real r.v.s $\mathbf{x}_r$ and $\mathbf{x}_i$.

A complex r.v. can be characterized using the symmetry properties of its distribution. The most commonly made symmetry assumption in the statistical signal processing literature is that of *circular symmetry* (Picinbono, 1994). A complex r.v. $\mathbf{x}$ is said to have a *circularly symmetric* distribution about $\boldsymbol{\mu} \in \mathbb{C}^p$, called the *symmetry center*, if

$$(\mathbf{x} - \boldsymbol{\mu}) \stackrel{\mathrm{d}}{=} e^{J\theta}(\mathbf{x} - \boldsymbol{\mu}), \quad \forall \theta \in \mathbb{R}, \tag{4.4}$$

where the notation $\stackrel{\mathrm{d}}{=}$ should read "has same distribution as." If the r.v. $\mathbf{x}$ has finite 1st-order moments, then the symmetry center is equivalent with its *mean vector*, that is,

$$\boldsymbol{\mu} = \mathsf{E}[\mathbf{x}] = \mathsf{E}[\mathbf{x}_r] + J\mathsf{E}[\mathbf{x}_i].$$

We say that a r.v. $\mathbf{x}$ is *circular* if it is circularly symmetric about the origin $\mathbf{0}$ in which case its distribution remains invariant under multiplication by any (complex) number on the complex unit circle. A circular r.v. $\mathbf{x}$, in general, does not necessarily possess a density. However, if it does, then its pdf $f(\mathbf{x})$ satisfies $f(e^{J\theta}\mathbf{x}) = f(\mathbf{x})$, $\forall \theta \in \mathbb{R}$.

The *covariance matrix* of a complex r.v. $\mathbf{x} = \mathbf{x}_r + J\mathbf{x}_i$ is defined as

$$\begin{aligned}
\mathsf{cov}(\mathbf{x}) &= \mathsf{E}\big[(\mathbf{x} - \mathsf{E}[\mathbf{x}])(\mathbf{x} - \mathsf{E}[\mathbf{x}])^H\big] \\
&= \mathsf{cov}(\mathbf{x}_r) + \mathsf{cov}(\mathbf{x}_i) + J\{\mathsf{cov}(\mathbf{x}_i, \mathbf{x}_r) - \mathsf{cov}(\mathbf{x}_r, \mathbf{x}_i)\}.
\end{aligned}$$

Recall that the class of (real or complex) covariance matrices is the same as the class of all positive semidefinite matrices, so $\mathsf{cov}(\mathbf{x}) \succeq \mathbf{0}$. Also note that the diagonal elements of $\mathsf{cov}(\mathbf{x})$ collect the variances, that is, $[\mathsf{cov}(\mathbf{x})]_{ii} = \mathsf{var}(x_i) = \mathsf{E}[|x_i - \mathsf{E}[x_i]|^2]$, $i = 1, \ldots, p$ and if $\mathbf{x}$ is nondegenerate in any subspace of $\mathbb{C}^p$, then $\mathsf{cov}(\mathbf{x}) \succ \mathbf{0}$.

Circularity obviously implies some structure on the covariances between the elements of $\mathbf{x}$. If $\mathbf{x}$ has a circular distribution, and possesses finite 2nd-order moments, then $\boldsymbol{\mu} = \mathsf{E}[\mathbf{x}]$ and the covariance matrix $\mathsf{cov}(\mathbf{x})$ has the so-called *2nd-order circularity* (Picinbono, 1994) property

$$\mathsf{cov}(\mathbf{x}_r) = \mathsf{cov}(\mathbf{x}_i) \quad \text{and} \quad \mathsf{cov}(\mathbf{x}_r, \mathbf{x}_i) = -\mathsf{cov}(\mathbf{x}_i, \mathbf{x}_r), \tag{4.5}$$

which is equivalent to stating that the real $2p \times 2p$ covariance matrix of $[\mathbf{x}]_{\mathbb{R}}$ is of complex form:

$$\text{cov}([\mathbf{x}]_{\mathbb{R}}) = (1/2)\{\text{cov}(\mathbf{x})\}_{\mathbb{R}}. \tag{4.6}$$

The first property states that the covariance matrices of the real and complex parts are equal and the covariance matrix between the real and complex parts is skew-symmetric.

In this chapter, we study an important subclass of circularly symmetric distributions, referred to as complex elliptically symmetric (CES) distributions (Ollila et al., 2012). These form an important extension of multivariate Gaussian distributions, and include many of the commonly used distributions such as the complex multivariate t-distribution, the generalized Gaussian (also referred to as the power exponential distribution), the K-distribution, and all compound Gaussian distributions. The class of ES distributions are parametrized by the symmetry center $\boldsymbol{\mu} \in \mathbb{C}^p$ and the scatter matrix parameter $\boldsymbol{\Sigma} \succ 0$, which (when the 2nd-order moment exists) can be taken to be equivalent to the covariance matrix $\text{cov}(\mathbf{x})$ of the r.v. $\mathbf{x} \in \mathbb{C}^p$.

4.2   Elliptically Symmetric Distributions

In this chapter, we assume that the continuous r.v. $\mathbf{x} \in \mathbb{F}^p$ $(= \mathbb{C}$, or, $\mathbb{R})$ has a non-singular distribution, that is, its pdf exists. ES distributions can be defined, however, more generally without the existence of a density. In this section, we review the main properties of ES distributions. For a thorough treatment of real elliptical distributions and their generalizations, see for example, Fang et al. (1990), Frahm (2004), and Ollila et al. (2012) for the complex-valued case.

4.2.1   Real Elliptically Symmetric Distributions

DEFINITION 14   *A r.v. $\mathbf{x} \in \mathbb{R}^p$ is said to have a real elliptically symmetric distribution (RES) with symmetry center $\boldsymbol{\mu} \in \mathbb{R}^p$, and a scatter matrix parameter $\boldsymbol{\Sigma} \in \mathbb{S}_{++}^p$, if it has a pdf of the form*

$$f(\mathbf{x}|\boldsymbol{\mu}, \boldsymbol{\Sigma}) = A_{p,h} \cdot |\boldsymbol{\Sigma}|^{-1/2} h\{[\mathbf{x} - \boldsymbol{\mu}]^{\top} \boldsymbol{\Sigma}^{-1} [\mathbf{x} - \boldsymbol{\mu}]\}, \tag{4.7}$$

where $A_{p,h}$ is a normalizing constant (ensuring that f integrates to 1) and $h : \mathbb{R}_0^+ \to \mathbb{R}^+$ is called the density generator. The function h satisfies the relationship

$$\delta_{p/2,h} = \int_0^{\infty} t^{p/2-1} h(t) \mathrm{d}t < \infty, \tag{4.8}$$

which ensures the integrability of $f(\cdot)$. We denote this case by $\mathbf{x} \sim \mathcal{E}_p(\boldsymbol{\mu}, \boldsymbol{\Sigma}, h)$.

The p-variate normal (Gaussian) distribution, denoted $\mathbf{x} \sim \mathcal{N}_p(\boldsymbol{\mu}, \boldsymbol{\Sigma})$, belongs to the class of RES distributions and is obtained when $h(t) = \exp(-t/2)$. Note that RES distributions are a subclass of *symmetric* distributions, that is, $(\mathbf{x} - \boldsymbol{\mu}) \overset{\mathrm{d}}{=} -(\mathbf{x} - \boldsymbol{\mu})$, and the name "elliptically" stems from the fact that the contours of equal density are ellipsoids in $\mathbb{R}^p$.

A fundamental theorem that characterizes RES distributions is stated next.

THEOREM 3 (stochastic representation theorem)   *A r.v.* $\mathbf{x} \sim \mathcal{E}_p(\boldsymbol{\mu}, \boldsymbol{\Sigma}, h)$ *if and only if there exists a real random variable $R > 0$ possessing a density $f_R(r)$, called the modular variate, such that $\mathbf{x}$ admits a representation*

$$\mathbf{x} \overset{d}{=} \boldsymbol{\mu} + R\mathbf{A}\mathbf{u},$$

where the r.v. $\mathbf{u} \in \mathbb{R}^p$ is independent of R and is uniformly distributed on the unit p-sphere $S^p = \{\mathbf{x} \in \mathbb{R}^p : \|\mathbf{x}\|^2 = 1\}$, and $\boldsymbol{\Sigma} = \mathbf{A}\mathbf{A}^\top$.

An important consequence of Theorem 3 is that the distribution of $Q = R^2$, referred to as a *2nd-order modular variate*, gives the distribution of the quadratic form:

$$\mathrm{QF}(\mathbf{x}) = (\mathbf{x} - \boldsymbol{\mu})^\top \boldsymbol{\Sigma}^{-1} (\mathbf{x} - \boldsymbol{\mu}) \overset{d}{=} Q.$$

The pdf of Q (and hence of a quadratic form) is determined by the density generator $h(\cdot)$:

$$f_Q(t) = t^{p/2-1} h(t) \delta_{p/2,h}^{-1}, \tag{4.9}$$

where (the normalizing constant) $\delta_{p/2,h}$ is defined in (4.8). Note that there is a one-to-one relationship between h and f_R, that is, specifying h determines f_R and vice versa. Another important consequence of Theorem 3 is that the pth-order moment of $\mathbf{x}$ exists if and only if the pth-order moment of R exists. For example, RES distributions have finite second-order moments iff $\mathsf{E}[Q] < \infty$ which by (4.9) can be rephrased as

$$\int_0^\infty t^{p/2} h(t) < \infty. \tag{4.10}$$

The density generator of the Gaussian distribution is $h(t) = \exp(-t/2)$ in which case (4.9) coincides with the pdf of a chi square random variable with p degrees of freedom. This verifies the well-known result that

$$\mathrm{QF}(\mathbf{x}) = \|\boldsymbol{\Sigma}^{-1/2}(\mathbf{x} - \boldsymbol{\mu})\|_2^2 \sim \chi_p^2 \quad \text{when} \quad \mathbf{x} \sim \mathcal{N}_p(\mathbf{0}, \boldsymbol{\Sigma}).$$

A stochastic representation of $\mathbf{x}$ also provides an easy means to compute the mean vector $\mathsf{E}[\mathbf{x}]$ and the covariance matrix $\mathsf{cov}(\mathbf{x})$. The mean vector is

$$\mathsf{E}[\mathbf{x}] = \boldsymbol{\mu} + \mathsf{E}[R]\mathbf{A}\mathsf{E}[\mathbf{u}] = \boldsymbol{\mu}$$

where we used the fact that $\mathsf{E}[\mathbf{u}] = \mathbf{0}$ (which follows from symmetry, $\mathbf{u} \overset{d}{=} -\mathbf{u}$). The mean vector exists if $\mathsf{E}[R] < \infty$. Recalling that $\boldsymbol{\Sigma} = \mathbf{A}\mathbf{A}^\top$, it follows that

$$\mathsf{cov}(\mathbf{x}) = \mathsf{cov}(R\mathbf{A}\mathbf{u}) = \mathsf{E}[R^2]\mathbf{A}\mathsf{cov}(\mathbf{u})\mathbf{A}^\top = (\mathsf{E}[Q]/p)\boldsymbol{\Sigma} = \sigma_h^2 \boldsymbol{\Sigma}, \tag{4.11}$$

where $\sigma_h^2 = \mathsf{E}[Q]/p$ depends on the specific elliptical distribution only through its density generator $h(\cdot)$ (or equivalently through the pdf $f_Q(\cdot)$ in (4.9)). In obtaining (4.11), we used the result that $\mathsf{E}[\mathbf{u}\mathbf{u}^\top] = (1/p)\mathbf{I}$. When $\mathbf{x} \sim \mathcal{N}_p(\boldsymbol{\mu}, \boldsymbol{\Sigma})$, we have that $\sigma_h^2 = 1$ as $Q \sim \chi_p^2$ and hence $\mathsf{E}[Q] = p$.

Finally, we note the following slight ambiguity in the definition of RES distributions: The information about the scale of $\boldsymbol{\Sigma}$ is confounded in $h(\cdot)$ as $h(t)$ can be replaced by

$h_0(t) = a^{-p/2}h(t/a)$ and Σ by $\Sigma_0 = (1/a)\Sigma$ without altering the value of the pdf $f(\mathbf{x})$. This ambiguity is easily avoided by restricting the scale of Σ, or the density generator $h(\cdot)$, in a suitable way. Commonly, one presumes that Σ is equal to the covariance matrix $\text{cov}(\mathbf{x})$, which is equivalent to requiring that $h(t)$ satisfies

$$\int_0^\infty t^{p/2}h(t) = p \tag{4.12}$$

in which case $\sigma_h^2 = 1$ and hence $\Sigma = \text{cov}(\mathbf{x})$. Therefore, when $\mathbf{x}$ has finite 2nd-order moments, we may, without loss of generality, identify the scatter matrix parameter Σ with the covariance matrix $\text{cov}(\mathbf{x})$ by assuming that h satisfies the condition (4.12).

The class of RES distributions is closed under affine transformations. Thus, for $\mathbf{x} \sim \mathcal{E}_p(\boldsymbol{\mu}, \Sigma, h)$, it holds that $\mathbf{B}\mathbf{x} + \mathbf{b} \sim \mathcal{E}_p(\mathbf{B}\boldsymbol{\mu} + \mathbf{b}, \mathbf{B}\Sigma\mathbf{B}^\top, h)$ for any nonsingular $\mathbf{B} \in \mathbb{R}^{p \times p}$ and $\mathbf{b} \in \mathbb{R}^p$. This result follows at once by using the stochastic representation theorem because $\mathbf{x} \overset{\text{d}}{=} \boldsymbol{\mu} + R\mathbf{A}\mathbf{u}$ and hence

$$\mathbf{B}\mathbf{x} + \mathbf{b} \overset{\text{d}}{=} (\mathbf{B}\boldsymbol{\mu} + \mathbf{b}) + R(\mathbf{B}\mathbf{A})\mathbf{u}.$$

Reapplying the stochastic representation theorem implies that $\mathbf{B}\mathbf{x} + \mathbf{b}$ has an RES distribution with a density generator h, a symmetry center $\mathbf{B}\boldsymbol{\mu} + \mathbf{b}$, and a scatter matrix $(\mathbf{B}\mathbf{A})(\mathbf{B}\mathbf{A})^\top = \mathbf{B}\Sigma\mathbf{B}^\top$.

4.2.2   Complex Elliptically Symmetric Distributions

DEFINITION 15   *A r.v $\mathbf{x} \in \mathbb{C}^p$ is said to have a noncircular CES distribution if $[\mathbf{x}]_\mathbb{R}$ has a 2p-variate RES distribution. Furthermore, $\mathbf{x}$ is said to have a circular CES distribution if $\mathbf{x}$ is circularly symmetric in the sense of (4.4).*

Noncircular CES distributions have been studied in Ollila and Koivunen (2004) and Ollila et al. (2011a). In this chapter, we study circular CES distributions (Krishnaiah and Lin, 1986; Ollila et al., 2012). We drop the term *circular* from the prefix, which is consistent with the current terminology used in signal processing, for example, the complex normal (CN) distribution commonly refers to the circular CN distribution, yet the prefix "circular" is often omitted.

Let $\mathbf{x} \in \mathbb{C}^p$ have a CES distribution and let $\boldsymbol{\mu} \in \mathbb{C}^p$ be the center of (circular) symmetry. Then, by construction, $[\boldsymbol{\mu}]_\mathbb{R}$ needs to be also the symmetry center of $[\mathbf{x}]_\mathbb{R}$ and, hence, by definition $[\mathbf{x}]_\mathbb{R} \sim \mathcal{E}_{2p}([\boldsymbol{\mu}]_\mathbb{R}, \boldsymbol{\Omega}, h)$ or, equivalently, $[\mathbf{x} - \boldsymbol{\mu}]_\mathbb{R} \sim \mathcal{E}_{2p}(\mathbf{0}, \boldsymbol{\Omega}, h)$ for some real-valued scatter matrix parameter $\boldsymbol{\Omega} \in \mathbb{S}_{++}^{2p}$. Circularity of $\mathbf{x}$ implies the following:

THEOREM 4   *If $\mathbf{x}$ has a CES distribution, then $[\mathbf{x}]_\mathbb{R} \sim \mathcal{E}_{2p}([\boldsymbol{\mu}]_\mathbb{R}, \boldsymbol{\Omega}, h)$ for some (real) scatter matrix parameter $\boldsymbol{\Omega} \in \mathbb{S}_{++}^{2p}$ of complex form (4.2), meaning that $\boldsymbol{\Omega} = \{\Sigma_0\}_\mathbb{R}$ for some complex matrix $\Sigma_0 \in \mathbb{S}_{++}^p$.*

Proof This result follows from the equivalence of the distribution of $[(\mathbf{x} - \boldsymbol{\mu})]_{\mathbb{R}}$ and $[e^{J\theta}(\mathbf{x} - \boldsymbol{\mu})]_{\mathbb{R}} \, \forall \theta \in \mathbb{R}$. First, note that

$$[e^{J\theta}(\mathbf{x} - \boldsymbol{\mu})]_{\mathbb{R}} = \{e^{J\theta}\mathbf{I}\}_{\mathbb{R}}[(\mathbf{x} - \boldsymbol{\mu})]_{\mathbb{R}} = \begin{pmatrix} \cos(\theta)\mathbf{I} & -\sin(\theta)\mathbf{I} \\ \sin(\theta)\mathbf{I} & \cos(\theta)\mathbf{I} \end{pmatrix} \begin{pmatrix} \mathbf{x}_r \\ \mathbf{x}_i \end{pmatrix}.$$

Then, due to the properties of elliptical distributions, it follows that $\mathbf{B}[\mathbf{x}]_{\mathbb{R}} \sim \mathcal{E}_{2p}(\mathbf{0}, \mathbf{B}\boldsymbol{\Omega}\mathbf{B}^\top, h)$ for any nonsingular $2p \times 2p$ matrix $\mathbf{B}$. Hence the distributions of $[\mathbf{x} - \boldsymbol{\mu}]_{\mathbb{R}}$ and $[e^{J\theta}(\mathbf{x} - \boldsymbol{\mu})]_{\mathbb{R}}$ are equivalent if and only if

$$\underbrace{\begin{pmatrix} \boldsymbol{\Omega}_{11} & \boldsymbol{\Omega}_{21}^\top \\ \boldsymbol{\Omega}_{21} & \boldsymbol{\Omega}_{22} \end{pmatrix}}_{= \boldsymbol{\Omega}} = \begin{pmatrix} \cos(\theta)\mathbf{I} & -\sin(\theta)\mathbf{I} \\ \sin(\theta)\mathbf{I} & \cos(\theta)\mathbf{I} \end{pmatrix} \begin{pmatrix} \boldsymbol{\Omega}_{11} & \boldsymbol{\Omega}_{21}^\top \\ \boldsymbol{\Omega}_{21} & \boldsymbol{\Omega}_{22} \end{pmatrix} \begin{pmatrix} \cos(\theta)\mathbf{I} & -\sin(\theta)\mathbf{I} \\ \sin(\theta)\mathbf{I} & \cos(\theta)\mathbf{I} \end{pmatrix}^\top.$$

This identity holds if and only if $\boldsymbol{\Omega}$ is of complex form (4.2), that is, $\boldsymbol{\Omega}_{11} = \boldsymbol{\Omega}_{22}$ and $\boldsymbol{\Omega}_{21} = -\boldsymbol{\Omega}_{21}^\top$. Thus,

$$\{\boldsymbol{\Sigma}_0\}_{\mathbb{R}} = \boldsymbol{\Omega}, \quad \text{where} \quad \boldsymbol{\Sigma}_0 = \boldsymbol{\Omega}_{11} + J\boldsymbol{\Omega}_{21} \in \mathbb{S}_{++}^p. \tag{4.13}$$

The fact that $\boldsymbol{\Sigma}_0$ is a positive definite Hermitian matrix follows from Theorem 2(h). □

So far we have shown that circular CES distributions are parametrized by the symmetry center $\boldsymbol{\mu} \in \mathbb{C}^p$ and a complex $p \times p$ scatter matrix $\boldsymbol{\Sigma}_0 \succ 0$. With the help of Theorem 4, we may rewrite the pdf of $\mathbf{x}$ in a form that resembles the real-valued case. Using Theorem 2(d), (g), (h), and (4.7), we may write the pdf of $[\mathbf{x}]_{\mathbb{R}} \sim \mathcal{E}_{2p}([\boldsymbol{\mu}]_{\mathbb{R}}, \{\boldsymbol{\Sigma}_0\}_{\mathbb{R}}, h)$ as

$$f\big([\mathbf{x}]_{\mathbb{R}} \,\big|\, [\boldsymbol{\mu}]_{\mathbb{R}}, \{\boldsymbol{\Sigma}_0\}_{\mathbb{R}}\big) = A_{2p,h} \cdot |\{\boldsymbol{\Sigma}_0\}_{\mathbb{R}}|^{-1/2} h([\mathbf{x} - \boldsymbol{\mu}]_{\mathbb{R}}^\top \{\boldsymbol{\Sigma}_0\}_{\mathbb{R}}^{-1} [\mathbf{x} - \boldsymbol{\mu}]_{\mathbb{R}})$$

$$= A_{2p,h} \cdot |\boldsymbol{\Sigma}_0|^{-1} h\{(\mathbf{x} - \boldsymbol{\mu})^\mathsf{H} \boldsymbol{\Sigma}_0^{-1} (\mathbf{x} - \boldsymbol{\mu})\}. \tag{4.14}$$

It will be more convenient to define $\boldsymbol{\Sigma} = 2\boldsymbol{\Sigma}_0$ as the scatter matrix parameter of the CES distribution. This can be done easily by recalling the slight ambiguity in the definition of RES distributions. Thus, $h(t)$ can be replaced with $h_0(t) = a^{-p}h(t/a)$ and $\boldsymbol{\Sigma}_0$ by $\boldsymbol{\Sigma} = (1/a)\boldsymbol{\Sigma}_0$ for any $a > 0$, yet the value of the pdf in (4.14) remains unchanged for any $\mathbf{x} \in \mathbb{C}^p$. Let us define $h_0(t) = 2^p h(2t)$ and $\boldsymbol{\Sigma} = 2\boldsymbol{\Sigma}_0$ to write the pdf in (4.14) in the form

$$f(\mathbf{x}|\boldsymbol{\mu}, \boldsymbol{\Sigma}) = C_{p,g}|\boldsymbol{\Sigma}|^{-1} g\{(\mathbf{x} - \boldsymbol{\mu})^\mathsf{H} \boldsymbol{\Sigma}^{-1} (\mathbf{x} - \boldsymbol{\mu})\}, \tag{4.15}$$

where $C_{p,g} = 2^p A_{2p,h}$ is the normalizing constant, $g(t) = h(2t)$ denotes the density generator of the CES distribution, and $\boldsymbol{\Sigma} \succ 0$ denotes its scatter matrix parameter. We write $\mathbf{x} \sim \mathcal{E}_p(\boldsymbol{\mu}, \boldsymbol{\Sigma}, g)$ or $\mathbf{x} \sim \mathbb{C}\mathcal{E}_p(\boldsymbol{\mu}, \boldsymbol{\Sigma}, g)$ to denote this case.

Example 10 (**CN distribution**) Recall that the $2p$-variate Gaussian r.v. has a density generator $h(t) = \exp(-t/2)$ and the normalizing constant $A_{2p,h}$ of the density is $A_{2p,h} = (2\pi)^{-p}$. Thus, the density generator of the CN distribution is

$$g(t) = h(2t) = \exp(-t)$$

and the normalizing constant is $C_{p,g} = 2^p A_{2p,h} = \pi^{-p}$. Hence, from (4.15), it follows that the pdf of the p-variate CN distribution is

$$f(\mathbf{x}|\boldsymbol{\mu}, \boldsymbol{\Sigma}) = \pi^{-p}|\boldsymbol{\Sigma}|^{-1}\exp\{-(\mathbf{x}-\boldsymbol{\mu})^H\boldsymbol{\Sigma}^{-1}(\mathbf{x}-\boldsymbol{\mu})\}.$$

We denote this case by $\mathbf{x} \sim \mathcal{N}_p(\boldsymbol{\mu}, \boldsymbol{\Sigma})$ or $\mathbf{x} \sim \mathbb{C}\mathcal{N}_p(\boldsymbol{\mu}, \boldsymbol{\Sigma})$. The distribution of the (circular) p-variate Gaussian distribution was first derived by Goodman (1963).

Example 11 (**Complex t_ν distribution**) Another important elliptical distribution is the multivariate t-distribution with $\nu > 0$ degrees of freedom. We recall that the density generator of the $2p$-variate real t_ν-distribution is $h(t) = (1 + t/\nu)^{-(\nu+2p)/2}$ and the normalizing constant $A_{2p,h}$ of the density is $A_{2p,h} = \Gamma(\frac{\nu+2p}{2})/[\Gamma(\frac{\nu}{2})(\nu\pi)^p]$. See, for example, Fang et al. (1990). Thus, the density generator of the p-variate complex t_ν-distribution is

$$g(t) = h(2t) = (1 + t/\nu)^{-(\nu+2p)/2}$$

and the normalizing constant is $C_{p,g} = 2^p A_{2p,h}$, which can then be substituted into (4.15) to obtain the full pdf $f(\mathbf{x}|\boldsymbol{\mu}, \boldsymbol{\Sigma})$ in the complex case. We denote this case by $\mathbf{x} \sim t_{p,\nu}(\boldsymbol{\mu}, \boldsymbol{\Sigma})$ or $\mathbf{x} \sim \mathbb{C}t_{p,\nu}(\boldsymbol{\mu}, \boldsymbol{\Sigma})$.

Next, we show that the CES distributions are characterized by the equivalent stochastic representation theorem that was stated in Theorem 3. First, observe that if $\mathbf{x} \sim \mathcal{E}_p(\boldsymbol{\mu}, \boldsymbol{\Sigma}, g)$, then $[\mathbf{x}]_{\mathbb{R}} \sim \mathcal{E}_{2p}([\boldsymbol{\mu}]_{\mathbb{R}}, (1/2)\{\boldsymbol{\Sigma}\}_{\mathbb{R}}, h)$, where we used Theorem 4 and the fact that $\boldsymbol{\Sigma}_0 = (1/2)\boldsymbol{\Sigma}$. Then, note that a complex matrix $\mathbf{M} \in \mathbb{S}_{++}^p$ has a decomposition $\mathbf{M} = \mathbf{A}\mathbf{A}^H$ for some nonsingular matrix $\mathbf{A} \in \mathbb{C}^{p\times p}$ (Horn and Johnson, 1985). Hence, by Theorem 2(i), it follows that $\{\boldsymbol{\Sigma}\}_{\mathbb{R}} = \{\mathbf{A}\}_{\mathbb{R}}\{\mathbf{A}\}_{\mathbb{R}}^{\top}$. Then, by Theorem 3, $[\mathbf{x}]_{\mathbb{R}}$ has the following stochastic representation

$$[\mathbf{x}]_{\mathbb{R}} \stackrel{d}{=} [\boldsymbol{\mu}]_{\mathbb{R}} + R\frac{1}{\sqrt{2}}\{\mathbf{A}\}_{\mathbb{R}}[\mathbf{u}]_{\mathbb{R}} \qquad (4.16)$$

$$= \left[\boldsymbol{\mu} + \mathcal{R}\mathbf{A}\mathbf{u}\right]_{\mathbb{R}} \qquad (4.17)$$

where $\mathcal{R} = R/\sqrt{2}$ is a scaled version of R and $\mathbf{u} = \mathbf{u}_r + j\mathbf{u}_i$ is a r.v. uniformly distributed on the complex unit p-sphere $\mathcal{S}^p = \{\mathbf{x} \in \mathbb{C}^p : \|\mathbf{x}\|^2 = 1\}$, which is equivalent to stating that $[\mathbf{u}]_{\mathbb{R}}$ is uniformly distributed on $\mathcal{S}^{2p} = \{\mathbf{x} \in \mathbb{R}^{2p} : \|\mathbf{x}\|^2 = 1\}$. Note that (4.16) follows from (4.17) by using (4.3). Thus, due to (4.17), and the isomorphism between $\mathbf{x}$ and $[\mathbf{x}]_{\mathbb{R}}$, an analogous stochastic representation theorem holds for the complex valued case:

THEOREM 5   *A r.v.* $\mathbf{x} \sim \mathbb{C}\mathcal{E}_p(\boldsymbol{\mu}, \boldsymbol{\Sigma}, g)$ *if and only if there exists a real random variable* $\mathcal{R} > 0$ *possessing a density* $f_{\mathcal{R}}(r)$, *called the modular variate, such that* $\mathbf{x}$ *admits a representation*

$$\mathbf{x} \stackrel{d}{=} \boldsymbol{\mu} + \mathcal{R}\mathbf{A}\mathbf{u},$$

where the r.v. $\mathbf{u} \in \mathbb{C}^p$ is independent of $\mathcal{R}$ and is uniformly distributed on the complex unit p-sphere $\mathcal{S}^p = \{\mathbf{x} \in \mathbb{R}^p : \|\mathbf{x}\|^2 = 1\}$, and $\boldsymbol{\Sigma} = \mathbf{AA}^{\mathsf{H}}$.

Because the CES distribution possesses a similar stochastic representation theorem, the following results, which are similar to the real case, can be shown:

- The quadratic form $\mathrm{QF}(\mathbf{x}) = \|\boldsymbol{\Sigma}^{-1/2}(\mathbf{x}-\boldsymbol{\mu})\|_2^2 =_d \mathcal{Q}^2$, where the random variable $\mathcal{R}^2 = \mathcal{Q}$ is referred to as 2nd-order modular variate. The pdf of $\mathcal{Q}$ is

$$f_{\mathcal{Q}}(t) = t^{p-1} g(t) \delta_{p,g}^{-1}. \tag{4.18}$$

  where $\delta_{p,g}$ is as defined in (4.8).

- $\mathsf{E}[\mathbf{x}] = \boldsymbol{\mu}$ and $\mathsf{cov}(\mathbf{x}) = \sigma_g^2 \boldsymbol{\Sigma}$, where $\sigma_g^2 = \mathsf{E}[\mathcal{Q}]/p$.
- $\mathbf{x}$ has finite qth-order moments if and only if $\mathcal{R}$ has finite qth-order moments.
- The class is closed under affine transformations: $\mathbf{Bx}+\mathbf{b} \sim \mathbb{C}\mathcal{E}_p(\mathbf{B}\boldsymbol{\mu}+\mathbf{b}, \mathbf{B}\boldsymbol{\Sigma}\mathbf{B}^{\mathsf{H}}, g)$ for any nonsingular $\mathbf{B} \in \mathbb{C}^{p \times p}$ and $\mathbf{b} \in \mathbb{C}^p$.

4.2.3   Related Model: The Angular Central Gaussian Distribution

If $\mathbf{x}$ is a (complex or real) r.v. with an ES distribution, then the distribution of its projection onto the unit p-sphere, $\mathbf{x}/\|\mathbf{x}\|$, is said to have a (complex or real) *angular elliptical distribution*. In particular, if the elliptical distribution is a normal (Gaussian) distribution, then the distribution of $\mathbf{x}/\|\mathbf{x}\|$ is said to have an *angular Gaussian (AG) distribution*. The central case ($\boldsymbol{\mu} = \mathbf{0}$) is referred to as an *angular central Gaussian (ACG)* distribution. The complex ACG distribution has been widely used in statistical shape analysis (Kent, 1997; Dryden and Mardia, 1998), for example.

DEFINITION 16   *A (real or complex) r.v. $\mathbf{x}_a$ is said to have an ACG distribution if it admits a stochastic representation $\mathbf{x}_a \overset{d}{=} \mathbf{x}/\|\mathbf{x}\|$, where $\mathbf{x} \sim \mathcal{N}_p(\mathbf{0}, \boldsymbol{\Sigma})$. We write $\mathbf{x}_a \sim \mathrm{AG}_p(\mathbf{0}, \boldsymbol{\Sigma})$ to denote this case.*

For a nonsingular $\boldsymbol{\Sigma}$, the pdf of the distribution is given by

$$f_{\mathbf{x}_a}(\mathbf{x}) = \begin{cases} s_{2p}^{-1} |\boldsymbol{\Sigma}|^{-1} (\mathbf{x}^{\mathsf{H}} \boldsymbol{\Sigma}^{-1} \mathbf{x})^{-p}, & \text{complex-valued case} \\ s_p^{-1} |\boldsymbol{\Sigma}|^{-1/2} (\mathbf{x}^{\mathsf{T}} \boldsymbol{\Sigma}^{-1} \mathbf{x})^{-p/2}, & \text{real-valued case} \end{cases} \tag{4.19}$$

where s_p denotes the surface area of the (real) unit p-sphere $\mathcal{S}^p$, that is, $s_p = \frac{2\pi^{p/2}}{\Gamma(p/2)}$. It can be noted, from either its characterizing definition or its density, that the parameter $\boldsymbol{\Sigma}$ can only be identified up to a scale factor as $\boldsymbol{\Sigma}$ and $c\boldsymbol{\Sigma}$ yield the same distribution. It turns out, for the central case, that the term *angular central Gaussian* is a slight misnomer because the central Gaussian distribution can be replaced by any central ES distribution and the resulting angular distribution is the same. That is, if $\mathbf{x} \sim \mathcal{E}_p(\mathbf{0}, \boldsymbol{\Sigma}, g)$, then $\mathbf{x}/\|\mathbf{x}\|$ follows an $\mathrm{AG}_p(\mathbf{0}, \boldsymbol{\Sigma})$ distribution.

Although the density of a (real or complex) $\mathbf{x}_a$, has the apparent form of the density of an elliptical distribution, it is, in fact, not an ES distribution. The difference is that the densities specified by (4.7) and (4.15) are defined with respect to the Lebesgue measure on $\mathbb{R}^p$ and $\mathbb{R}^{2p}$, respectively, whereas the density (4.19) is with respect to the uniform measure on S^p (or S^{2p} in the complex case). Consequently, the ACG distribution does not possess the characterizing stochastic representation stated earlier for the ES distribution. However, the (real or complex) ACG distribution can be generated from the (real or complex) r.v. $\mathbf{u}$ possessing a uniform distribution on the (real or complex) unit sphere according to $\mathbf{Au}/\|\mathbf{Au}\| \sim \mathrm{AG}_p(\mathbf{0}, \boldsymbol{\Sigma})$ for $\boldsymbol{\Sigma} = \mathbf{AA}^H$ and a nonsingular $\mathbf{A}$. Also, the class of ACG distributions is closed under *standardized linear transformations*, that is, if $\mathbf{x}_a \sim \mathrm{AG}_p(\mathbf{0}, \boldsymbol{\Sigma})$, then $\mathbf{Bx}_a/\|\mathbf{Bx}_a\| \sim \mathrm{AG}_p(\mathbf{0}, \mathbf{B}\boldsymbol{\Sigma}\mathbf{B}^H)$ for any nonsingular $p \times p$ matrix $\mathbf{B}$.

4.3 ML- and *M*-estimation of the Scatter Matrix

Due to the fact that the pdf of the RES and CES distribution have similar forms, it is possible to develop the theory for the ML-estimators of their parameters in parallel. From this point onward, we assume that the symmetry center $\boldsymbol{\mu} \in \mathbb{F}^p$ ($\mathbb{F} = \mathbb{C}$, or $\mathbb{R}$) of the elliptical distribution is known or fixed and, without loss of generality, we assume that $\boldsymbol{\mu} = \mathbf{0}$. Thus, we only need to estimate the scatter matrix parameter $\boldsymbol{\Sigma} \succ 0$. We assume that we have a set of N i.i.d. p-variate observations, $\mathbf{x}_1, \ldots, \mathbf{x}_N$, distributed as $\mathcal{E}_p(\mathbf{0}, \boldsymbol{\Sigma}, g)$. We further assume that $N > p$, that is, the number of available observations exceeds the dimensionality of the data. As was shown in the previous section, the pdf of real or complex r.v. $\mathbf{x}_i \sim \mathcal{E}_p(\mathbf{0}, \boldsymbol{\Sigma}, g)$ has the form

$$f(\mathbf{x}|\mathbf{0}, \boldsymbol{\Sigma}) \propto |\boldsymbol{\Sigma}|^{-1/\gamma} g(\mathbf{x}^H \boldsymbol{\Sigma}^{-1} \mathbf{x}),$$

where the constant γ is as defined in (2.43), that is, $\gamma = 2$ in the real-valued case ($\mathbb{F} = \mathbb{R}$) and $\gamma = 1$ in the complex-valued case ($\mathbb{F} = \mathbb{C}$).

Then, the MLE of the scatter matrix $\boldsymbol{\Sigma} \succ 0$, denoted by $\hat{\boldsymbol{\Sigma}} \succ 0$, minimizes the negative log-likelihood function

$$L(\boldsymbol{\Sigma}) = -\prod_{i=1}^{N} \ln\{|\boldsymbol{\Sigma}|^{-1/\gamma} g(\mathbf{x}_i^H \boldsymbol{\Sigma}^{-1} \mathbf{x}_i)\} = \sum_{i=1}^{N} \rho_{\mathrm{ML}}(\mathbf{x}_i^H \boldsymbol{\Sigma}^{-1} \mathbf{x}_i) - \frac{N}{\gamma} \ln |\boldsymbol{\Sigma}^{-1}| \quad (4.20)$$

where the function $\rho_{\mathrm{ML}} : \mathbb{R}_0^+ \to \mathbb{R}^+$ is defined as

$$\rho_{\mathrm{ML}}(t) = -\ln g(t) \quad (4.21)$$

and is referred to as the *ML loss function* to follow the convention used in regression analysis.

Example 12 (**SCM as the MLE**) We solve the MLE of $\boldsymbol{\Sigma}$ when the samples are from a p-variate (real or complex) Gaussian distribution $\mathcal{N}_p(\mathbf{0}, \boldsymbol{\Sigma})$. As discussed in Example 10, the density generator is $g(t) = \exp(-t/\gamma)$, so the pdf in (4.7) has the familiar form

$$f(\mathbf{x}|\mathbf{0}, \boldsymbol{\Sigma}) \propto |\boldsymbol{\Sigma}|^{-1/\gamma} \exp\left(-\frac{1}{\gamma}\mathbf{x}^H \boldsymbol{\Sigma}^{-1} \mathbf{x}\right).$$

The Gaussian ML loss function is $\rho_G(t) = -\ln g(t) = t/\gamma$ and, hence, the negative log-likelihood function in (4.20) becomes

$$
\begin{aligned}
L_G(\boldsymbol{\Sigma}) &= \sum_{i=1}^{N} \rho_G(\mathbf{x}_i^H \boldsymbol{\Sigma}^{-1} \mathbf{x}_i) - \frac{N}{\gamma} \ln |\boldsymbol{\Sigma}^{-1}| \\
&= \frac{1}{\gamma} \sum_{i=1}^{N} \mathbf{x}_i^H \boldsymbol{\Sigma}^{-1} \mathbf{x}_i - \frac{N}{\gamma} \ln |\boldsymbol{\Sigma}^{-1}| \\
&= \frac{N}{\gamma} \left\{ \mathsf{Tr}\left(\boldsymbol{\Sigma}^{-1} \frac{1}{N} \sum_{i=1}^{N} \mathbf{x}_i \mathbf{x}_i^H \right) - \ln |\boldsymbol{\Sigma}^{-1}| \right\} \\
&= \frac{N}{\gamma} \left\{ \mathsf{Tr}(\boldsymbol{\Sigma}^{-1} \mathbf{S}) - \ln |\boldsymbol{\Sigma}^{-1}| \right\}
\end{aligned}
$$

where $\mathbf{S} = \frac{1}{N} \sum_{i=1}^{N} \mathbf{x}_i \mathbf{x}_i^H$ is the SCM. Here, the relation $\mathbf{x}_i^H \boldsymbol{\Sigma}^{-1} \mathbf{x}_i = \mathsf{Tr}(\boldsymbol{\Sigma}^{-1} \mathbf{x}_i \mathbf{x}_i^H)$ has been used. Taking the matrix derivative w.r.t $\boldsymbol{\Sigma}^{-1}$, yields

$$
\begin{aligned}
\frac{\gamma}{N} \times \frac{\partial}{\partial \boldsymbol{\Sigma}^{-1}} L_G(\boldsymbol{\Sigma}) &= \frac{\partial}{\partial \boldsymbol{\Sigma}^{-1}} \left\{ \sum_{i=1}^{N} \mathsf{Tr}(\boldsymbol{\Sigma}^{-1} \mathbf{S}) - \ln |\boldsymbol{\Sigma}^{-1}| \right\} \\
&= \mathbf{S} - \boldsymbol{\Sigma}
\end{aligned}
$$

where the results $\frac{\partial}{\partial \mathbf{X}} \mathsf{Tr}(\mathbf{XA}) = \mathbf{A}^\top$ and $\frac{\partial}{\partial \mathbf{X}} \ln |\mathbf{X}| = (\mathbf{X}^{-1})^\top$ have been used. Setting the derivative to $\mathbf{0}$ gives $\hat{\boldsymbol{\Sigma}} = \mathbf{S}$ as the critical point. The found local solution is well known to be a global minimum and, hence, the MLE. This follows from the fact that $L_G(\boldsymbol{\Sigma})$ is strictly convex[1] in $\boldsymbol{\Sigma}^{-1}$, or that $L_G(\boldsymbol{\Sigma})$ is geodesically strictly convex both in $\boldsymbol{\Sigma}$ and $\boldsymbol{\Sigma}^{-1}$ (Zhang et al., 2013), and, hence, the found solution is the unique minimum.

The ML-estimator of a scatter matrix can be embedded in a larger family of estimators referred to as the *M*-estimators of a scatter matrix and are developed in the seminal paper for real-valued observations by Maronna (1976). The complex-valued case can be developed similarly. In this chapter, we define *M*-estimators as solutions of an optimization problem that is equivalent to the negative log-likelihood function defined in (4.20), but we allow a more general loss function ρ that does not need to be related to an elliptical density, that is, it does not need to be a ML loss function as defined in (4.21).

An *M*-estimator of a scatter matrix is defined as the solution to the optimization problem

$$
\hat{\boldsymbol{\Sigma}} = \arg \min_{\boldsymbol{\Sigma} > 0} \left\{ L(\boldsymbol{\Sigma}) = \sum_{i=1}^{N} \rho(\mathbf{x}_i^H \boldsymbol{\Sigma}^{-1} \mathbf{x}_i) - \frac{N}{\gamma} \ln |\boldsymbol{\Sigma}^{-1}| \right\} \tag{4.22}
$$

where $\rho : \mathbb{R}_0^+ \to \mathbb{R}^+$, referred to as the *loss function*, is assumed to verify Condition 2 that is stated in the following text.

---

[1] Note, however, that the Gaussian likelihood $L_G(\boldsymbol{\Sigma})$ is not convex in $\boldsymbol{\Sigma}$.

CONDITION 2 **(Uniqueness)** *The loss function $\rho(t)$ is nondecreasing (or strictly increasing) and continuous for $0 < t < \infty$. The function $r(x) = \rho(e^x)$ is convex (or strictly convex) for $-\infty < x < \infty$.*

It was shown in Zhang et al. (2013) that Condition 2 is sufficient for $L(\Sigma)$ to be a geodesically convex (or strictly convex) function in $\Sigma \in \mathbb{S}_{++}^p$. In particular, when $\rho(t)$ is strictly increasing and $r(x) = \rho(e^x)$ is strictly convex, the optimization problem (4.22) has a unique global minimizer (Zhang et al., 2013, theorem 1). Because $L(\Sigma)$ is geodesically convex (or strictly convex), then any local minimizer of $L(\Sigma)$ is also its global minimizer (Wiesel and Zhang, 2015). This means that if the loss function $\rho(t)$ is differentiable, then a global minimum $\hat{\Sigma}$ of (4.22) can always be found by solving the *M-estimating equation*, which is obtained by setting the matrix derivative of $L(\Sigma)$, with respect to Σ^{-1}, to zero, that is, $\nabla_{\Sigma^{-1}} L(\Sigma) = \mathbf{0}$. Solving this equation yields

$$\hat{\Sigma} = \frac{1}{N} \sum_{i=1}^{N} u(\mathbf{x}_i^H \hat{\Sigma}^{-1} \mathbf{x}_i) \mathbf{x}_i \mathbf{x}_i^H \tag{4.23}$$

where the function $u(t) = \gamma \cdot \rho'(t)$ is referred to as the *weight function*. Hence, an *M*-estimator of a scatter matrix can be viewed as "an adaptively weighted sample covariance matrix."

All the commonly used loss functions $\rho(t)$ used in the literature are differentiable and, hence, finding the solution to (4.22) is equivalent to solving the *M*-estimating equation (4.23). For a differentiable ρ function, the population counterpart, the *M*-functional of a scatter matrix, can be defined as the solution to

$$\mathbf{C}(F) = \mathsf{E}_F\left[u(\mathbf{x}^H \mathbf{C}(F)\mathbf{x})\mathbf{x}\mathbf{x}^H\right] \tag{4.24}$$

where F denotes the cumulative distribution function (cdf) of the p-variate (real or complex) r.v. $\mathbf{x}$. When F is the ES distribution, $F = \mathcal{E}_p(\mathbf{0}, \Sigma, g)$, it is easy to verify that the *M*-functional of a scatter matrix is proportional to the scatter matrix parameter Σ of the elliptical distribution, that is, $\mathbf{C}(F) = \sigma_{u,g}^2 \Sigma$, where the constant $\sigma_{u,g}$ is a solution to

$$\mathsf{E}\left[\psi(\|\mathbf{x}\|^2/\sigma_{u,g}^2)\right] = p, \quad \mathbf{x} \sim \mathcal{E}_p(\mathbf{0}, \mathbf{I}, g). \tag{4.25}$$

Here, $\psi(t) = tu(t)$ and $u(t) = \gamma \cdot \rho'(t)$. Often the loss function $\rho(t)$ is standardized so that the *M*-estimator $\hat{\Sigma}$, obtained by minimizing (4.22), is Fisher consistent when sampling from a p-variate (real or complex) Gaussian distribution $\mathcal{N}_p(\mathbf{0}, \Sigma)$, that is, $\mathbf{C}(F) = \Sigma$, when $F = \mathcal{N}_p(\mathbf{0}, \Sigma)$. This holds true if and only if (4.25) holds for $\mathbf{x} \sim \mathcal{N}_p(\mathbf{0}, \mathbf{I})$.

A technical condition on the sample is needed in order for the minimizer of $L(\Sigma)$ to exist. The condition for existence of the minimizer of $L(\Sigma)$, for the real case, has been solved by Kent and Tyler (1991) and is given in the following text.

CONDITION 3 **(Existence)** *A minimizer of $L(\Sigma)$ exists, for the real case, if $\rho(t)$ is a continuous function in $[0, \infty)$ and*

$$a_1 = \sup\{a \mid t^{a/2} \exp\{-\rho(t)\} \to 0 \text{ as } t \to \infty\}$$

is positive ($a_1 > 0$). A further requirement is that the observations $\mathcal{X} = \{\mathbf{x}_i\}_{i=1}^{N}$ in $\mathbb{R}^p$ satisfy

$$\frac{|\mathcal{X} \cap V|}{N} < 1 - \frac{p - \dim(V)}{a_1} \tag{4.26}$$

for any linear subspace $V \subset \mathbb{R}^p$.

Loosely speaking, these existence conditions require that only a small proportion of the data $\mathcal{X} = \{\mathbf{x}_i\}_{i=1}^{N}$ lie in any set subspace. For example, if $a_1 = \infty$, then (4.26) simply implies that $\mathcal{X}$ spans $\mathbb{R}^p$.

The M-estimator $\hat{\boldsymbol{\Sigma}}$ can be computed by a fixed-point algorithm as follows: given an arbitrary initial start value $\boldsymbol{\Sigma}_0 \in \mathbb{S}_{++}^{p}$, iterate

$$\hat{\boldsymbol{\Sigma}}_{k+1} = \frac{1}{N} \sum_{i=1}^{N} u(\mathbf{x}_i^H \hat{\boldsymbol{\Sigma}}_k^{-1} \mathbf{x}_i) \mathbf{x}_i \mathbf{x}_i^H \tag{4.27}$$

for $k = 0, 1, \ldots$ until convergence. Under some mild regularity conditions, the algorithm converges to the solution of (4.23). The technicalities are detailed in Huber and Ronchetti (2009), Maronna et al. (2006), and Ollila et al. (2012).

4.4   Examples of *M*- and ML-estimators

In this section, we review some of the most commonly used loss functions $\rho(t)$ and their respective M-estimators of scatter. We also specify the corresponding weight functions $u(t) = \gamma \cdot \rho'(t)$. All loss functions are assumed to satisfy Condition 2. Thus, the objective function $L(\boldsymbol{\Sigma})$ is assumed to be geodesically convex and the global minimizer $\hat{\boldsymbol{\Sigma}}$ can be found by simply solving the estimating equation (4.23) using the fixed point algorithm. Each (complex or real) M-estimator introduced in this section can be computed by using the RobustSP toolbox MATLAB© function Mscat.

4.4.1 $t_\nu M$-estimator

$t_\nu M$-estimators have been studied by Kent and Tyler (1991) for the real-valued case and by Ollila and Koivunen (2003b) and Ollila et al. (2012) for the complex-valued case. As discussed in Example 11, the density generator $g(t)$ of a p-variate real (or complex) t_ν distribution with $\nu > 0$ degrees of freedom is $g(t) = (1 + t/\nu)^{-\frac{1}{2}(\nu+2p)}$ in the complex-valued case and $g(t) = (1 + t/\nu)^{-\frac{1}{2}(\nu+p)}$ in the real-valued case. Note that $g(t)$ can be replaced by $\equiv g(t) \cdot C$ for any constant $C > 0$ without affecting the ML-estimation problem. The M-estimator of scatter, based on the induced ML loss function (4.21),

$$\rho_\nu(t) = \begin{cases} \frac{(\nu+p)}{2} \ln(\nu + t), & \text{real-valued case} \\ \frac{(\nu+2p)}{2} \ln(\nu + 2t), & \text{complex-valued case} \end{cases}, \tag{4.28}$$

is referred to as the $t_\nu M$-*estimator* and corresponds to the MLE when $\{\mathbf{x}_i\}_{i=1}^N$ are i.i.d. from $t_{p,\nu}(\mathbf{0}, \boldsymbol{\Sigma})$. The corresponding weight function $u_\nu(t) = \gamma \cdot \rho_\nu'(t)$ is

$$u_\nu(t) = \gamma \times \rho_\nu'(t) = \begin{cases} \dfrac{\nu + p}{\nu + t}, & \text{real-valued case} \\[2mm] \dfrac{\nu + 2p}{\nu + 2t}, & \text{complex-valued case} \end{cases}$$

The resulting M-estimator of scatter is not Fisher-consistent for the multivariate Gaussian distribution case. However, such a Fisher consistent version of the $t_\nu M$-estimators can be obtained by using a scaled loss function

$$\rho_\nu^*(t) = \frac{1}{b}\rho_\nu(t), \quad b = \mathsf{E}[\psi_\nu(\|\mathbf{x}\|^2)]/p, \quad \mathbf{x} \sim \mathcal{N}_p(\mathbf{0}, \mathbf{I})$$

where $\psi_\nu(t) = tu_\nu(t)$. Similarly a scaled weight function can be defined according to $u_\nu^*(t) = u_\nu(t)/b$. Figure 4.1 illustrates the (standardized) loss and weight functions for the real-valued case and for some specific values of ν.

Example 13 **(Effect of an Outlier)** We generate a data set $\{\mathbf{x}_i\}_{i=1}^N$ from a real-valued bivariate Gaussian distribution, $\mathcal{N}_2(\mathbf{0}, \boldsymbol{\Sigma})$, where the sample size is $N = 100$ and the covariance matrix $\boldsymbol{\Sigma}$ has the eigenvector decomposition $\boldsymbol{\Sigma} = \mathbf{E}\boldsymbol{\Lambda}\mathbf{E}^\top$, with $\mathbf{E}$ and $\boldsymbol{\Lambda}$ being defined according to

$$\mathbf{E} = \begin{pmatrix} \cos(\frac{\pi}{4}) & -\sin(\frac{\pi}{4}) \\ \sin(\frac{\pi}{4}) & \cos(\frac{\pi}{4}) \end{pmatrix} \quad \text{and} \quad \boldsymbol{\Lambda} = \mathrm{diag}(5, 1).$$

We then plot the 95 percent *tolerance ellipses*

$$\{\mathbf{x} \in \mathbb{R}^2 : \mathbf{x}^\top \hat{\boldsymbol{\Sigma}}^{-1} \mathbf{x} = F_{\chi_2^2}^{-1}(0.95)\}$$

based on an estimated covariance matrix $\hat{\boldsymbol{\Sigma}}$, where $F_{\chi_2^2}(x)$ denotes the cdf of a chi-squared distribution with 2 degrees of freedom. Note, because $\mathbf{x}^\top \boldsymbol{\Sigma}^{-1} \mathbf{x} \sim \chi_2^2$ for a Gaussian r.v. $\mathbf{x} \sim \mathcal{N}_2(\mathbf{0}, \boldsymbol{\Sigma})$, one can expect that approximately 95 percent of the sampled observations will lie inside the ellipse.

Figure 4.2a shows the obtained 95 percent tolerance ellipse using the SCM and the $t_3 M$-estimator. As can be seen, both the $t_3 M$-estimator (solid ellipse) and the SCM **S** produce essentially overlapping tolerance ellipses, meaning that they give closely similar estimates of the eigenvectors and eigenvalues. Figure 4.2b and 4.2c illustrate the cases when one of the observations in the original data set is replaced with an outlier in the *southeast direction*, which is gradually increasing in magnitude. As can be seen, the eigenvector of the SCM gets pulled toward the outlier whereas the tolerance ellipses of the $t_3 M$-estimator remain unaffected.

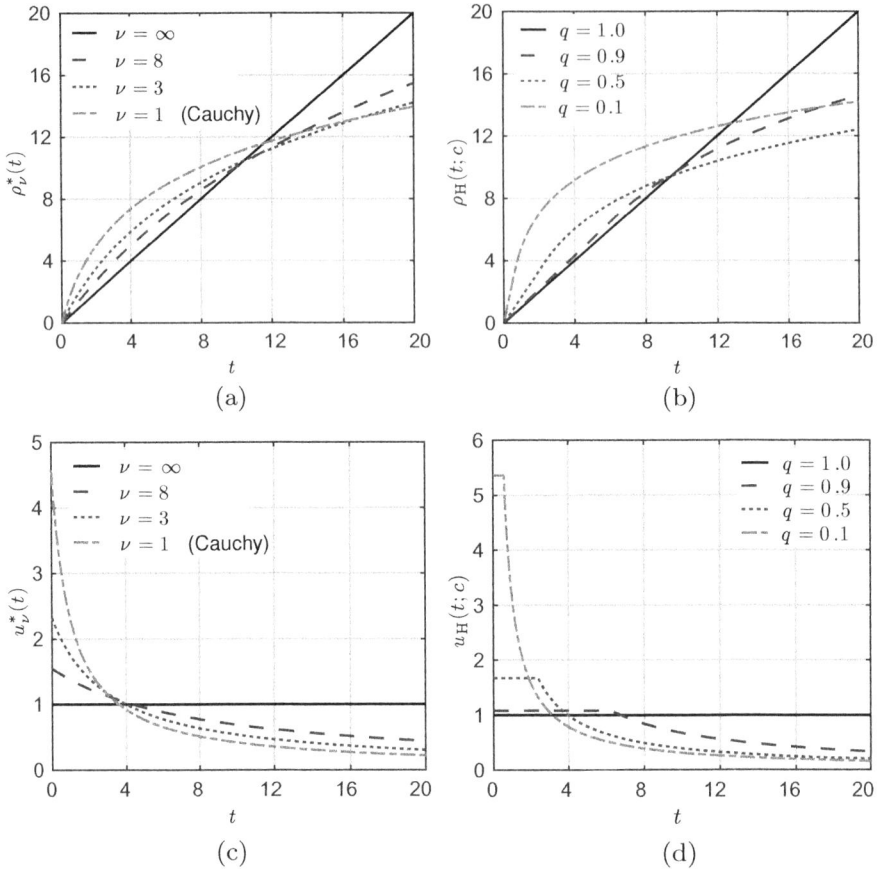

Figure 4.1 Loss and weight function of the tM-estimator (left panel) and Huber's M-estimator (right panel) for various choices of the degree of freedom ν and the threshold c^2. The dimension used is $p = 3$.

4.4.2   Huber's Loss Function

In his seminal work, Huber (1964) proposed a location and scale family of univariate heavy-tailed distributions often referred to as "least favorable distributions" (LFDs). A LFD corresponds to a symmetric unimodal distribution that follows a Gaussian distribution in the middle and a double exponential distribution in the tails. The corresponding MLE of the parameters are then referred to as Huber's M-estimators. The extension of Huber's M-estimators to the multivariate setting, is usually defined as a generalization of the corresponding univariate M-estimating equations to the multivariate setting, see for example, Maronna (1976), or Ollila and Koivunen (2003b) and Ollila et al. (2012) for the complex-valued case.

It was shown recently by Ollila et al. (2016), however, that Huber's M-estimator of scatter can also be viewed as the MLE of a family of heavy-tailed p-variate

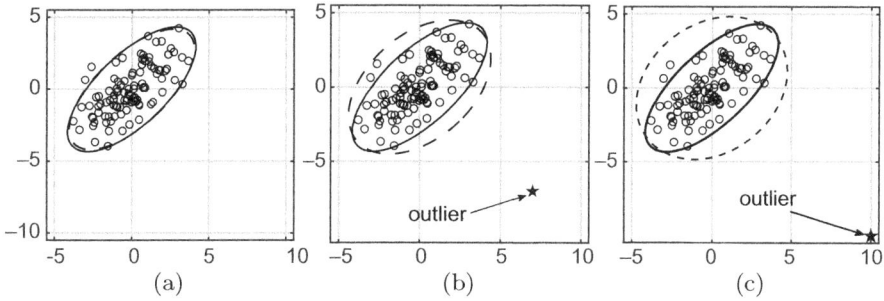

Figure 4.2 A random sample of length $N = 100$ from $\mathcal{N}_2(\mathbf{0}, \mathbf{\Sigma})$ and estimated tolerance ellipses based on the $t_3 M$-estimator $\hat{\mathbf{\Sigma}}$ (solid line) and the SCM **S** (dotted line). The left panel shows the results for the original data. In the middle and right panels, one of the datapoints is replaced by an outlier (marked by a star) in the *south-east direction*. As the magnitude of the outlier increases, the tolerance ellipses of the SCM start to inflate but those based on the $t_3 M$ estimator remain unaffected.

elliptical distributions, namely those with a density generator of the form $g_H(t; c) = \exp\{-\rho_H(t; c)\}$, where

$$\rho_H(t; c) = \frac{1}{\gamma} \times \begin{cases} t/b & \text{for } t \leqslant c^2, \\ (c^2/b)\big(\log(t/c^2) + 1\big) & \text{for } t > c^2. \end{cases} \tag{4.29}$$

These distributions follow a multivariate Gaussian distribution in the middle, but have tails that decrease according to an inverse polynomial. The distribution is a valid distribution for $c > 0$, and for the corresponding MLE of scatter, that is, the Huber M-estimator of multivariate scatter, the threshold c represents a user-defined tuning constant that determines the robustness and efficiency of the estimator. The constant $b > 0$ is a scaling factor. This can be seen by noting that if $\hat{\mathbf{\Sigma}}$ is the Huber M-estimator of scatter when $b = 1$, then the Huber M-estimator of scatter when $b = b_o$ is simply $b_o \hat{\mathbf{\Sigma}}$. The scaling constant b is usually chosen so that the resulting scatter estimator is Fisher consistent for the covariance matrix at the p-variate Gaussian distribution. Given a value of c, the value of b needed to obtain Fisher consistency, $\mathbf{C}(F) = \mathbf{\Sigma}$ at $F = \mathcal{N}_p(\mathbf{0}, \mathbf{\Sigma})$, is

$$b = \begin{cases} F_{\chi^2_{p+2}}(c^2) + c^2(1 - F_{\chi^2_p}(c^2))/p, & \text{real-valued case} \\ F_{\chi^2_{2(p+1)}}(2c^2) + c^2(1 - F_{\chi^2_{2p}}(2c^2))/p, & \text{complex-valued case} \end{cases}$$

We refer to $\rho_H(t; c)$ as Huber's loss function because it gives rise to Huber's weight function, namely

$$u_H(t; c) = \gamma \times \rho'_H(t; c) = \begin{cases} 1/b, & \text{for } t \leqslant c^2 \\ c^2/(tb), & \text{for } t > c^2 \end{cases}.$$

Thus, an observation $\mathbf{x}$ with squared Mahalanobis distance (MD) $t = \mathbf{x}^H \mathbf{\Sigma}^{-1} \mathbf{x}$ smaller than c^2 receives constant weight, while observations with a larger MD are heavily downweighted.

The threshold c^2 can be chosen, for example, as the qth upper quantile of the distribution having a quadratic form QF($\mathbf{x}$) and based on the reference distribution $\mathcal{N}_p(\mathbf{0}, \boldsymbol{\Sigma})$. For example, because QF($\mathbf{x}$) $\sim \chi_p^2$ when $\mathbf{x} \sim \mathcal{N}_p(\mathbf{0}, \boldsymbol{\Sigma})$, c^2 may be chosen as $c^2 = F_{\chi_p^2}^{-1}(q)$ for some $0 < q < 1$. Figure 4.1 illustrates the loss and weight functions for the real-valued case and for some specific values of q when the data dimension is $p = 3$.

4.4.3   Tyler's Loss Function

The M-estimator proposed by Tyler (1987) has become a very popular robust covariance estimator in the signal-processing literature. Tyler's M-estimator, or its regularized version, has been well studied and the following papers are useful references: Gini and Greco (2002), Conte et al. (2002), Ollila and Koivunen (2003b), Pascal et al. (2008), Abramovich and Spencer (2007), Chen et al. (2011), Ollila and Tyler (2012), Wiesel (2012), Sun et al. (2014), Ollila and Tyler (2014), Pascal et al. (2014), Wiesel and Zhang (2015), and Soloveychik and Wiesel (2015).

The Gaussian loss function can be viewed as a limiting case of either a t loss function or Huber's loss function by considering $\nu \to \infty$ or $c \to \infty$, respectively. At the other extreme, that is, as $\nu \to 0$ or $c \to 0$, Tyler's loss function

$$\rho_{\mathrm{T}}(t) = \frac{p}{\gamma} \log t, \tag{4.30}$$

whose corresponding weight function is $u_{\mathrm{T}}(t) = \gamma \times \rho_{\mathrm{T}}'(t) = p/t$, is obtained. To obtain this limit using Huber's loss function, first note that the Huber's M-estimator is not affected by replacing $\rho_{\mathrm{H}}(t; c)$ with $\rho_{\mathrm{H}}^*(t; c) = \rho_{\mathrm{H}}(t; c) - h(c, b)$, as $h(c, b) = c^2\{1 - \log(c^2)\}/b$ is constant with respect to t. Then, because $c^2/b \to p$ as $c \to 0$, it follows that $\rho_{\mathrm{H}}^*(t; c) \to \rho_{\mathrm{T}}(t)$. Using this loss function, the corresponding (scaled by γ) objective function (4.22) becomes

$$\mathcal{L}_{\mathrm{T}}(\boldsymbol{\Sigma}) = p \sum_{i=1}^{N} \log(\mathbf{x}_i^{\mathsf{H}} \boldsymbol{\Sigma}^{-1} \mathbf{x}_i) - N \log |\boldsymbol{\Sigma}^{-1}|. \tag{4.31}$$

This is the negative log-likelihood function for N i.i.d. observations from an ACG (cf. Section 4.2.3) distribution AG$_p(\mathbf{0}, \boldsymbol{\Sigma})$, which has shown to be minimized (Kent and Tyler, 1988; Kent, 1997) by Tyler's distribution-free M-estimator of scatter. Note that (4.31) does not have a unique minimum because if $\hat{\boldsymbol{\Sigma}}$ is a minimum then so is $b\hat{\boldsymbol{\Sigma}}$ for any $b > 0$. That is, Tyler's M-estimator estimates the shape of $\boldsymbol{\Sigma}$ only.

Note that replacing $\mathbf{x}_i$ with $\mathbf{x}_i/\|\mathbf{x}_i\|$ in (4.31) does not affect its minimizer because it is equivalent to subtracting the term $\frac{p}{N}\sum_{i=1}^{N} \log(\mathbf{x}_i^{\mathsf{H}}\mathbf{x}_i)$, which does not depend on $\boldsymbol{\Sigma}$. Consequently, the finite sample and asymptotic distribution of Tyler's M-estimator of scatter is the same under any elliptical distribution. This is the reason why Tyler (1987) referred to his estimator as a *distribution-free M-estimator*.

Besides being the MLE for an ACG distribution, Tyler's M-estimator has another ML interpretation that is stated in Theorem 6:

THEOREM 6 (Ollila and Tyler, 2012)   *Suppose* $\mathbf{x}_1, \mathbf{x}_2, \ldots, \mathbf{x}_N \in \mathbb{F}^p$ *($\mathbb{F} = \mathbb{R}$, or $\mathbb{C}$) are independent observations from (possibly different) elliptical distributions of proportional covariance matrices, so* $\mathbf{x}_i \sim \mathcal{E}_p(\mathbf{0}, \tau_i \Sigma, g_i)$, *where* $\tau_i > 0$, $i = 1, \ldots, N$ *and* $\Sigma \in \mathbb{S}_{++}^p$ *are unknown parameters. Then, the maximum likelihood estimate for* Σ *corresponds to Tyler's M-estimate of scatter.*

It is also important to realize that in Theorem 6, the samples $\mathbf{x}_i$-s can be from different ES distributions (i.e., the density generators $g_i(\cdot)$ are not specified). This fact allows the noise conditions to fluctuate from sample to sample. This powerful MLE property of Tyler's *M*-estimator may explain why the estimator often performs well in diverse applications.

4.5   Regularized *M*-estimators of the Scatter Matrix

M-estimators of scatter, as described in the previous section, are only applicable in the $N > p$ case. Low sample support (i.e., p is of the same magnitude as N) is a commonly occurring problem in diverse data analysis problems. In the case of insufficient sample support, that is, $p > N$, the inverse of the SCM cannot be computed as the estimator is not full rank. Thus, for example, classic beamforming techniques such as minimum variance distortionless response (MVDR) beamforming or the adaptive normalized matched filter cannot be realized because they require an estimate of the inverse covariance matrix.

A common ad-hoc solution has been to use diagonal loading, that is, to use regularized (shrinkage) SCM (RSCM) of the form,

$$\mathbf{S}_{\alpha,\beta} = \beta \mathbf{S} + \alpha \mathbf{I}, \tag{4.32}$$

where $\beta > 0$ and $\alpha \geq 0$ are regularization parameters. Regularized SCM has a long history in signal processing and related fields. For example, in array processing, such estimators are often used in adaptive beamforming; see for example, Carlson (1988), Li et al. (2003), and Vorobyov et al. (2003). In finance, an estimator of the form (4.32) was popularized by Ledoit and Wolf (2004) who proposed consistent estimators of the regularization parameter pair (α, β) which are optimal in the minimum mean squared error (MMSE) sense. Estimators that minimize the MMSE were also proposed by Chen et al. (2010) assuming that the data set is from a Gaussian distribution and by Ollila (2017) under the (less restrictive) assumption that the samples are from an unspecified elliptical distribution.

One of the first contributions for constructing a robust regularized covariance matrix estimator was by Abramovich and Spencer (2007). Since then, robust regularized *M*-estimators have been an active research topic; see for example, Chen et al. (2011), Wiesel (2012), Ollila and Tyler (2014), Sun et al. (2014), Couillet and McKay (2014), Pascal et al. (2014), and Zhang and Wiesel (2016), as well as the recent textbook by Wiesel and Zhang (2015).

Here we describe the large class of regularized M-estimators developed by Ollila and Tyler (2014). First, for a given ρ-function, define a class of *tuned ρ-functions* as

$$\rho_\beta(t) = \beta\rho(t) \quad \text{for } \beta > 0,$$

where β represents an additional tuning constant that can be used to tune the estimator toward some desirable property. Using $\rho_\beta(t) = \beta\rho(t)$ as a loss function in (4.22), we can define a respective tuned optimization function $L_\beta(\boldsymbol{\Sigma})$ as

$$L_\beta(\boldsymbol{\Sigma}) = \beta \sum_{i=1}^{N} \rho(\mathbf{x}_i^H \boldsymbol{\Sigma}^{-1} \mathbf{x}_i) - \frac{N}{\gamma} \ln |\boldsymbol{\Sigma}^{-1}|.$$

Notice that for $\beta = 1$, the conventional M-estimation criterion function in (4.22) is obtained. We then introduce an additive penalty function $P(\boldsymbol{\Sigma}) = \text{Tr}(\boldsymbol{\Sigma}^{-1})$ to the preceding criterion, which yields

$$L_{\alpha,\beta}(\boldsymbol{\Sigma}) = L_\beta(\boldsymbol{\Sigma}) + (\alpha N)P(\boldsymbol{\Sigma})$$

$$= \beta \sum_{i=1}^{N} \rho(\mathbf{x}_i^H \boldsymbol{\Sigma}^{-1} \mathbf{x}_i) - \frac{N}{\gamma} \ln |\boldsymbol{\Sigma}^{-1}| + \alpha N \text{Tr}(\boldsymbol{\Sigma}^{-1}). \tag{4.33}$$

Naturally, when $\alpha = 0$ and $\beta = 1$, the conventional nonregularized (and nontuned) M-estimation criterion function is obtained. The penalty function $\text{Tr}(\boldsymbol{\Sigma}^{-1})$ penalizes candidate solutions that are close to being singular. This is because $\text{Tr}(\boldsymbol{\Sigma}^{-1}) = \sum_{j=1}^{p} \frac{1}{\lambda_j(\boldsymbol{\Sigma})}$, where $\lambda_1(\boldsymbol{\Sigma}), \ldots, \lambda_p(\boldsymbol{\Sigma})$ are the ordered eigenvalues of $\boldsymbol{\Sigma}$. Thus, any candidate $\boldsymbol{\Sigma} \in \mathbb{S}_{++}^p$, having $\lambda_j(\boldsymbol{\Sigma})$ close to a zero, is heavily penalized. This feature is necessary in the ill-conditioned insufficient sample support case ($N < p$).

Ollila and Tyler (2014) then defined the regularized M-estimator of scatter as the minimizer of (4.33), that is,

$$\hat{\boldsymbol{\Sigma}} = \underset{\boldsymbol{\Sigma} \succ 0}{\arg \min} \, L_\beta(\boldsymbol{\Sigma}) + \alpha N P(\boldsymbol{\Sigma}). \tag{4.34}$$

The authors then showed that if $\rho(t)$ satisfies Condition 2, and is bounded below, then the solution always exists and is unique. Furthermore, if $\rho(t)$ is also differentiable, then the minimum corresponds to the unique solution of the *regularized M-estimating equation*,

$$\hat{\boldsymbol{\Sigma}} = \frac{\beta}{N} \sum_{i=1}^{N} u(\mathbf{x}_i^H \hat{\boldsymbol{\Sigma}}^{-1} \mathbf{x}_i)\mathbf{x}_i\mathbf{x}_i^H + \alpha \mathbf{I}. \tag{4.35}$$

For $\alpha > 0$, existence of the solution to (4.34) does not require any conditions on the sample $\{\mathbf{x}_i\}_{i=1}^N$. This is not the case in the nonpenalized case ($\alpha = 0$), which requires Condition 3. Also note, for $\alpha = 0$ (and $\beta = 1$), that equation (4.35) reduces to the conventional M-estimating equation in (4.23). If $\rho(t)$ satisfies Condition 2, is differentiable, and $u(t) = \rho'(t)$ is nonincreasing, then the solution to the regularized M-estimating equation (4.35) can be found using the iterative algorithm

$$\hat{\Sigma}_k = \frac{\beta}{N} \sum_{i=1}^{N} u(\mathbf{x}_i^{\mathsf{H}} \hat{\Sigma}_k^{-1} \mathbf{x}_i) \mathbf{x}_i \mathbf{x}_i^{\mathsf{H}} + \alpha \mathbf{I}, \qquad (4.36)$$

$(k = 0, 1, \ldots)$, which converges to the solution of (4.35) for any initial start $\hat{\Sigma}_0 \in \mathbb{S}_{++}^p$; see Ollila and Tyler (2014, theorem 2).

The preceding results hold for the Gaussian, Huber's, and t_ν ($\nu > 0$) loss function cases. If $\rho(t)$ is the Gaussian loss function $\rho_G(t) = t/\gamma$, then the unique minimum of (4.34) is obtained by the regularized SCM $\mathbf{S}_{\alpha,\beta}$ given in (4.32). Because Huber's loss function $\rho_H(t; c)$ in (4.29) is differentiable, bounded from below, and satisfies Condition 2, the optimization problem (4.34) has a unique solution that can be found simply by solving the regularized M-estimating equation (4.35) using the algorithm (4.36). The same holds true for the t-loss function case when $\nu > 0$.

Next, we consider the tuned Tyler's loss function $\rho_\beta(t) = p\beta \log t$ for fixed $0 < \beta < 1$. The regularized objective function, based on Tyler's loss function, is then

$$L_{\alpha,\beta}(\Sigma) = \beta p \sum_{i=1}^{N} \log(\mathbf{x}_i^{\mathsf{H}} \Sigma^{-1} \mathbf{x}_i) - N \ln |\Sigma^{-1}| + \alpha N \mathsf{Tr}(\Sigma^{-1}),$$

and the respective regularized M-estimating equation is

$$\hat{\Sigma} = \beta \frac{p}{N} \sum_{i=1}^{N} \frac{\mathbf{x}_i \mathbf{x}_i^{\mathsf{H}}}{\mathbf{x}_i^{\mathsf{H}} \hat{\Sigma}^{-1} \mathbf{x}_i} + \alpha \mathbf{I} \qquad (4.37)$$

The regularized Tyler's M-estimator has been independently studied, for example, by Ollila and Tyler preceding (2014), Pascal et al. (2014), and Sun et al. (2014). In the case of Tyler's loss function, we often use $\alpha = 1 - \beta$, $\beta \in (0, 1]$, as in this case the solution verifies $\mathsf{Tr}(\hat{\Sigma}^{-1}) = p$. Because Tyler's loss function $\rho_T(t) = p \log t$ is not bounded below, the earlier stated result for existence of the solution does not hold, and it is necessary to impose some conditions on the sample; see Ollila and Tyler (2014) for details. The solution to (4.37) can again be computed using (4.36).

4.6   Signal Detection Application

We address the problem of detecting a known complex *signal vector* $\mathbf{s} \in \mathbb{C}^p$ in *received data* $\mathbf{x} = \alpha \mathbf{s} + \mathbf{v} \in \mathbb{C}^p$, where $\mathbf{v} \in \mathbb{C}^p$ represents the unobserved complex *noise* r.v. and $\alpha \in \mathbb{C}$ is a signal parameter modeled as an unknown deterministic parameter or as a random variable depending on the application at hand. The signal $\mathbf{s}$ is considered to be normalized s.t. $\|\mathbf{s}\|_2^2 = p$ holds. This avoids the inherent ambiguity in the definition of $\mathbf{x}$ as the scale of $\mathbf{s}$ can always be absorbed into α. In radar applications, for example, α is a complex unknown parameter accounting for both channel propagation effects and target backscattering and $\mathbf{s}$ is the transmitted known (normalized) radar pulse vector. The signal-absent versus signal-present problem can then be expressed as a test between the hypotheses

$$H_0 : |\alpha| = 0 \quad \text{vs} \quad H_1 : |\alpha| > 0. \tag{4.38}$$

Assuming that the noise $\mathbf{v}$ follows a centered CES distribution with a scatter matrix $\boldsymbol{\Sigma} = \sigma^2 \mathbf{V}$, where $\sigma > 0$ represents the unknown *noise level* (average noise power) and $\mathbf{V}$ denotes the known *shape matrix* (or normalized covariance matrix) verifying $\text{Tr}(\mathbf{V}) = p$, we investigate the properties of the detector

$$\Lambda \equiv \Lambda(\mathbf{x}, \mathbf{s}) = \frac{|\mathbf{s}^H \boldsymbol{\Sigma}^{-1} \mathbf{x}|^2}{(\mathbf{x}^H \boldsymbol{\Sigma}^{-1} \mathbf{x})(\mathbf{s}^H \boldsymbol{\Sigma}^{-1} \mathbf{s})} \underset{H_0}{\overset{H_1}{\gtrless}} \lambda, \tag{4.39}$$

which has been derived independently by several different authors. The detector has many names such as the constant false alarm rate (CFAR), the matched subspace detector (MSD) (Scharf and Friedlander, 1994; Scharf and McWhorter, 1996), the normalized matched filter (NMF) (Conte et al., 2002), the linear quadratic generalized likelihood ratio test (Gini, 1997; Gini and Greco, 2002), to mention a few. The adaptive version of the detector, that is, when the unknown $\boldsymbol{\Sigma}$ is replaced by its estimate $\hat{\boldsymbol{\Sigma}}$, also has many names, including the adaptive coherence estimator (ACE) (Scharf and McWhorter, 1996; Kraut et al., 2005), and the adaptive NMF (Conte et al., 2002).

An important feature of the detector is its invariance under scalar multiples of $\mathbf{x}$. This implies, owing to the stochastic representation (cf. Theorem 5) of CES distributions, that its nulldistribution is independent of the noise level σ^2 or the functional form of the CES distribution, that is, it is distribution free (Kraut et al., 2001; Ollila and Tyler, 2012). Specifically, we have the following result, the proof of which can be found in Ollila and Tyler (2012):

THEOREM 7 *If the null (noise only) hypothesis H_0 holds, so $\mathbf{x} = \mathbf{v} \sim \mathcal{E}_p(\mathbf{0}, \sigma^2 \mathbf{V}, g)$, where $\mathbf{V} \in \mathbb{S}_{++}^p$ is known and verifies $\text{Tr}(\mathbf{V}) = p$, then Λ in (4.39) has a $\text{Beta}(1, p-1)$ distribution.*

This property of the detector is elemental as it holds true regardless of the functional form of the density generator g or the noise level σ^2. Hence, when the average noise power is unknown, the detector has a CFAR property when the noise follows an unspecified elliptical distribution. The rejection threshold λ in (4.38) can be set as the $(1 - P_{FA})$th quantile of the $\text{Beta}(1, p-1)$ distribution to guarantee a desired level $P_{FA} \in [0, 1]$ for the probability of false alarm (PFA). This one-to-one relationship between PFA and the threshold λ of the detector can be expressed in closed form as

$$P_{FA} = \text{Prob}(\Lambda > \lambda | H_0) = (1 - \lambda)^{p-1}, \text{ or, } \lambda = 1 - \sqrt[p-1]{P_{FA}}. \tag{4.40}$$

An *adaptive detector* is obtained by computing an estimate $\hat{\boldsymbol{\Sigma}}$ of $\boldsymbol{\Sigma}$ from signal-free secondary data $\mathbf{x}_1, \ldots, \mathbf{x}_N$ and substituting the estimate in place of $\boldsymbol{\Sigma}$ in (4.39); the resulting test statistic is then denoted by $\hat{\Lambda}$. An important feature of the detector (4.39) is that it requires $\boldsymbol{\Sigma}$ only up to a constant scalar because one may replace $\boldsymbol{\Sigma}$ by $c\boldsymbol{\Sigma}$ in (4.39) without altering the value of Λ. Because any M-estimator $\hat{\boldsymbol{\Sigma}}$ estimates $\boldsymbol{\Sigma}$ up to a scalar $\sigma_{u,g}^2$ (cf. Section 4.3), we may use any M-estimator in place of $\boldsymbol{\Sigma}$. In radar applications, for example, clutter is commonly heavy-tailed non-Gaussian, and, hence, robustness of the selected M-estimator is perhaps the most important design criterion. Due to consistency of $\hat{\boldsymbol{\Sigma}}$ toward $a\boldsymbol{\Sigma}$ for $a > 0$ (Ollila et al., 2012), the resulting $\hat{\Lambda}$ will have an approximate Beta$(1, p - 1)$ distribution for sufficiently large sample sizes N. Naturally, the goodness of the Beta$(1, p - 1)$ approximation depends on the selected estimator, the sample length, and the reference CES distribution of the secondary data. Next, we address some of these issues using a simulation study.

4.6.1   Simulation Study

We first investigate the accuracy of the Beta$(1, p - 1)$ approximation for the null distribution of the adaptive detector $\hat{\Lambda}$ under short data records of the secondary data. We generate secondary data $\mathbf{x}_1, \ldots, \mathbf{x}_N$ as i.i.d. random samples from the $p = 8$ variate K-distribution $K_{p,\nu}(\mathbf{0}, \boldsymbol{\Sigma})$ with $\nu = 0.5$ (i.e., a heavy-tailed K-distribution) as well as 10 data observations from the same distribution that are classified using the adaptive form of the detector (4.39). For 10,000 trials, we calculated the empirical P_{FA} (the proportion of incorrect rejections) for a fixed rejection threshold λ. For each trial, the noise covariance matrix $\boldsymbol{\Sigma}$ was generated randomly as $\boldsymbol{\Sigma} = \mathbf{E}\mathbf{L}\mathbf{E}^H$, where $\mathbf{E}$ is a random orthonormal matrix and $\mathbf{L} = \mathsf{diag}(l_1, \ldots, l_p)$ contains the eigenvalues that are generated from the $\mathcal{U}(0, 1)$ distribution and postnormalized so that $\sum_{i=1}^{p} l_i = p$ holds (i.e., $\mathsf{Tr}(\boldsymbol{\Sigma}) = p$). Thus, we obtained $10,000 \cdot 10$ decisions of rejection/acceptance for the H_0 hypothesis. Figure 4.3a and 4.3b depict the results for sample lengths $N = 32, 64$ and $N = 96$ for the adaptive detector based on Tyler's M-estimator and Huber's M-estimator of scatter (using $q = 0.8$ to determine the threshold c^2). The solid curve $(N = \infty)$ depicts the theoretical PFA stated in (4.40). For $P_{FA} \approx 0.01$, only a small mismatch is observed between the empirical PFA and the nominal PFA. In this setting, the adaptive detector based on Tyler's M-estimator performs better than Huber's M-estimator based on $q = 0.8$. Recall from Section 4.4 that Tyler's M-estimator corresponds to Huber's M-estimator in the limiting case $(q \to 0)$.

Figure 4.3c shows the empirical pdf plots (histogram) of the obtained test statistic $\hat{\Lambda}$ based on Tyler's M-estimator for the case of $N = 96$, as well as the density of a Beta$(1, p - 1)$ distribution (solid curve). As can be seen, the adaptive detector, based on Tyler's M-estimator, provides a good fit to the Beta$(1, p - 1)$ distribution, especially in the tail region of the distribution. This is also illustrated in the quantile-quantile (QQ-) plot of Figure 4.3d. In the ideal case, the QQ-plot resembles a straight line through the origin having slope 1. The dashed vertical (resp. horizontal) line indicates the value for the theoretical (resp. empirical) 0.01 and 0.05 (or 1 percent and 0.1 percent) upper quantiles.

Next, we assess the probability of detection (PD) of the adaptive detector. For this case, the test data is generated under the H_1 model, and consistent with $\mathbf{x} = \alpha\mathbf{s} + \mathbf{v}$, where we assume that $|\alpha|$ is a Rayleigh fluctuating amplitude with scale $\sigma_{|\alpha|}$ and the noise $\mathbf{v}$ has a K-distribution with parameter $\nu = 0.5$ and covariance matrix $\boldsymbol{\Sigma} = \sigma^2\mathbf{I}$. The signal-to-noise power ratio (SNR) ratio is then defined as $\mathrm{SNR} = \mathsf{E}[\|\alpha\mathbf{s}\|_2^2]/\mathsf{E}[\|\mathbf{v}\|_2^2] = \sigma_{|\alpha|}^2/\sigma^2$. In this case, the theoretical probability of detection

(a) Tyler's M-estimator

(b) Huber's M-estimator with $q = 0.8$

(c) Histogram, Tyler's M-estimator

(d) QQ-plot, Tyler's M-estimator

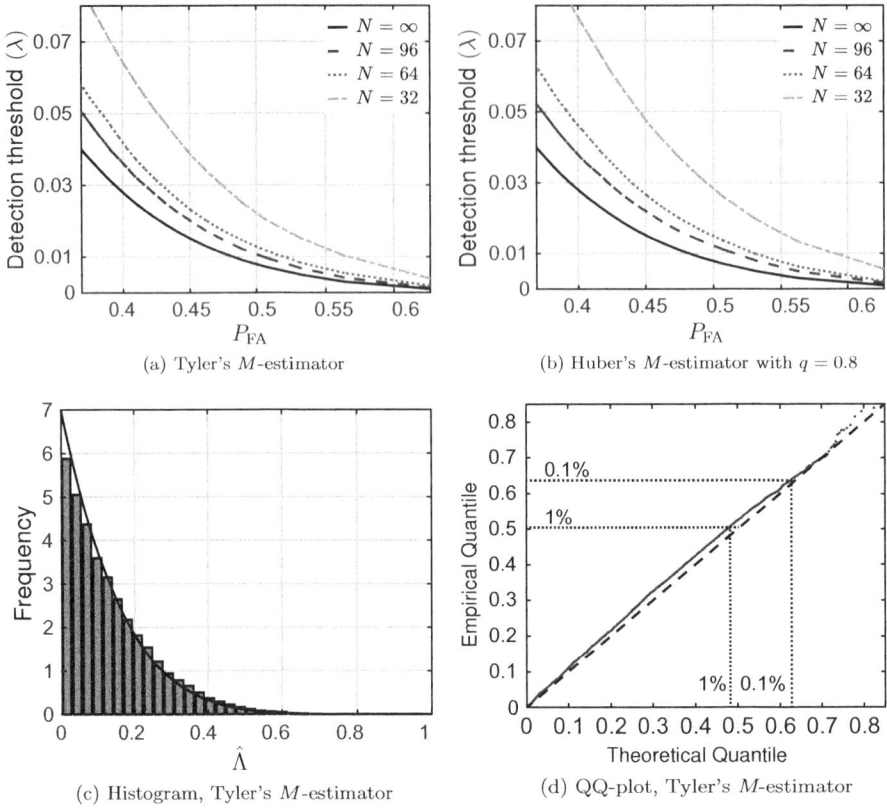

Figure 4.3 Empirical P_{FA} for adaptive detectors employing (a) Tyler's M-estimator and (b) Huber's M-estimator with $q = 0.8$ under K-distributed clutter and different sample sizes N of the secondary data. Figure (c) shows the corresponding empirical pdf for $N = 96$ for $\hat{\Lambda}$ based on Tyler's M-estimator and (d) depicts the QQ-plot of the empirical quantiles and theoretical quantiles of the Beta$(1, p - 1)$ distribution. Simulation parameters are $p = 8$, $\nu = 0.5$, and the clutter covariance matrix Σ was generated randomly for each of the 10,000 trials.

$P_{\mathrm{D}} = \mathsf{Prob}(\Lambda > \lambda | H_1)$ is easy to compute as a function of the SNR (Ollila and Tyler, 2012):

$$P_{\mathrm{D}}(\gamma) = 1 - \int_0^\infty F_{2;2(p-1)}\left(\lambda'; 4xp^2\gamma\right) f_{2;2\nu}(x)\mathrm{d}x \qquad (4.41)$$

where $\gamma = \sigma_{|\alpha|}^2 / \sigma^2$ denotes the SNR, $\lambda' = (p - 1)\lambda/(1 - \lambda)$, $f_{a,b}(\cdot)$ denotes the pdf of the F-distribution with a and b degrees of freedom and $F_{a,b}(\cdot; c)$ denotes the cdf of the F-distribution with a and b degrees of freedom and a noncentrality parameter c. We note that P_{D} depends on the SNR, the data dimension p, and the parameter ν of the K-distribution. We set the threshold λ for the adaptive detector so that the theoretical (for known Σ) level is $P_{\mathrm{FA}} = 0.01$.

Figure 4.4 plots the theoretical PD curve given in (4.41) as a function of the SNR parameter γ. Also plotted are the observed PD (the proportion of correct rejections

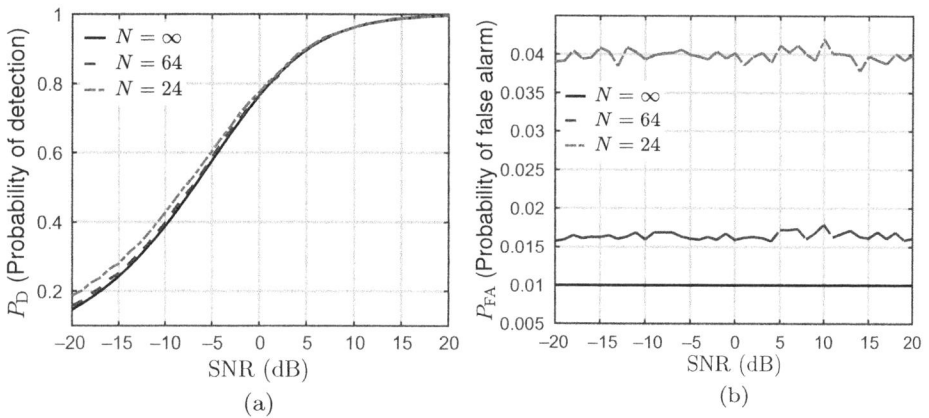

Figure 4.4 The left panel shows the observed PD, as a function of the SNR $\sigma_{|\alpha|}^2/\sigma^2$ (dB), for the adaptive detector based on Tyler's M-estimator under K-distributed clutter. The right panel shows the observed probability of false alarm rate as a function of the SNR. Simulation parameters are $p = 8$, $\mathbf{\Sigma} = \sigma^2\mathbf{I}$, and $\nu = 4.5$ and the detection threshold λ was set to give a PFA equal to 1 percent.

of the null hypothesis) for the adaptive detector with test statistic $\hat{\Lambda}$ that uses Tyler's M-estimator. The results are based on 5,000 independent trials for each set SNR level from the set $\{-20, -19, \ldots, 19, 20\}$ (dB). In each Monte Carlo trial, we have drawn a set of length $N = 3p = 24$ and $N = 8p = 64$ (signal free) secondary data from which $\mathbf{\Sigma}$ was estimated and then 10 test data $\mathbf{x}$ were generated from a H_1 model and classified using the detector. For each SNR level considered, $5,000 \cdot 10$ decisions were obtained and, from these, the empirical (averaged) proportions of correct rejections were computed. As Figure 4.4a demonstrates, for $N = 24$, the empirical PD deviates from the theoretical PD ($N = \infty$) especially at low SNRs, but for $N = 64$, the agreement with the theoretical PD is very good. This deviation can also be seen in Figure 4.4b, which shows the empirical probability of the false alarm rate for the case in which the test data is generated from the null hypothesis H_0. Indeed, for $N = 24$, the empirical PFA is fluctuating around 0.04 and, thus, the adaptive detector based on $N = 24$ secondary data is less conservative than what was desired. The detector based on $N = 64$ yields results that are much closer to the nominal PFA level of 0.01.

4.7 Concluding Remarks

In this chapter, robust M-estimators of the scatter (covariance) matrix were considered. We reviewed the properties of the class of ES distributions, which include the most commonly used multivariate distributions as special cases, for example the multivariate normal (Gaussian), the t-distribution, or the generalized Gaussian distribution. The class of ES distributions is parametrized by a positive definite Hermitian scatter matrix parameter, which can be taken to be equal to the covariance matrix when the second-order moments exist. We reviewed ML estimation and M-estimation of the scatter matrix parameter of the ES distribution. The most commonly used robust loss functions for scatter estimation, such as Huber's, Tyler's, or the t-loss and their properties were discussed in detail. We also discussed the conditions on the loss function under which the M-estimation criterion function is geodesically convex. We illustrated how the minimization of the ML or M-estimation criterion function of scatter leads to fixed point estimating equations and to the related fixed point algorithm to compute the solution. We further addressed the case of regularized M-estimators of the scatter matrix, which can be used also in the cases in which the sample size, N, is smaller than the dimensionality, p. Finally, usefulness of the robust M-estimators of scatter was illustrated in a signal detection application.

5   Robustness in Sensor Array Processing

5.1   Introduction

In sensor array signal processing, one is typically interested in characterizing, synthesizing, enhancing, or attenuating certain aspects of propagating wavefields by employing a collection of sensors in distinct locations, known as a *sensor array*. Interest is typically in detecting signal sources or estimating parameters such as target or source location, velocity and bearing. Moreover, sensor arrays are used for different imaging tasks in medical and radar applications. Signals are processed in the spatial domain and the explaining variables are the locations of the sensors and time. Characterizing an observed wavefield refers to determining its *spatial spectrum*, that is, the angular distribution of energy. From such a spectrum, information regarding the location of the sources generating the wavefield can be acquired. Synthesizing a wavefield refers to generating a propagating wavefield with a desired spatial spectrum to focus the transmitted energy toward certain directions or locations in space. Finally, attenuating or enhancing a received wavefield, based on its spatial spectrum, refers to the ability of resolving sources, in the spatial domain, canceling interfering sources, or enhancing the signal by improving the SNR and maximizing the energy received from certain directions. Examples of characterization, synthesis, and enhancement of propagating wavefields include direction-of-arrival (DOA) estimation, source localization, channel estimation, interference cancellation, as well as transmit and receive beamforming.

Traditionally, array signal processing has found applications in radar and sonar systems, acoustics, geophysics, signal intelligence and surveillance, as well as medical imaging applications. Radioastronomy also extensively employs array processing techniques. More recently, it has been used in wireless communication systems, where various aspects, including interference control and spectral efficiency increase, are of interest.

Signals in sensor arrays are modeled either as deterministic sequences or random processes, and unobservable noise is typically modeled as a random process. In case of stochastic models, statistical methods are needed in characterizing signal properties, performance of the estimation procedures, and uncertainties present in the system. For the case of deterministic signals, classical signal theory and matrix perturbation theory are used. The actual sensors and their properties also play a major role in array signal models. The standard approach to array signal processing assumes rather idealistic array models. In particular, all array elements are assumed to have similar ideal gain patterns, and the employed sensor array is assumed to have a regular geometry such as a uniform linear array (ULA), uniform circular array (UCA), or a uniform rectangular array (URA). Further, it is assumed that there is no mutual coupling among the sensors and the sensor phase centers are exactly in their desired locations. In real-world sensor arrays, these idealized assumptions are not valid and arrays are subject to a variety of nonidealities. Sensor array imperfections can lead to significant performance losses if appropriate calibration procedures are not employed. Moreover, there is a growing interest to use nonuniform and sparse array geometries. They have the benefit that fewer sensors and related radio-frequency intermediate-frequency (RF-IF) front ends are needed for the desired performance, which reduces the price of the sensor array. At the same time, a larger displacement between the sensors leads to reduced mutual coupling and reduces the impact of array impairments. Examples of such array configurations include minimum redundancy arrays (MRA) and passive and active arrays employing difference and sum co-arrays (Hoctor and Kassam, 1990).

A key design goal for signal processing algorithms in sensor array systems is to achieve optimal performance in a given application. Statistical methods are also required to define the criteria for optimality, for example, minimum mean square error, minimum variance, or the maximum of the likelihood function. The optimization is typically performed based on strict assumptions on the propagation environment, statistical or structural properties of the source signals, the sensor array configuration, the sensor characteristics, as well as the interference and noise probability models. As discussed extensively in previous chapters, a shortcoming of optimal estimation procedures is that they are extremely sensitive even to small deviations from the assumed model. In reality, the assumptions underpinning the signal model, as well as the statistical models for the noise and interference, may not be valid and a significant degradation from the optimal performance is experienced. Procedures that perform highly reliably under uncertainty are commonly called robust (Kassam and Poor, 1985; Swami and Sadler, 2002; Godara, 2004; Hua et al., 2004; Hampel et al., 2011). Robust methods typically trade-off optimality for highly reliable performance for the case of departures from the underlying signal and noise models, as well as nonideal sensor array configurations and propagation environments.

In this chapter, we will consider the robustness of sensor array processing when the ideal assumptions that are made for the signal and noise models are not necessarily valid. In practice, the uncertainties are caused by errors in the array response and array calibration, impairments in the signal waveforms caused by nonidealities in the RF-IF front ends, propagation environment, lack of data, or the highly time-varying nature of the propagation environment, interferers, or a misspecified noise model. The main focus is on the statistical robustness needed to deal with misspecified noise distribution models, for example, heavy-tailed, highly impulsive noise may be experienced instead of nominal Gaussian noise. In such cases, statistically optimal array processing techniques, such as the maximum likelihood method, derived assuming a Gaussian noise model will experience a significant performance degradation. As an example application, high-resolution direction-finding methods are considered. Such methods use the array output covariance matrix, the subspaces spanned by its eigenvectors and associated eigenvalues, and the projection matrices to these subspaces. This is justified because the subspaces of the array covariance matrix are related to the subspaces spanned by the array steering vectors that contain information about the unknown parameters of interest, such as the angles of arrival. Hence, we pay special attention to the robust estimation of the array covariance matrix and its eigenvalues and eigenvectors. Subspace and maximum likelihood methods are prominent examples of techniques that rely on subspaces and projection matrices. Robust estimation of covariance or scatter matrices was considered in Sections 4.3 and 4.4. In array processing algorithms, the data is typically complex valued.

This chapter is organized as follows. The basic array processing signal model for ideal arrays and the underlying assumptions used in estimating the angles of arrival in such arrays are presented. The uncertainties in the signal model include uncertainties arising from the modeling of the emitted signals, the array configuration, and the propagation environment. Some array configurations, and DOA estimation methods, that provide robustness in the face of such uncertainties are briefly discussed. The uncertainties in modeling the noise are considered next. The main focus is on statistical robustness and is the subject of Section 5.4. Robust procedures, which are close to optimal in the presence of nominal Gaussian noise and highly robust in non-Gaussian noise are presented. The qualitative robustness of these procedures is described using the concept of influence functions. Finally, some examples of robust angle-of-arrival estimation are given.

5.2   Basic Array Signal Model

In the following, we describe the basic signal model and the underlying assumptions that are used in angle-of-arrival estimation. An ideal ULA model, as illustrated in Figure 5.1, is assumed for the sake of clarity. We assume the standard narrowband and low rank signal model of K incoherent source signals impinging on an array of M sensors ($K < M$). The sources originate from a far-field point source. Consequently, plane waves are assumed to be impinging on the array. At a given discrete time instant, the received signal vector $\mathbf{x}$ is an $M \times 1$ complex vector given by

$$\mathbf{x} = \mathbf{As} + \mathbf{w}, \tag{5.1}$$

where $\mathbf{A}$ is an $M \times K$ matrix such that

$$\mathbf{A} = [\mathbf{a}(\theta_1), \mathbf{a}(\theta_2), \ldots, \mathbf{a}(\theta_K)],$$

with $\mathbf{a}(\theta_k)$ denoting the $M \times 1$ array steering vector corresponding to the DOA θ_k of the kth signal. The K-vector

$$\mathbf{s} = [s_1, s_2, \ldots, s_K]^\top$$

is the vector of incident signals and $\mathbf{w}$ is the $M \times 1$ complex noise vector.

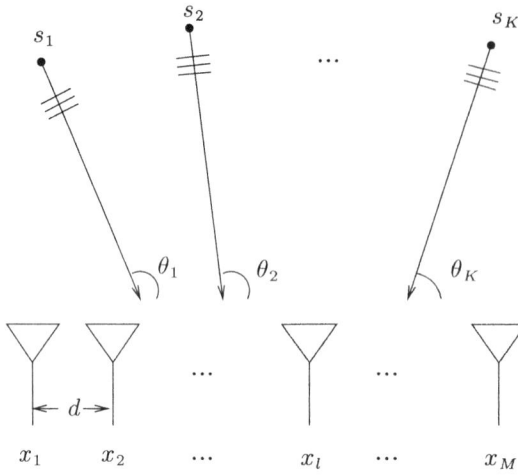

Figure 5.1 An ideal ULA of M sensors receiving plane waves from K far-field and narrow-band point sources.

It is assumed, for any collection of K distinct θ_i's, that the matrix **A** has full rank. When the array is a ULA, the array steering vector is given by

$$\mathbf{a}(\theta) = [1, e^{2\pi J(d/\lambda)\cos(\theta)}, \ldots, e^{2\pi J(M-1)(d/\lambda)\cos(\theta)}]^\top,$$

where d denotes the element spacing and λ denotes the wavelength. The angle is here expressed with respect to the line connecting the array elements, as shown in Figure 5.1. For the case in which the angle is measured with respect to the broadside of the array, $-\sin(\theta)$ should be used instead of $\cos(\theta)$. The elements of the steering vector depend on the array structure. For the case of a ULA with identical array elements, the matrix **A** becomes a Vandermonde matrix. The array steering vector model would be more realistic if the following array nonidealities are considered: mutual coupling between array elements, mounting platform reflections, cross-polarization effects, element mis-placement errors, and the individual directional beam patterns associated with each array element. These nonidealities are discussed in more detail in the subsection dealing with errors in the array signal model. However, for the purpose of studying statistical robustness of array processing algorithms, the ideal array model illustrated in Figure 5.1 is suitable.

Two-dimensional sensor arrays are needed to do estimation and beamsteering both in azimuth and elevation. Elevation information is relevant in many applications, for example, radar and 5G communications. The UCA is a widely used 2-D array geometry and with such an array the elements of the steering vector are of the form

$$\mathbf{a}(\theta, \phi) = \begin{pmatrix} e^{J\omega\frac{r}{c}\sin\phi\cos(\theta-\gamma_0)} \\ e^{J\omega\frac{r}{c}\sin\phi\cos(\theta-\gamma_1)} \\ \vdots \\ e^{J\omega\frac{r}{c}\sin\phi\cos(\theta-\gamma_{M-1})} \end{pmatrix},$$

where $\omega = 2\pi f$ is the angular frequency, r is the radius of the array, $\gamma_i = \frac{2\pi i}{M}$ is the angular position of the element (counted in a counterclockwise manner from the x-axis), and c is the speed of light. The azimuth and elevation angles are denoted by θ_l and ϕ_l.

Assume that the narrowband signals impinging on the array are zero-mean wide-sense stationary stochastic signals and the noise is spatially and temporally white and uncorrelated with the signals. For a treatment of robust array processing in the nonstationary case, the reader is referred to Sharif et al. (2013). At this point, we focus only on spatial processing of the data. For this case, the covariance matrix of the signals $\mathbf{x}$ (if it exists) that is obtained from the array is given by

$$\Sigma = E[\mathbf{x}\mathbf{x}^H] = \mathbf{A}\Sigma_s\mathbf{A}^H + \sigma^2\mathbf{I},$$

where $\Sigma_s = E[\mathbf{s}\mathbf{s}^H]$ is the $K \times K$ signal covariance matrix of full rank, that is, the signals are not coherent, σ^2 is the noise variance, and $\mathbf{I}$ denotes the $K \times K$ identity matrix. The eigenvalue decomposition (EVD) of the array covariance matrix is given by

$$\Sigma = \mathbf{U}\Lambda\mathbf{U}^H,$$

where Λ is a diagonal matrix of eigenvalues and matrix $\mathbf{U}$ contains the corresponding eigenvectors. If $M > K$, the signal model is low rank and the $M - K$ smallest eigenvalues of Σ are smaller than the signal eigenvalues and equal to σ^2, that is, the noise variance. The eigenvectors corresponding to the smallest eigenvalues are orthogonal to the columns of the steering vector matrix $\mathbf{A}$. These $M - K$ eigenvectors $\mathbf{U}_n$ span the *noise subspace* and the remaining K eigenvectors $\mathbf{U}_s$, corresponding to the K largest eigenvalues, span the *signal subspace*. The signal subspace eigenvectors span the same subspace as the columns of the steering vector matrix $\mathbf{A}$. This property is explicitly exploited in high-resolution methods such as MUSIC (Multiple Signal Classification) and ESPRIT (Estimation of Signal Parameters by Rotational Invariance Techniques), as well as in maximum likelihood methods where projection matrices to signal or noise subspaces are employed. For example, the MUSIC method is based on the property that the signals lie in the signal subspace and they are orthogonal to the entire noise subspace spanned by $\mathbf{U}_n$. In terms of matrix notation this may be expressed by the MUSIC pseudospectrum

$$V_M(\theta) = \frac{1}{\mathbf{a}^H(\theta_i)\mathbf{U}_n\mathbf{U}_n^H\mathbf{a}(\theta_i)}$$

at the correct DOAs θ_i, $i = 1, \ldots, K$ for the K sources. The denominator may be seen as an index of orthogonality because it quantifies how orthogonal the steering vectors are to the entire noise subspace. When θ_i corresponds to a true angle-of-arrival, the steering vector lies in a subspace that is orthogonal to the noise subspace and the inner product in the denominator should be zero. For other angles $\theta \neq \theta_i$, the quadratic expression in the denominator is positive. Estimates of the K angles of arrival can be found from the K highest peaks of the pseudospectrum, that is, the K angles that produce the highest index of orthogonality. Hence, the height of a peak in a MUSIC pseudospectrum is not associated with the power of the source signal, but with the orthogonality of the steering vector with the noise subspace.

In case of broadband signals, a space-time (ST) array model is used. There are two basic approaches. First, the problem can be changed into a set of narrowband problems by employing fast Fourier transform (FFT) processing to subdivide the broadband signal into multiple narrowband signals. Then narrowband methods can be used for each subband. The results from each subband can then be combined to obtain the output for the broadband case. Alternatively, ST processing can be done at the receiver. For the case of a uniform linear ST arrays of M elements, a tapped delay line of length L is associated with each element. The tap delays are chosen as $T \leq 1/(2\omega_c)$, where ω_c is the center frequency, to avoid temporal aliasing. The element spacing is typically $\lambda/2$ to ensure that no spatial aliasing takes place. For the case of delay lines with L taps, the ST snapshot is an $ML \times 1$ observation vector, denoted by $\mathbf{x}$, formed by stacking the spatial snapshots. The signal model for ST observations, at a given discrete time-instant, is

$$\mathbf{x} = \sum_{k=1}^{K} \mathbf{v}(\theta_k, \omega_k)s_k + \mathbf{w},$$

where $\mathbf{x}$ is the $ML \times 1$ ST array output vector, $s_1, \ldots, s_K$ are the K source signals, $\mathbf{w}$ is the $NM \times 1$ noise-plus-interference vector, and the ST steering vector

$$\mathbf{v}(\theta, \omega) = \mathbf{b}(\omega) \otimes \mathbf{a}(\theta, \omega)$$

is the Kronecker product of well-known spatial and temporal steering vectors denoted by $\mathbf{a}(\theta, \omega)$ and $\mathbf{b}(\omega)$, respectively. The angle of arrival is denoted by θ_k and ω_k is associated with the temporal spectrum of the kth signal, $k = 1, \ldots, K$, impinging on the array, for example, the Doppler frequency. The signal and noise-plus-interference components are assumed to be mutually uncorrelated.

Most ST processing methods employ the ST covariance matrix, which assuming that $\mathbf{x}$ is zero-mean, is

$$\boldsymbol{\Sigma}_{\mathrm{ST}} = \mathsf{E}[\mathbf{x}\mathbf{x}^{\mathsf{H}}].$$

It is conventionally estimated using the sample covariance matrix

$$\mathbf{S} = \frac{1}{N} \sum_{i=1}^{N} \mathbf{x}_i \mathbf{x}_i^{\mathsf{H}}$$

as given in (4.1) with $\bar{\mathbf{x}} = 0$, where $\mathbf{x}_1, \ldots, \mathbf{x}_N$ are the sampled ST snapshots. The ULA that is used for acquiring ST snapshots is illustrated in Figure 5.2. The space-only array for narrowband signals is a special case in which no tapped-delay line is used.

Classical direction-finding techniques based on spatial spectrum estimation, such as the classical beamformer and Capon's minimum variance distortionless response (MVDR) method, as well as high-resolution methods, including different subspace and maximum likelihood methods, employ the array covariance matrix and its structure extensively. Even small departures from the underlying assumptions, both in the signal and the noise models, are significantly reflected in the array covariance matrix, its eigenvalue-eigenvector decomposition, and subspaces spanned by its eigenvectors. Hence, the estimation of the array covariance matrix $\boldsymbol{\Sigma}$ plays a crucial role in array processing algorithms as well as their performance and optimality. Various uncertainties

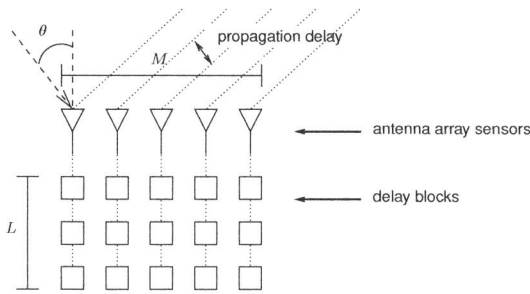

Figure 5.2 Schematic diagram of a ULA used for acquiring a ST snapshot.

in the signal and the noise models can cause additional variance in the DOA estimates or they may lead to highly biased estimates. Moreover, a reduced resolution of the array, that is, the capability to resolve closely spaced sources, may result. Further, some signals impinging on the array may remain completely undetected. Estimating the number of signals and the dimensions of the noise and signal subspaces is also an important issue because it describes the number of signals present in the observed data, as well as characterizes the rank of Σ. It is closely related to the problem of estimating the noise variance and eigenvalues reliably even though the underlying assumptions for the noise model do not necessarily hold. Commonly, methods such as the Bayesian information criterion (BIC), Akaike's information criterion (AIC), or minimum description length (MDL), see Krim and Viberg (1996) and Koivunen and Ollila (2014), are employed.

5.3   Uncertainties in the Array Signal Model

In this section, the impact of signal modeling errors, as well as departures from assumptions made with respect to the array configuration and propagation environment, are considered. These errors typically cause a systematic error (bias) in DOA estimates, excess variance of the estimates, and loss of the high-resolution property. This section addresses the problem briefly because the issue of performance in the presence of modeling errors has been considered in textbooks, see for example, Gershman (2004) and Godara (2004). In particular, the impact of modeling errors on beamforming and spatial spectrum estimation have been extensively studied and a variety of regularization techniques, which reduce the impact of such errors, have been proposed. In general, employing regularization techniques makes an array processing method statistically suboptimal.

5.3.1   Sources of Uncertainty

Some commonly occurring modeling errors faced in array signal processing are:

1. *Signal model errors*: signals may not be narrowband, hence different frequencies may experience different delays; the signals may not originate from a point

source; emitters may not be in the far field and, for this case, the plane wave assumption is not necessarily valid.

2. *Array model errors*: uncertainty in the gain and phase responses of individual array sensor elements; uncertainty or changes in the element positions; and mutual coupling among the elements and the platform.

3. *Propagation model errors*: the medium may not be homogeneous; the propagation conditions may vary (e.g., due to weather); the scattering environment may be rich or changes because of mobility; there may not be a line-of-sight component present (e.g., in biomedical measurements or through-the-wall radar imaging); there may be local scatterers causing near-field effects; and there may be correlated signals, caused by multipath or intentional jamming.

Some of the errors are angle dependent and some angle independent (Rao and Hari, 1988). Moreover, too few observations may be available, which is a problem because sample estimates of the array covariance matrix have to be used because the theoretical covariance matrix is typically not known. In particular, subspace methods may suffer from the small samples being too small because the signal covariance matrix and the array covariance matrix are assumed to be of full rank and a sufficient number of samples are needed to build up the rank and ensure that the matrices are positive definite. Each snapshot will increase the rank by one and an array of M elements needs at least M snapshots to have a full rank covariance matrix (RCM). In highly dynamic signal environments, one may not have enough time to build up the rank of the array covariance matrix (RCM). In sample-starved scenarios, it is common to use a regularized covariance matrix where diagonal loading is added to the sample covariance matrix to ensure that the matrix is not ill-conditioned. In practice, a scaled identity matrix is added to the array covariance matrix to improve its condition number. Determining the amount of diagonal loading needed is an interesting problem in its own right. Regularized robust covariance estimation was treated in Section 4.5. Other solutions include using Krylov-subspaces and Multistage Wiener Filtering (MSWF) (Goldstein et al., 1998; Werner et al., 2007) as well as sequential updating of the square root of the array covariance matrix or its inverse (Aittomaki and Koivunen, 2004). Such methods either completely avoid inverting the array covariance matrix or ensure a well-conditioned matrix because the product of square root matrices is a positive definite matrix; such matrices are desirable in most array processing algorithms.

All the preceding errors cause perturbations in the array covariance matrix and, consequently, to the subspaces spanned by its eigenvectors. These perturbations obviously lead to errors in subspace-based array processing methods such as DOA estimation. The impacts of these errors, in terms of bias and variance, are characterized in detail for different algorithms, for example, in Swindlehurst and Kailath (1992), Godara (2004), and Rao and Hari (1988). In these books, expressions for the model errors are also derived. For small enough errors, a first-order Taylor series formula is derived in Swindlehurst and Kailath (1992) along with the Cramér–Rao bound (CRB).

5.3.2   Robustness and Signal Model Errors

Some of the array and propagation model errors may be handled by appropriate calibration of the array (Costa et al., 2013). However, the conditions where the calibration was done may be significantly different from the conditions where the antenna system is deployed. Moreover, the calibration parameters may change if the array or its platform is moved, the cabling is changed, or the temperature or weather conditions change significantly. There exist also autocalibration procedures (see, e.g., Weiss and Friedlander, 1989) that allow calibrating the array while it is being deployed in an operational use. Such methods require the use of external signals to excite the sensors and may be subject to identifiability problems as well. It is well known that some array configurations are more sensitive to calibration errors than others. For example, the two-subarray configuration used in the ESPRIT algorithm requires only partial calibration and is more robust against calibration errors when compared to the MUSIC algorithm, which assumes an accurately calibrated array. Similarly, there are sparse array configurations, based on the co-array concept, where the impact of mutual coupling is reduced because of larger interelement spacing for the majority of elements. The minimum redundancy array (MRA) is a prime example of such a configuration. The array geometries are different for passive and active settings. An interesting technique called RARE, with less strict calibration requirements, has been proposed. It is more flexible than ESPRIT in terms of allowed subarray configurations and requires only partial array calibration (Pesavento et al., 2002).

For the case in which the signals are not narrowband, ambiguities in the estimates may result. This is due to the fact that different frequencies may experience different delays and the single-delay-processing done in most algorithms does not uniquely define the angle-of-arrival. As a result, an interval of potential angle estimates may be available instead of a point estimate. This problem can be avoided by processing the data in the frequency domain, subdividing the problem into multiple narrowband problems, or using ST processing where a tapped delay line of L coefficients is attached to each array element (Werner et al., 2007). Each ST snapshot is then an $ML \times 1$ vector.

The departures from the far-field and point source assumptions may force the array response to be removed from the assumed manifold. These errors are typically due to the lack of understanding the physical problem. Hence, correcting these error may require changes in the array configuration. One simple approach is to assume that these errors are random, independent, and zero-mean. Hence, they just decrease the SNR by increasing the noise variance. If it is known in advance that the sources are not point sources, or there is a significant angular spread in the signals, techniques used in channel sounding should be considered, see for example, Pedersen et al. (2000) and Ribeiro et al. (2005). A deterministic model considers a large number of discrete beams with associated parameter values (Pedersen et al., 2000), whereas a stochastic model approach uses angular distribution models, such as *von Mises distributions*, or a mixture of such distributions, and finds the maximum likelihood estimates of the mean angle and the scatter about the mean angle (Ribeiro et al., 2005).

Assuming random errors, the perturbations in element positions, errors in the gain and phase responses of the sensors and the mutual coupling among the sensors, may be expressed in matrix form using the array covariance matrix as follows (Swindlehurst and Kailath, 1992):

$$\hat{\boldsymbol{\Sigma}} = (\mathbf{I} + \Delta)[(\mathbf{A} + \tilde{\mathbf{A}})\boldsymbol{\Sigma}_s(\mathbf{A} + \tilde{\mathbf{A}})^{\mathsf{H}} + \sigma^2(\mathbf{I} + \tilde{\boldsymbol{\Sigma}}_v)](\mathbf{I} + \Delta)^{\mathsf{H}},$$

where the matrix Δ captures the errors that influence both the signal and noise components of the data. Departures from the nominal array response are included in the matrix $\tilde{\mathbf{A}}$. This matrix describes the perturbations in element positions, errors in gain and phase responses of the sensors, and mutual coupling among the sensors. The term $\tilde{\boldsymbol{\Sigma}}_v$ models the deviation of the noise covariance matrix from the nominal (identity) matrix $\mathbf{I}$. The effects of the perturbations to the signal and noise subspaces can be studied using the first-order analysis introduced by Swindlehurst and Kailath (1992), or by using tools from the matrix perturbation theory. One convenient way to describe perturbations of subspaces is obtained through the singular value decomposition of

$$\hat{\mathbf{U}}_s^{\mathsf{H}}\mathbf{U}_s,$$

where $\hat{\mathbf{U}}_s$ and $\mathbf{U}_s$ are the perturbed and true signal subspace eigenvectors. The canonical angles of the basis vectors in $\hat{\mathbf{U}}_s$ and $\mathbf{U}_s$ are obtained by $\arccos(\gamma_i)$, where γ_i is a singular value of $\hat{\mathbf{U}}_s^{\mathsf{H}}\mathbf{U}_s$. Obviously, the true and perturbed subspaces are close to each other if the largest canonical angle $\arccos(\gamma_1)$ is very small.

There are array processing techniques that are robust in the face of array and signal model errors. Beamspace processing (Zoltowski et al., 1993) makes the subspace methods less sensitive to modeling errors. The observed data is preprocessed by a beamformer before applying a high-resolution subspace method. A very powerful technique for modeling a variety of array nonidealities is based on the manifold separation principle (Belloni et al., 2007; Costa et al., 2010). Using this principle, the array model is decomposed into array and wavefield dependent parts. The array dependent part is found by using calibration measurements, and it includes all the array nonidealities. It depends only on the array properties, not on the observed wavefields. The wavefield dependent part facilitates using low-complexity algorithms developed for ideal regular array geometries. It also facilitates the derivation of optimal array processing methods that can attain the Cramér-Rao lower bound (CRLB) (Belloni et al., 2007; Costa et al., 2010) despite the array nonidealities. The manifold separation principle ensures that the performance depends on the wavefields only (Costa and Koivunen, 2014).

In case of coherent sources caused by multipath or jamming, many high-resolution methods fail. The spatial smoothing preprocessing technique (Pillai and Kwon, 1989) can be employed to build up the rank of the signal covariance matrix $\boldsymbol{\Sigma}_s$. The conditioning of the array covariance matrix can be improved by a regularization that uses diagonal loading, where a scaled identity matrix is added to $\hat{\boldsymbol{\Sigma}}$ to ensure it is positive semidefinite. This may be done at the cost of statistical optimality. Maximum likelihood (ML) methods can deal with coherent sources as well.

The array and signal modeling errors may be considered by using appropriate weighting factors in MUSIC and subspace fitting algorithms, for example. The weighted

MUSIC (Swindlehurst and Kailath, 1992) introduces a weighting matrix $\mathbf{W}$ in the pseudospectrum equation

$$V_M(\theta, \mathbf{W}) = \frac{1}{\mathbf{a}^H(\theta_i)\mathbf{U}_n\mathbf{W}\mathbf{U}_n^H\mathbf{a}(\theta_i)}$$

and the weights are chosen such that the variance of the estimates is minimized. Obviously, choosing $\mathbf{W} = \mathbf{I}$ results in the conventional MUSIC algorithm. In addition, the asymptotically optimal weighted subspace fitting (WSF) method

$$\hat{\theta} = \arg\min_{\theta} \mathsf{Tr}\{\Pi_A^\perp \hat{\mathbf{U}}_s \mathbf{W} \hat{\mathbf{U}}_s^H\},$$

where $\Pi_A^\perp = \mathbf{I} - \Pi_A$ is a projection matrix to the noise subspace and $\Pi_A = \mathbf{A}[\mathbf{A}^H\mathbf{A}]^{-1}\mathbf{A}^H$ can be employed to reduce the effects of modeling errors, see Jansson et al. (1998) and Swindlehurst and Kailath (1993) for details. The eigenvectors are again weighted to minimize the variance of the angle estimates. The optimum weighting matrix, based on estimated eigenvalues and eigenvectors, is

$$\hat{\mathbf{W}}_{\mathrm{opt}} = (\hat{\Lambda}_s - \hat{\sigma}^2\mathbf{I})^2\hat{\Lambda}_s^{-1},$$

where $\hat{\Lambda}_s$ are the eigenvalues of $\hat{\Sigma}$ corresponding to $\hat{\mathbf{U}}_s$. The weighting is impacted by the modeling errors, see, for example, Jansson et al. (1998) for a detailed derivation and further references.

5.4   Statistically Robust Methods

An array processing procedure is robust if it is not sensitive to small departures from the assumed signal and noise models. Typically, the departures occur in the form of outliers, that is, observations that deviate significantly from the majority of the data. Other causes for the lack of robustness include model class-selection errors and incorrect assumptions with respect to the noise distribution model or its parameters. Also the physical sensor system structure may not be sufficiently well specified or known. For example, in the context of array processing the array may be poorly calibrated and mutual coupling and other array nonidealities may have been ignored (Costa et al., 2013). Moreover, the array calibration parameters may change, for example, if the array or its platform is moved, cabling is changed, or the weather conditions change. All these issues emphasize the importance of validating all the underlying assumptions by physical measurements. After all, many assumptions are just made to facilitate the algorithm design for a sensor array receiver. For example, by assuming a Gaussian probability model, the derivation of the algorithms often leads to easy-to-compute linear structures (vector and matrix equations) because linear transformations of Gaussians are Gaussians.

   In this section, uncertainty in the noise model is considered in more detail. The influence of departures from the underlying noise model is studied quantitatively as well as qualitatively. Robust estimation procedures that have desirable robustness properties are presented as well. As discussed in Chapter 1, a signal processing procedure is statistically robust if it is not sensitive to small departures from the assumed noise models (Kassam and Poor, 1985; Hampel et al., 2011; Zoubir et al., 2012). Typically the departures occur in the form of outliers, that is, highly deviating observations that do not follow the pattern of the majority of the data. The noise may also come from a heavier-tailed distribution than assumed. Other causes of departures include noise model class-selection errors and incorrect assumptions with respect to the noise and interference environments.

The noise in sensor array processing applications may be heavy tailed for many reasons. Man-made interference typically has a heavy-tailed distribution. In outdoor radio channels, heavy-tailed noise is often encountered (Kozick and Sadler, 2000; Swami and Sadler, 2002). Middleton's class A and class B noise models are used to describe man-made and natural noise that may have a heavy-tailed component that, as known from previous chapters, causes severe performance losses, especially for MLEs derived under the Gaussian noise assumption (Middleton, 1996).

5.4.1   Characterizing Robustness in Array Processing

Before we describe the quantitative and qualitative measures of robustness for array processing, we introduce some notation. We use $\hat{\Sigma} \in \mathbb{S}_{++}^{M \times M}$ to denote a scatter matrix (covariance matrix, if it exists) estimate based on the snapshot data set $\mathbf{X}_N = (\mathbf{x}_1 \ldots \mathbf{x}_N)$ in $\mathbb{C}^M$ where $\mathbb{S}_{++}^{M \times M}$ denotes the set of all positive definite Hermitian $M \times M$ matrices. As discussed in Chapter 4, the scatter matrix is a generalization of the covariance matrix and is used to account for the noise distributions where second-order moments are not necessarily defined. For example, the well-known symmetric α-stable distribution and the Cauchy distribution belong to such distribution classes. We also write $\hat{\Sigma}(\mathbf{X}_N)$ to indicate that $\hat{\Sigma}$ is based on data set $\mathbf{X}_N$. We require that $\hat{\Sigma}$ is *affine equivariant*, that is, for any nonsingular $M \times M$-matrix $\mathbf{B}$, the estimate for the transformed data $\mathbf{B}\mathbf{X}_N = (\mathbf{B}\mathbf{x}_1 \ldots \mathbf{B}\mathbf{x}_N)$ is

$$\hat{\Sigma}(\mathbf{B}\mathbf{X}_N) = \mathbf{B}\hat{\Sigma}(\mathbf{X}_N)\mathbf{B}^H.$$

This equivariance allows for establishing various properties of an estimator in cases in which different components in the model may have different scales, data may experience different transformations or may be correlated. Assume that $F \in \mathcal{F}$ where $\mathcal{F}$ denotes a large subset of distributions on $\mathbb{C}^M$. By "large," we mean one that contains plausible models for the unknown population as well as the empirical distribution F_N associated with the data set $\mathbf{X}_N$. Then, a map $\mathbf{C} : \mathcal{F} \rightarrow \mathbb{S}_{++}^{M \times M}$ is a *statistical functional corresponding to* $\hat{\Sigma}$ whenever $\hat{\Sigma} = \mathbf{C}(F_N)$. Affine equivariance implies that, at a *centered complex elliptically symmetric (CES) distribution F*,

$$\mathbf{C}(F) = \sigma \Sigma,$$

for some positive σ, the value of which depends both on the functional $\mathbf{C}$ and the centered CES distribution F. CES distributions are a class of distributions that are parametrized by a scatter matrix, which is proportional to the covariance matrix of the CES distribution when it exists. The multivariate complex Gaussian and Cauchy distribution, for example, are prominent members in this class of distributions. Here "centered" means that the location or the symmetry center $\boldsymbol{\mu}$ of the CES distribution is assumed to be known, assumed to be fixed, or can be reliably estimated. Without loss of generality, we assume $\boldsymbol{\mu} = \mathbf{0}$. A detailed discussion of CES distributions was given in Section 4.2.

Quantitative Robustness

The concept of a breakdown point, as discussed in earlier chapters, is used here for quantifying the robustness of array covariance matrix estimates. Recall that the breakdown point is defined by the smallest fraction of outliers that can cause the estimator to *break down*. In the context of array covariance matrix estimation, this could mean the smallest number of observations, k out of N, that could make either the largest eigenvalue over all bounds or the smallest eigenvalue arbitrarily close to zero, such that the matrix is ill-conditioned. The ill-conditioning can be observed from the condition number of the covariance matrix, that is, the ratio of the smallest and the largest eigenvalue. The (finite-sample) breakdown point for a scatter matrix estimator may be defined as follows

$$\varepsilon^*(\hat{\Sigma}, \mathbf{X}_N) = \min_{1 \leq k \leq N} \left\{ \frac{k}{N} \mid \sup_{\mathbf{X}_{N,k}} D\big(\hat{\Sigma}(\mathbf{X}_N), \hat{\Sigma}(\mathbf{X}_{N,k}) \big) = \infty \right\},$$

where $\mathbf{X}_{N,k}$ is the corrupted sample obtained by replacing k of the observations of $\mathbf{X}_N$ with arbitrary values and the supremum is taken over all possible data sets $\mathbf{X}_{N,k}$. D is defined as

$$D(\mathbf{A}, \mathbf{B}) = \max\{|\lambda_1(\mathbf{A}) - \lambda_1(\mathbf{B})|, |\lambda_M(\mathbf{A})^{-1} - \lambda_M(\mathbf{B})^{-1}|\},$$

where $\lambda_1(\mathbf{A}) \geq \ldots \geq \lambda_M(\mathbf{A})$ are the ordered eigenvalues of the matrix $\mathbf{A} \in \mathbb{S}_{++}^{M \times M}$ and is a measure of the dissimilarity of two positive definite Hermitian matrices. This particular measure of dissimilarity was introduced by Lopuhaa and Rousseeuw (1991). The breakdown point for covariance matrices is here defined in terms of eigenvalues. However, similar concepts for eigenvectors could be constructed as well. As discussed in Section 1.3.2, a breakdown point is always below 50 percent, that is, the majority of the data should be good. It often depends on the dimensionality of the problem (the number of components in an observation vector, the number of parameters in a regression problem) so that the breakdown point decreases as the dimension of the problem increases. For additional definitions and detailed description of breakdown point, see Hampel et al. (2011), Lopuhaa and Rousseeuw (1991), and Kassam and Poor (1985).

Qualitative Robustness

The qualitative robustness of an estimator of an array covariance matrix can be defined in terms of the IF. A robust estimator should have a bounded and continuous IF. Intuitively, boundedness implies that a small amount of contamination at any point should not have an arbitrarily large influence on the obtained estimate whereas continuity implies that small changes in the data set should only cause a small change in the estimate. An IF may also be redescending such that outliers have no impact on the estimator at all.

An IF is essentially the first derivative of the functional version of an estimator. If we denote the point mass at $\mathbf{z}$ by $\delta_{\mathbf{z}} \in \mathbb{R}^{M \times N}$ and consider the contaminated distribution

$$F_{\varepsilon} = (1 - \varepsilon)F + \varepsilon\delta_{\mathbf{z}},$$

then, the IF of a functional $\mathbf{T}$ at F is

$$\mathsf{IF}(\mathbf{z}; \mathbf{T}(F), F) = \lim_{\varepsilon \downarrow 0} \frac{\mathbf{T}(F_{\varepsilon}) - \mathbf{T}(F)}{\varepsilon} = \left[\frac{\partial \mathbf{T}(F_{\varepsilon})}{\partial \varepsilon} \right]_{\varepsilon=0} \tag{5.2}$$

One may interpret the IF as describing the effect that an *infinitesimal* contamination at point $\mathbf{z}$ has on the estimator, normalized by the mass of the contamination. See Section 1.3.1 for an introduction to IFs and Hampel et al. (2011) for a comprehensive study.

For any affine equivariant scatter matrix functional $\mathbf{C}(F) \in \mathbb{S}_{++}^{M \times M}$, there exists functions $\alpha, \beta \colon \mathbb{R}^+ \to \mathbb{R}$ such that the IF of $\mathbf{C}(F)$, at a CES distribution F, is

$$\mathsf{IF}(\mathbf{z}; \mathbf{C}, F) = \alpha(r)\boldsymbol{\Sigma}^{1/2}(\mathbf{u}\mathbf{u}^{\mathsf{H}} - (1/M)\mathbf{I})\boldsymbol{\Sigma}^{1/2} + \beta(r)\boldsymbol{\Sigma}, \tag{5.3}$$

where $t = \mathbf{x}^{\mathsf{H}}\boldsymbol{\Sigma}^{-1}\mathbf{x}$ and $\mathbf{u} = \boldsymbol{\Sigma}^{-1/2}\mathbf{x}/\sqrt{t}$. The preceding implies that the IF of the scatter functional is bounded if and only if the corresponding "weight functions" α and β are bounded. See Ollila and Koivunen (2003a) for this result and for more details and examples.

The knowledge of the IF of the scatter matrix functional $\mathbf{C}(F)$ allows us to obtain the IFs of its eigenvector and eigenvalue functionals. Eigenvalues and eigenvectors are extensively used in multisensor systems and smart antennas. They are in the core of many high-resolution sensor array processing algorithms as well as subspace-based processing used in interference cancellation and signal separation. As an example, the IF for the eigenvector functional $\mathbf{g}_i(F)$ of $\mathbf{C}(F)$ corresponding to a simple eigenvalue λ_i is given by (Ollila and Koivunen, 2003a):

$$\mathsf{IF}(\mathbf{z}; \mathbf{g}_i, F) = \frac{\alpha(t)}{\sigma} \sum_{j=1, j \neq i}^{M} \frac{\sqrt{\lambda_j \lambda_i}}{\lambda_i - \lambda_j} u_j u_i^* \boldsymbol{\gamma}_j,$$

where $\boldsymbol{\gamma}_1, \ldots, \boldsymbol{\gamma}_M$ denote the eigenvectors of the scatter matrix $\boldsymbol{\Sigma}$. The IF and breakdown point allow establishing the qualitative and quantitative robustness of an array processing procedure in a rigorous manner. This is important as the word robust is often used informally to describe signal processing algorithms without any theoretical justification.

5.4.2   Robust Procedures

In this subsection, methods for angle-of-arrival estimation, which are both qualitatively and quantitatively robust, are detailed. The first class of methods stems from nonparametric statistics and in particular a multivariate generalization of the sign function. In addition, semiparametric *M*-estimation and stochastic ML (SML) estimation methods are considered. Deriving maximum likelihood methods for heavy-tailed distributions often leads to statistically robust procedures that still perform well in the standard Gaussian case. SML methods assume that the signal **s** is random. In a stochastic model, the signal part is practically always modeled as being Gaussian distributed, whereas the noise can be distributed according to some other distribution. See Section 4.3 for an example of the ML method for the heavy-tailed *t*-distribution case. Statistically robust methods can be employed to reduce the impact of modeling errors, for example, by employing a *minimax* criterion in the design. However, proper array calibration and understanding of the physical sensor is likely to lead to better performance. Robust estimates for the array covariance matrix and signal or noise subspaces can be utilized in beamspace methods, or in methods using partially calibrated arrays, for example, ESPRIT. Spatial smoothing of robustly estimated covariance matrices can be used to build up the rank, see for example, Visuri et al. (2001) and Kozick and Sadler (2000). In what follows, an example of statistically robust beamforming is provided.

Nonparametric Statistics

We begin by giving definitions for multivariate spatial sign and rank concepts. For an *M*-variate complex data set, $\mathbf{x}_1, \ldots, \mathbf{x}_N$ the *spatial rank function* is

$$\mathbf{r}(\mathbf{x}) = \frac{1}{N} \sum_{i=1}^{N} \mathbf{s}(\mathbf{x} - \mathbf{x}_i), \tag{5.4}$$

where **s** is the *spatial sign function*

$$\mathbf{s}(\mathbf{x}) = \begin{cases} \frac{\mathbf{x}}{||\mathbf{x}||}, & \mathbf{x} \neq \mathbf{0} \\ \mathbf{0} & \mathbf{x} = \mathbf{0} \end{cases} \tag{5.5}$$

with $||\mathbf{x}|| = (\mathbf{x}^H \mathbf{x})^{1/2}$. The spatial sign covariance matrix is defined as

$$\hat{\boldsymbol{\Sigma}}_{\text{SCM}} = \frac{1}{N} \sum_{i=1}^{N} \mathbf{s}(\mathbf{x}_i) \mathbf{s}^H(\mathbf{x}_i) \tag{5.6}$$

The spatial Kendall's Tau covariance matrix (TCM) and the spatial rank covariance matrix (RCM) are defined as

$$\hat{\boldsymbol{\Sigma}}_{\text{TCM}} = \frac{1}{N^2} \sum_{i=1}^{N} \sum_{j=1}^{N} \mathbf{s}(\mathbf{x}_i - \mathbf{x}_j) \mathbf{s}^H(\mathbf{x}_i - \mathbf{x}_j)$$

and

$$\hat{\boldsymbol{\Sigma}}_{\text{RCM}} = \frac{1}{N} \sum_{i=1}^{N} \mathbf{r}(\mathbf{x}_i) \mathbf{r}^H(\mathbf{x}_i),$$

respectively. It has been shown that these covariance matrix estimators produce convergent estimates of the eigenvectors and consequently of the subspaces (Visuri et al., 2001). Consequently, these estimators may be plugged in to any subspace estimator such as MUSIC. By performing spatial smoothing on these covariance matrices, correlated signals can be handled. Similar methods have been proposed in a seminal work by Gini and Greco (2002) but from a different viewpoint.

M-Estimation

As an example, consider the derivation of a robust *M*-estimator. As defined in (4.23), the *M*-estimator of scatter, based upon a sample $\mathbf{x}_1, \ldots, \mathbf{x}_N \in \mathbb{C}^M$, is

$$\hat{\boldsymbol{\Sigma}} = \frac{1}{N} \sum_{i=1}^{N} u\big(\mathbf{x}_i^H \hat{\boldsymbol{\Sigma}}^{-1} \mathbf{x}_i\big) \mathbf{x}_i \mathbf{x}_i^H, \tag{5.7}$$

where the function $u(t) = \gamma \cdot \rho'(t)$ is the *weight function*. Consistent with the discussion in Section 4.3, this estimator is affine equivariant. The statistical functional $\mathbf{C}(F) \in \mathbb{S}_{++}^{M \times M}$, corresponding to the *M*-estimator $\hat{\boldsymbol{\Sigma}}$, is defined in an analogous manner as a solution of

$$\mathbf{C}(F) = \mathsf{E}[u(\mathbf{x}^H \mathbf{C}(F)^{-1}\mathbf{x})\mathbf{x}\mathbf{x}^H].$$

Note that $\hat{\boldsymbol{\Sigma}} = C(F_N)$, where F_N denotes the empirical distribution.

The IF of the scatter *M*-functionals $\mathbf{C}(F)$, at a CES distribution F, is as given in (5.3) with (Ollila and Koivunen, 2003a):

$$\alpha(t) \propto u(t/\sigma)t \quad \text{and} \quad \beta(t) \propto u(t/\sigma)t - M\sigma.$$

Thus, the IF of the *M*-functional $\mathbf{C}(F)$ is continuous and bounded if $u(t)t$ is continuous and bounded.

As an interesting special case, consider the *complex t_1M-estimator* (Ollila and Koivunen, 2003a,b; Ollila et al., 2003c). As discussed in detail in Section 4.4, it is an *M*-estimator obtained using the following weight function:

$$u(t) = \frac{1 + 2M}{1 + 2t}.$$

This choice of weight function yields the MLE of the scatter matrix $\boldsymbol{\Sigma}$ if the underlying distribution is a complex multivariate Cauchy distribution. It turns out that it is also a highly efficient estimator for the case of multivariate Gaussian data. The estimates can be conveniently computed using the RobustSP toolbox MATLAB© function Mscat, which uses the fixed point algorithm specified (4.27). Many of the robust estimators for covariance matrices lend themselves to low complexity fixed-point estimation equations, and this makes them attractive choices in practice.

In Ollila and Koivunen (2003b), the t_1M-estimator is used in various sensor array signal processing applications. The IF of the t_1M-estimator is smooth and bounded (because $u(t)t$ is smooth and bounded). Consequently, its eigenvector and eigenvalue functionals are bounded as well. This facilitates robust subspace estimation as well as maintaining high-resolution properties in heavy-tailed noise environments. In Figure 5.3, the IF of the eigenvector of the sample covariance matrix estimator (4.1) and the t_1M-estimator are plotted. The sample covariance matrix clearly has an unbounded IF, whereas the IF of the t_1M-estimator is smooth and bounded, as shown earlier.

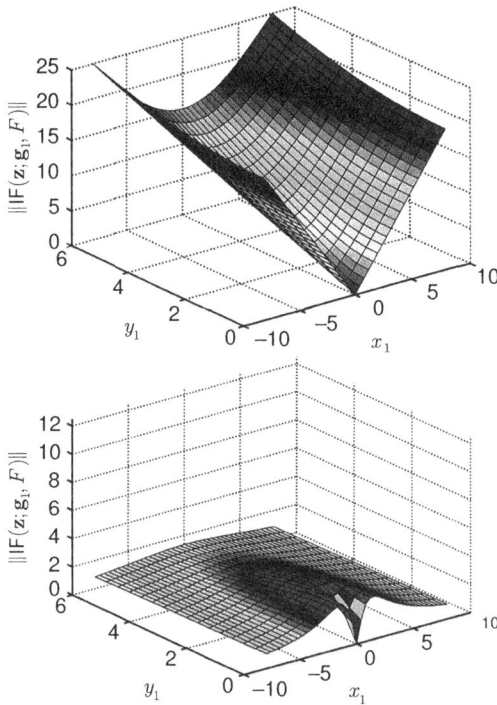

Figure 5.3 Graph of the IF $\|\mathsf{IF}(\mathbf{z}; \mathbf{g}_1, F)\|$ for the sample covariance estimator (upper) and $t_1 M$-estimator (lower) at the bivariate complex normal distribution ($\lambda_1 = 1$, $\lambda_2 = 0.6$). Here, $\mathbf{z} = (z_1, z_2)^T$ is such that z_2 is fixed and z_1 varies.

A further interesting M-estimator is obtained by using the complex analog of *Tyler's* M-estimator. As discussed in detail in Section 4.4, it is constructed with the weight function

$$u(t) = M/t.$$

Thus, Tyler's M-estimator $\hat{\Sigma}$ satisfies

$$\hat{\Sigma} = \frac{M}{N} \sum_{i=1}^{N} \frac{\mathbf{x}_i \mathbf{x}_i^{\mathsf{H}}}{\mathbf{x}_i^{\mathsf{H}} \hat{\Sigma}^{-1} \mathbf{x}_i},$$

provided $\mathbf{x}_i \neq \mathbf{0}$.

In sample-starved scenarios, the preceding estimators may not be able to produce positive definite covariance matrices because each $\mathbf{x}_i \mathbf{x}_i^{\mathsf{H}}$ is of rank one. A sufficient number of rank-1 terms are needed in the summation to build up a full rank matrix. A natural solution is to use a regularized estimator by adding a diagonal loading $\lambda \mathbf{I}$ to the covariance matrix, where $\mathbf{I}$ is identity matrix of the same dimension as the covariance matrix. This ensures that the matrix is well conditioned. However, it will trade-off optimality for reliability and the extent of the trade-off is determined by the multiplier λ. Regularized estimators, and how to choose λ, are discussed in detail in Section 4.5.

Deriving MLEs of a covariance matrix for heavy-tailed distributions often leads to highly robust estimators. For example, the MLEs for the complex t-distribution with different degrees of freedom have the desirable property that they are highly robust and with reasonable sized arrays are highly efficient, close to optimal estimators even in the multivariate complex Gaussian case. The efficiency of the estimator quickly increases as a function of the array size.

Stochastic Maximum Likelihood

In the widely used SML approach for sensor arrays, see Krim and Viberg (1996), the noise and the signal distributions are modeled as complex circular Gaussian, in which case $\mathbf{x}_i \sim \mathbb{C}N_M(\mathbf{0}, \boldsymbol{\Sigma})$ and the signal parameters $\theta = (\theta_1, ..., \theta_K)$, signal covariance matrix $\boldsymbol{\Sigma}_s \in \text{PDH}(K)$, and noise variance $\sigma^2 \in \mathbb{R}^+$ are found by solving

$$\{\hat{\theta}, \hat{\boldsymbol{\Sigma}}_s, \hat{\sigma}^2\} = \arg \min_{\theta, \boldsymbol{\Sigma}_s, \sigma^2} \{\log[\det(\boldsymbol{\Sigma})] + \text{Tr}[\boldsymbol{\Sigma}^{-1}\mathbf{S}]\},$$

where $\mathbf{S} = \frac{1}{N}\sum_{i=1}^{N} \mathbf{x}_i \mathbf{x}_i^H$. Although optimal (if the specified model is correct), the drawback of the SML method is that it leads to a difficult multidimensional nonlinear optimization problem with many local minima. Moreover, it is highly sensitive to heavy-tailed noise. To obtain robust DOA estimates, the complex multivariate t-distribution can be used as an array output model distribution. A highly robust, and surprisingly efficient estimator, is found by deriving an MLE for such a heavy-tailed model. Thus, if we model $\mathbf{x}_i \sim t_{M,\nu}(\mathbf{0}, \boldsymbol{\Sigma})$, the ML estimates of the signal parameters are found from

$$\{\hat{\theta}, \hat{\boldsymbol{\Sigma}}_s, \hat{\sigma}^2\} = \arg \min_{\theta, \boldsymbol{\Sigma}_s, \sigma^2} \left\{ \log(\det(\boldsymbol{\Sigma})) + \frac{2M + \nu}{2N} \sum_{i=1}^{N} \{\log(1 + 2\mathbf{x}_i^H \boldsymbol{\Sigma}^{-1}\mathbf{x}_i/\nu)\} \right\}.$$

Note that the scatter matrix $\boldsymbol{\Sigma}$ is used in this expression, instead of the covariance matrix, to account for distributions whose second-order moments are not defined, for example, the Cauchy distribution. SML estimation, in the context of heavy-tailed models, is also considered in Williams and Johnson (1993) and Kozick and Sadler (2000).

5.5   Array Processing Examples

We now demonstrate the robustness of the MUSIC method in DOA estimation based on robust estimates of covariance and consequently noise and signal subspaces. The basis for the example is an 8-element ($M = 8$) ULA with an interelement spacing equal to $\lambda/2$. Two uncorrelated signals ($K = 2$) with SNR 20 dB are assumed to be impinging on the array from DOA's $\theta_1 = -2°$ and $\theta_2 = 2°$ (broadside). In our study, $N = 300$ snapshots were generated from multivariate complex Gaussian and Cauchy distributions. The sample covariance matrix and $t_1 M$-estimators were used to estimate the noise subspace. Figure 5.4 depicts the MUSIC pseudospectrum associated with the two estimators of the scatter matrix for the simulated complex Gaussian and Cauchy snapshots. Note that both the sample covariance matrix $\hat{\boldsymbol{\Sigma}}$ and the $t_1 M$-estimator of the scatter matrix are able to resolve the two sources in the Gaussian case, and that in

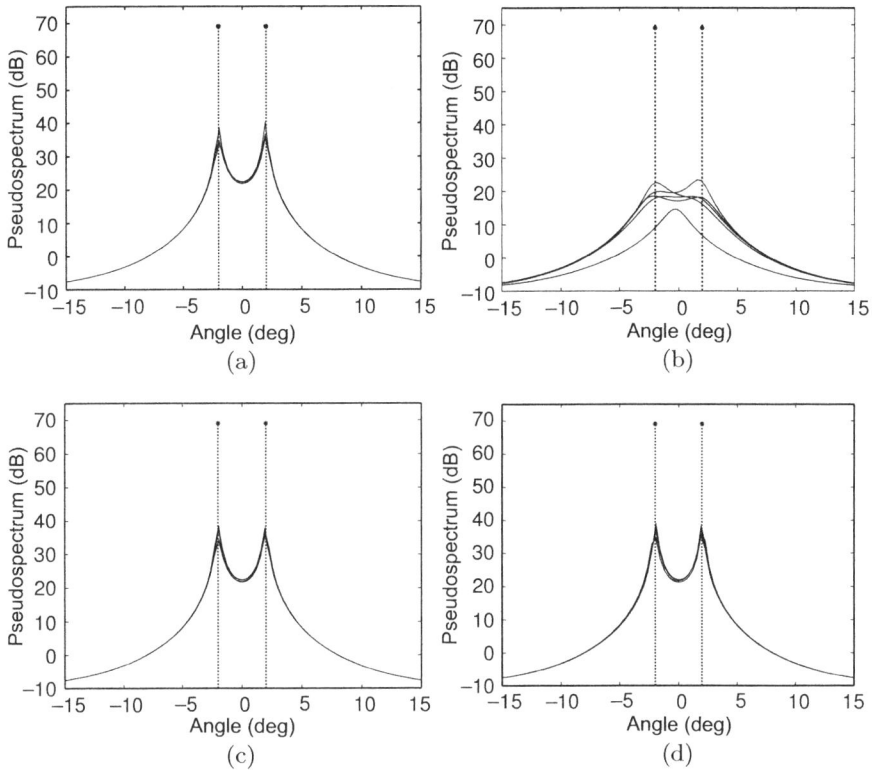

Figure 5.4 MUSIC pseudospectrum based on the sample covariance matrix (upper row) and the t_1M-estimator of the scatter matrix (lower row), for five simulated data sets generated from a complex normal (first column) and a Cauchy distribution (second column).

the Cauchy case the sample covariance matrix is not able to resolve the sources, that is, a loss of resolution is experienced. The robust t_1M-estimator of the scatter matrix, however, yields reliable estimates of the DOAs with high resolution.

A second example is based on a UCA geometry. Beamspace transformation is employed leading to a virtual array with a Vandermonde structure for the steering vector matrix. The unitary root-MUSIC algorithm derived for circular arrays in Belloni and Koivunen (2003) was used. As a by-product of the transformation, spatial smoothing was performed that led to improved robustness for the case of coherent signal sources. Additional robustness in the face of heavy-tailed noise was achieved by estimating the sign covariance matrix presented earlier in this chapter instead of the conventional sample covariance matrix. To demonstrate the robustness of the approach, additive noise, based on a complex Cauchy distribution, was introduced. For the Gaussian case, this would correspond to an SNR of 10 dB. This is a very demanding scenario because the moments of the noise distribution are not even defined for the Cauchy noise case. Hence, the definition of SNR needs to be given differently as well. The zeros of the unitary root-MUSIC for the UCA are plotted in Figure 5.5 using the conventional

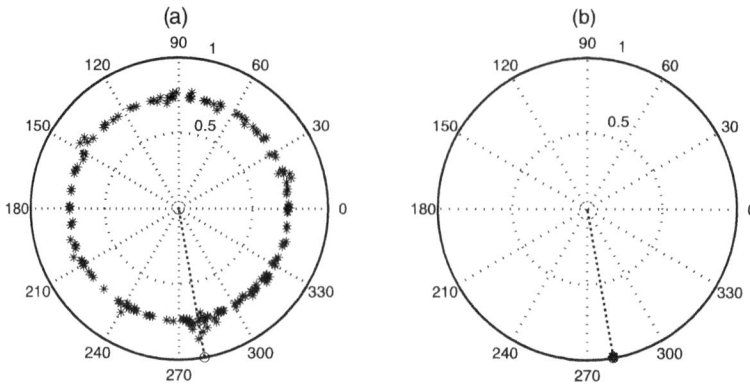

Figure 5.5 Zeros of the unitary root-MUSIC algorithm for UCA in Cauchy noise environment: (a) results using a conventional sample covariance matrix estimator and (b) result using a robust spatial sign covariance matrix estimator. The robust method finds the angle-of-arrival reliably and the roots are clustered about the true angle-of-arrival. The conventional covariance matrix based estimator completely fails because the roots are scattered all over the unit circle. In total, $N = 200$ snapshots are used, $r = \lambda$, $M = 19$ UCA elements.

sample covariance matrix and the robust spatial sign covariance matrix estimate. The DOA is $\phi = 280°$ and the robust method finds it very reliably whereas the conventional estimator completely fails.

Next, consider an example where coherent sources and heavy-tailed (symmetric α-stable noise with characteristic exponent α) noise are present simultaneously. Figure 5.6 shows five estimation results for the cases $\alpha = 2$ and $\alpha = 1$. In the case of Gaussian noise, the performance of both algorithms is almost identical. The algorithm based on using a robust RCM estimate of the covariance for each subarray with forward-backward spatial smoothing, followed by the conventional MUSIC algorithm, estimates the angles of arrival reliably even in extremely heavy-tailed noise conditions, whereas the standard spatial smoothing MUSIC algorithm typically fails when the noise is heavy tailed, that is, $\alpha < 2$.

Finally, an example of robust space-time receive beamforming exploiting robust estimates of the array covariance matrix is provided. Two different M-estimators of covariance are used. ST receivers can efficiently cancel several simultaneous jammers because they have more degrees of freedom (LM, where M is the number of sensors and L the number of elements in the tap delay line) than conventional arrays. We derived a robust version of the MMSE receiver algorithm (Fante and Vaccaro, 2000) for antijamming in a GPS receiver by using the $t_1 M$-estimator $\hat{\Sigma}$ of scatter instead of the conventional sample covariance matrix for a ST array. In the simulation example, a P–code navigation signal, on the L1 carrier ($f = 1575.42$ MHz) and with a bandwidth of 20 MHz, is received by a ST processor with $M = 8$ linear antenna elements, half wavelength spacing, and $L = 5$ taps delay associated to each element. The GPS satellites are assumed to be located at $-45°$, $-25°$, $15°$, and $55°$ azimuth and the SNR is set to -20 dB. Moreover, the antenna

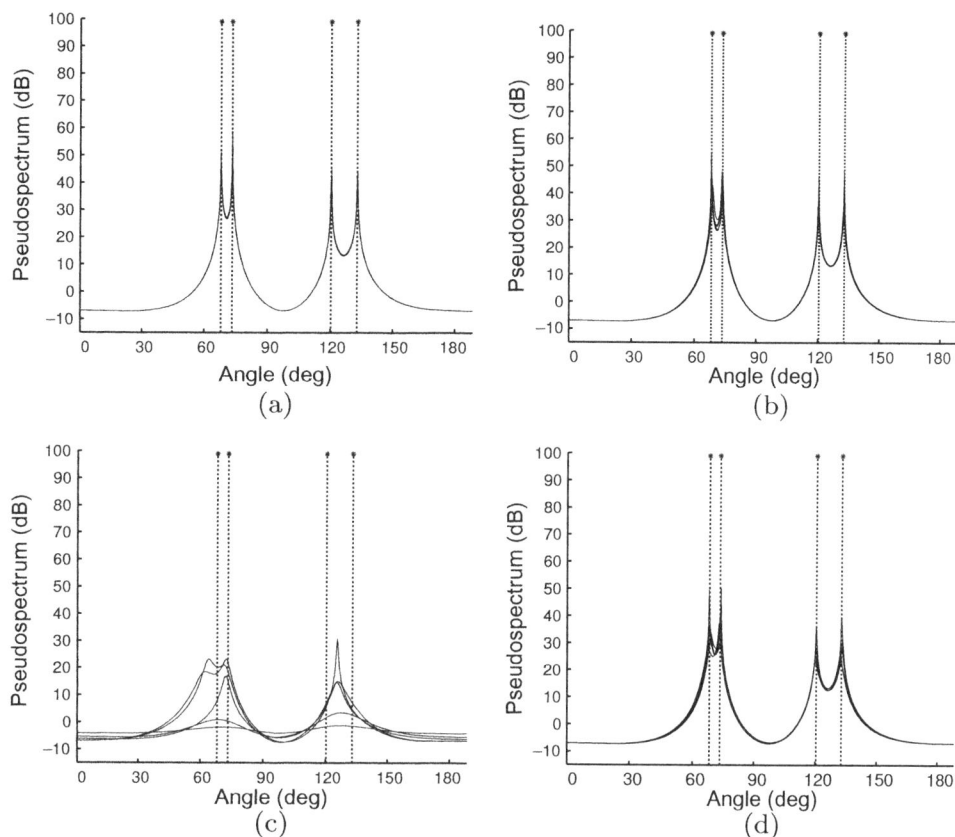

Figure 5.6 DOA estimation results for α-stable noise conditions: (a) spatial smoothing MUSIC, $\alpha = 2$; (b) RCM based spatial smoothing MUSIC, $\alpha = 2$; (c) spatial smoothing MUSIC, $\alpha = 1$; and (d) RCM based spatial smoothing MUSIC, $\alpha = 1$. The size of the ULA is 8 and the subarray size is 6. The DOAs are $65°$, $70°$, $115°$, and $127°$.

array is illuminated by one broadband intentional interferer with a jammer-to-noise ratio $= 60$ dB at $30°$ azimuth (Figure 5.7). The noise is heavy tailed with a Cauchy distribution. The beamformer, employing a robust estimate of the covariance, correctly places a deep null in the direction of the jammer and generally has a better overall beam pattern shape. The jammer-to-signal ratio, based on the sample covariance matrix and the $t_1 M$-estimator $\hat{\Sigma}$ was, respectively, -15 dB and -49 dB after the antijamming preprocessing stage. The jammer is effectively canceled and the GPS receiver operates reliably as if it was in a typical desired signal in noise environment. Transmit beamforming is typically based on the channel feedback from the desired user and the beam is then steered toward that user. Errors in feedback, as well as calibration errors in the array, can lead to erroneous results. Similarly, in MIMO systems, eigenvectors of the estimated channel matrix are used for spatial multiplexing. Non-robust estimation of channels and of the receive covariance matrix can lead to drastic performance degradation.

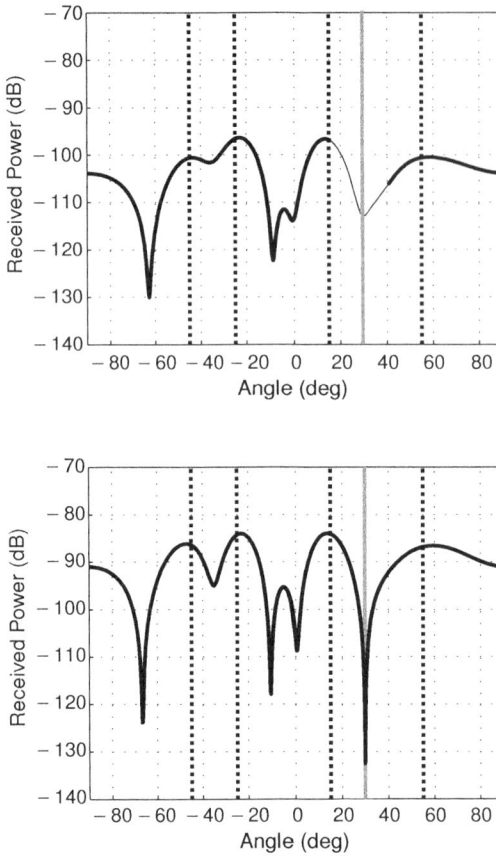

Figure 5.7 Robust antijamming of a GPS signal. Satellites at $-45°$, $-25°$, $15°$, and $55°$ azimuth (SNR $= -20$ dB), jammer at $30°$ (JNR $= 60$ dB). Robust method (beampattern in lower plot) steers a very deep null toward the jammer whereas the MMSE receiver using conventional sample covariance matrix estimation of the array covariance matrix (beampattern in top plot) cannot cancel the jammers effectively.

5.6   Concluding Remarks

In this section, robust estimation methods for sensor arrays were considered. Uncertainties in the signal model, noise model, array model, and propagation model were addressed. The emphasis was on statistical robustness and heavy-tailed noise models. The array covariance matrix is a key statistic in developing statistically robust methods for sensor arrays because the array covariance matrix and the subspaces spanned by its eigenvectors are used in most techniques. The qualitative and quantitative robustness of array processing algorithms were studied. Finally, signal processing examples for DOA estimation and beamforming, using different array configurations, were provided, including the case of a very demanding noise environment.

6   Tensor Models and Robust Statistics

6.1   Introduction

Tensor representations are a natural approach to model high-dimensional data. Such data may have a large volume, it may have been acquired using multiple sensing modalities, and there may be a variety of different couplings among its components. Tensors are basically a higher dimensional generalization of matrices. Tensor techniques have traditionally found applications in psychometrics, sociology, chromatography, and chemometrics. More recently, tensor models have found their way into many signal processing, wireless communication, and radar problems as well as neuroimaging, remote sensing, surveillance, social media, data analysis, and Internet applications. As an example, the radio channel data processed in multiantenna (multiple-input, multiple-output, MIMO) communication systems (employed, e.g., in 4G and 5G wireless and WiFi systems) may be represented as tensor-valued data because the samples of the channel transfer function are obtained for a set of frequencies, a number of transmitter and receiver antennas, and over time or code (Salmi et al., 2009). Similarly, in data mining, blog writings from the Internet can be represented using a 3-way tensor where the components are associated with authors, keywords, and time. Tensors accommodate high-dimensional data, processed in various applications, naturally as multiway arrays. For excellent tutorial and reference articles on tensors, and related signal processing techniques, see for example, Kolda and Bader (2009), Cichocki et al. (2015), Sidiropoulos et al. (2017), and references therein.

Tensor decompositions of multilinear models provide a unifying framework for multidimensional data analysis with simplified notation and algebra. This is important because intuition is usually of limited use in higher dimensions and having simple notation and algebra allows the manipulation of tensors and data analysis in a well-defined and transparent way.

Adopting a tensor representation for high-dimensional data facilitates using different tensor decomposition techniques for revealing relevant information, model identification, dimensionality reduction, or data compression, as well as identifying low-rank or sparse structures in the data. Tensor decompositions are often used, for example, in data mining, as well as finding a low rank approximation to the data. Sparsity or low-rank constraints can be used for accurate signal recovery, for example, in compressed sensing or to eliminate unnecessary redundant features of large-scale data sets. Practical examples of applications, where low-rank models and sparsity are employed, include

MIMO channels with specular and diffuse propagation components, financial data, multimodal observations made by smartphones, DNA micro-arrays, network traffic flows, and functional MRI medical imaging systems.

Tensor decompositions can be considered as higher-order generalizations of the well-known matrix decomposition methods such as the singular value decomposition method (SVD) or the principal component analysis method (PCA). Unfortunately, in higher dimensions the decompositions may not be unique. Two commonly used decomposition techniques are the Canonical Decomposition/PARAFAC (CP) (Carroll and Chang, 1970; Harshman, 1970) and the Tucker decomposition, which decompose a tensor into factor matrices. The Tucker decomposition also includes a core tensor in addition to factor matrices. More recently, tensor networks, especially tensor trains have been investigated. They are based on hierarchical representations of tensors using factor matrices and tensors. A brief overview can be found in Cichocki et al. (2015). The representation for a tensor and its elements may be written as a cascaded multiplication of appropriate matrices.

Tensor decompositions are often found by unfolding tensors into matrices along its modes and estimating factor matrices, one factor matrix at a time, using the alternating least square (ALS), or related techniques. Unfolding opens up the tensor into matrices along different modes. There is not only one way of unfolding a tensor, but many. This is fine as long as the unfoldings are done in a systematic way. The least square error criterion ensures that a high-fidelity approximation of the data tensor is obtained. However, if tensor-valued data are physical sensor measurements, the observations are contaminated by noise. In many applications, such as medical imaging and wireless communication, the noise may be decidedly non-Gaussian. Statistical robustness of tensor decompositions is necessary to ensure the reliability of the decompositions in the face of heavy-tailed errors or outliers that are common in such high-dimensional data, as well as to improve data analysis and to allow simple visualization, and exploration, of the data. Outliers may be difficult to identify because they do not necessarily appear in one component, or dimension, of the data alone.

This chapter is organized as follows. A brief overview of tensor representations is provided along with the commonly used notation. The basic ideas of CP tensor decomposition and Tucker decomposition are then described. A statistically robust method for finding a tensor decomposition is presented. Tensors often contain a massive amount of data and exploiting special structures in tensors allows for a more compact data representation. Low-rank structure, sparsity, and nonnegativity are among such special structures and finding a sparse representation is considered. Of particular interest is to find a statistically robust way of decomposing tensors while promoting sparsity. Finally, a simulation example of a statistically robust and sparse tensor decomposition is provided.

6.2   Tensor Notation and Basic Operations

A tensor $\mathcal{A} \in \mathbb{R}^{n_1 \times \cdots \times n_d}$, of order d, is a d-way array, with d indices $i_1, \ldots, i_d$, that represents a multilinear operator with coordinates $\mathcal{A}_{i_1, \ldots, i_d}$ (or $\mathcal{A}(i_1, \cdots, i_d)$). The order of a tensor gives the number of its modes. In the context of tensors, *mode* means the different dimensions, also known as *ways* or *orders*. We are familiar with low-order tensors: the scalar a is an order-0 tensor, the vector $\mathbf{a}$ is an order-1 tensor, and the matrix $\mathbf{A}$ is an order-2 tensor. The concept of a mode of a tensor also naturally arises. Consider, for example, brain data acquired through an electroencephalogram (EEG). The data, for example, can be represented according to time, frequency, and spatial information and such a model would lead to three modes in a tensor representation of the data.

There are many ways to describe parts of a tensor. A tensor can be described using *fibers* that are the higher-order analog of matrix rows and columns. Fibers are obtained simply by fixing all but one index in a tensor, for example, $\mathcal{A}(:, j, k)$, $\mathcal{A}(i, :, k)$, $\mathcal{A}(i, j, :)$ for a 3-way tensor $\mathcal{A}$. For such a tensor, mode-1 fibers are specified using MATLAB© notation, by $\mathcal{A}_{:jk}$, mode-2 fibers by $\mathcal{A}_{i:k}$, and mode-3 fibers by $\mathcal{A}_{ij:}$, respectively. Similarly, a two-dimensional component of a tensor, called a slice, is defined by fixing all but two indices, for example, $\mathcal{A}(i, :, :)$, $\mathcal{A}(:, j, :)$, $\mathcal{A}(:, :, k)$ for a 3-way tensor $\mathcal{A}$. Consequently, horizontal slices of the tensor are obtained by $\mathcal{A}_{i,:,:}$, lateral slices by $\mathcal{A}_{:,j,:}$, and frontal slices $\mathcal{A}_{:,:,k}$, respectively.

Most computational methods developed for tensors convert tensors into matrices when processing data. This operation is called *unfolding* or matricizing a tensor. Alternatively, one could represent tensors using block matrices, for example, by using Kronecker products. Mode-k unfolding for tensor $\mathcal{A}$ is obtained when mode-k fibers of $\mathcal{A} \in \mathbb{R}^{n_1 \times \cdots \times n_d}$ are assembled to produce an n_k by N/n_k matrix where $N = n_1, \ldots, n_d$ and $k = 1, \ldots, d$. As stated earlier, there are many ways to unfold a tensor because one can use a different ordering for the columns in the unfolding operation. One just needs to do it systematically to avoid any problems in a calculation.

A tensor unfolding of $\mathcal{A} \in \mathbb{R}^{n_1 \times \cdots \times n_d}$ is obtained by assembling the entries of tensor $\mathcal{A}$ into a matrix $\mathbf{A} \in \mathbb{R}^{N_1 \times N_2}$ where $N_1 N_2 = n_1, \ldots, n_d$.

For the tensor $\mathcal{A}$ shown in Figure 6.1, the three-mode-k unfoldings are

$$\mathbf{A}_{(1)} = \begin{bmatrix} 1 & 3 & 5 & 7 \\ 2 & 4 & 6 & 8 \end{bmatrix}$$

$$\mathbf{A}_{(2)} = \begin{bmatrix} 1 & 2 & 5 & 6 \\ 3 & 4 & 7 & 8 \end{bmatrix}$$

$$\mathbf{A}_{(3)} = \begin{bmatrix} 1 & 2 & 3 & 4 \\ 5 & 6 & 7 & 8 \end{bmatrix}$$

Different ways of unfolding a tensor arise from the different orderings of the columns for the mode-k unfoldings. In general, the specific permutation of the columns that is used is not important, but the same permutation must be used for all related calculations.

Similarly, tensors can be converted into vectors as is done with matrices in many signal processing algorithms. The vectorization operation vec turns a matrix into a vector by stacking its columns into a single column vector. Similarly a tensor may be

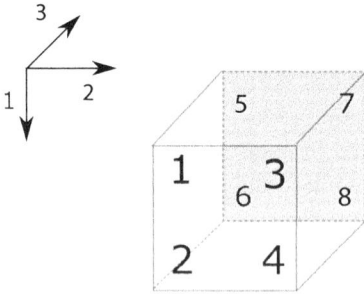

Figure 6.1 Example of the unfolding of a three-mode tensor $\mathcal{A}$.

converted into a vector by stacking mode-1 fibers. For example, for the tensor $\mathcal{A}$ in Figure 6.1:

$$
\mathrm{vec}(\mathcal{A}) = \begin{bmatrix} 1 \\ 2 \\ 3 \\ 4 \\ 5 \\ 6 \\ 7 \\ 8 \end{bmatrix}.
$$

Some basic mathematical operations of tensors follow:

- The inner product of $\mathcal{A}, \mathcal{B} \in \mathbb{R}^{n_1 \times n_2 \times n_3}$ is

$$
< \mathcal{A}, \mathcal{B} > = \sum_{i=1}^{n_1} \sum_{j=1}^{n_2} \sum_{k=1}^{n_3} \mathbf{a}_{ijk} \mathbf{b}_{ijk} = \mathrm{vec}(\mathcal{A})^\top \mathrm{vec}(\mathcal{B})
$$

- The Frobenius norm of a tensor is:

$$
\|\mathcal{A}\|_F = \sqrt{< \mathcal{A}, \mathcal{A} >}
$$

- Multiplication of a tensor and a matrix is defined as follows for $\mathcal{A} \in \mathbb{R}^{I \times J \times K}$, $\mathbf{B} \in \mathbb{R}^{M \times J}$, and $\mathbf{c} \in \mathbb{R}^I$. Consider, for example, the case in which multiplication is done along mode-2 of the tensor and the matrix has a compatible dimension:

$$
\mathcal{X} = \mathcal{A} \times_2 \mathbf{B} \quad (\in \mathbb{R}^{I \times M \times K})
$$

where each mode-2 fiber is multiplied by matrix $\mathbf{B}$. The following notation is used:

$$
\mathbf{x}_{imk} = \sum_j \mathbf{a}_{ijk} b_{mj}
$$

$$
\mathbf{X}_{(2)} = \mathbf{B} \mathbf{A}_{(2)}
$$

Similarly, a tensor can be multiplied by a vector. Such a multiplication along mode-1 of the tensor is defined as follows:

$$\mathbf{x} = \mathcal{A} \,\overline{\times}_1\, \mathbf{c} \quad (\in \mathbb{R}^{j \times k})$$

$$\mathbf{x}_{jk} = \sum_i \mathbf{a}_{ijk} c_i$$

where a dot product (inner product) between I-element vector $\mathbf{c}$ and each mode-1 fiber of $\mathcal{A}$ is computed.

Different matrix products are extensively used in processing tensor-valued data, finding factor matrices in tensor decompositions as well as representing tensors as matrices. An outer product is denoted by $\circ$, and it is defined as $\mathbf{a} \circ \mathbf{b} = \mathbf{a}\mathbf{b}^{\top}$ for $\mathbf{a} \in \mathbb{R}^I$, $\mathbf{b} \in \mathbb{R}^J$. Outer products are utilized extensively when representing a data tensor as a sum of rank-one tensors. This idea is used in finding a lower rank approximation of a tensor. For example, $\mathbf{a} \circ \mathbf{b} \circ \mathbf{c}$ $(\in \mathbb{R}^{I \times J \times K})$, where $\mathbf{c} \in \mathbb{R}^K$, is defined by $(\mathbf{a} \circ \mathbf{b} \circ \mathbf{c})_{ijk} = a_i b_j c_k$. Kronecker products are needed when tensors are represented as matrices, in particular block matrices. Different tensor decomposition methods, which unfold a tensor along each of its modes so that factor matrices can be estimated, also utilize Kronecker products. The Kronecker Product of the matrices $\mathbf{A} \in \mathbb{R}^{I \times J}$ and $\mathbf{B} \in \mathbb{R}^{K \times L}$ is defined as follows:

$$\mathbf{A} \otimes \mathbf{B} = \begin{bmatrix} a_{11}\mathbf{B} & a_{12}\mathbf{B} & \cdots & a_{1J}\mathbf{B} \\ a_{21}\mathbf{B} & a_{22}\mathbf{B} & \cdots & a_{2J}\mathbf{B} \\ \vdots & \vdots & \ddots & \vdots \\ a_{I1}\mathbf{B} & a_{I2}\mathbf{B} & \cdots & a_{IJ}\mathbf{B} \end{bmatrix}$$

The Khatri-Rao product may be defined as a columnwise Kronecker product of matrices $\mathbf{A} \in \mathbb{R}^{I \times R}$ and $\mathbf{B} \in \mathbb{R}^{J \times R}$. This results in an $IJ \times R$ matrix: $\mathbf{A} \odot \mathbf{B} = \begin{bmatrix} \mathbf{a}_1 \otimes \mathbf{b}_1 & \cdots & \mathbf{a}_R \otimes \mathbf{b}_R \end{bmatrix}$. In the vector case, it is given by $\mathbf{a} \otimes \mathbf{b} = \mathsf{vec}(\mathbf{b}\mathbf{a}^{\top})$.

The rank of a tensor, $\mathsf{rank}(\mathcal{X})$, is defined as the smallest number of rank-one tensors that sum to $\mathcal{X}$. The rank-one tensors in this definition are typically formed using an outer product of vectors from different factor matrices of the tensor. For example, if $\mathbf{a} \in \mathbb{R}^I, \mathbf{b} \in \mathbb{R}^J, \mathbf{c} \in \mathbb{R}^K$, then $\mathbf{a} \otimes \mathbf{b}$ is a rank-one matrix and $\mathbf{a} \circ \mathbf{b} \circ \mathbf{c}$ is a rank-1 tensor. If a tensor $\mathcal{X}$ has a minimal representation as

$$\mathsf{vec}(\mathcal{X}) = \sum_{i=1}^{r_1} \sum_{j=1}^{r_2} \sum_{k=1}^{r_3} \sigma_{ijk} (\mathbf{c}_k \otimes \mathbf{b}_j \otimes \mathbf{a}_i)$$

then $\mathsf{rank}(\mathcal{X}) = r_1 r_2 r_3$. Unfortunately, there is no known method to determine the rank of a tensor as there is for the case of matrices. There are also some ambiguities in defining the rank. For example, the rank of a particular tensor over the real field may be different from its rank over the complex field.

To provide an intuitive description of statistical robustness of tensor decompositions, we will focus on tensors with 3 modes, that is, 3-way tensors. A widely used structured

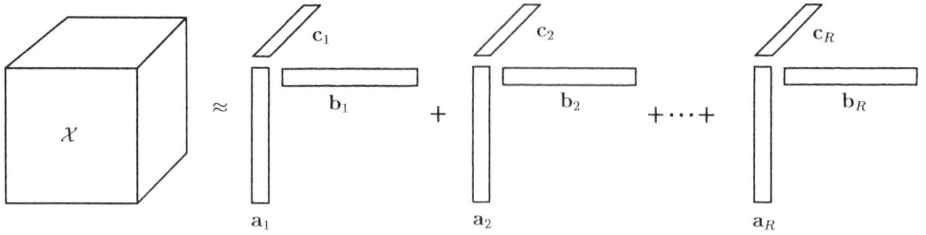

Figure 6.2 Kruskal tensor is defined using a decomposition of a tensor to sum of rank-one tensors.

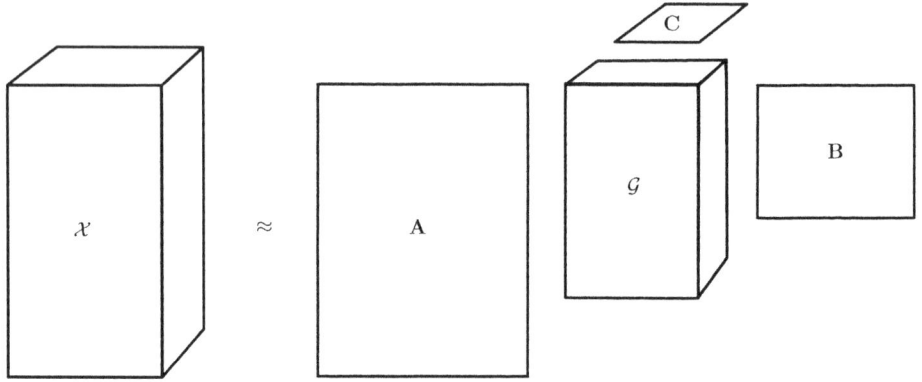

Figure 6.3 Illustration of a 3-way Tucker tensor, which is defined in terms of a core tensor and orthogonal factor matrices.

tensor model is called the Kruskal tensor, which is illustrated in Figure 6.2. It is defined as follows: $\mathcal{X} \in \mathbb{R}^{I \times J \times K}$, $\boldsymbol{\gamma} \in \mathbb{R}^R$,

$$\mathcal{X} \equiv [\![\boldsymbol{\gamma}; \mathbf{A}, \mathbf{B}, \mathbf{C}]\!] = \sum_{r=1}^{R} \gamma_r \mathbf{a}_r \circ \mathbf{b}_r \circ \mathbf{c}_r.$$

where $\mathbf{a} \in \mathbb{R}^I$, $\mathbf{b} \in \mathbb{R}^J$ and $\mathbf{c} \in \mathbb{R}^K$ $r = 1, \ldots, R$ form the unit-norm column vectors of the factor matrices $\mathbf{A} \in \mathbb{R}^{I \times R}$, $\mathbf{B} \in \mathbb{R}^{J \times R}$, and $\mathbf{C} \in \mathbb{R}^{K \times R}$.

Another widely used tensor model is the Tucker Tensor, which can be seen as a higher-order generalization of singular value decomposition (SVD). The Tucker tensor model is illustrated in Figure 6.3. In this generalization, a tensor is expressed in terms of orthogonal factor matrices, as well as a core tensor. For a tensor $\mathcal{X} \in \mathbb{R}^{I \times J \times K}$, the definition is

$$\mathcal{X} \equiv [\![\mathcal{G}; \mathbf{A}, \mathbf{B}, \mathbf{C}]\!] \equiv \mathcal{G} \times_1 \mathbf{A} \times_2 \mathbf{B} \times_3 \mathbf{C} = \sum_{p=1}^{P} \sum_{q=1}^{Q} \sum_{r=1}^{R} g_{pqr} \mathbf{a}_p \circ \mathbf{b}_q \circ \mathbf{c}_r,$$

where $\mathbf{A} \in \mathbb{R}^{I \times P}$, $\mathbf{B} \in \mathbb{R}^{J \times Q}$, and $\mathbf{C} \in \mathbb{R}^{K \times R}$ denote the orthogonal factor matrices $(\mathbf{A}^\top \mathbf{A} = \mathbf{I}, \mathbf{B}^\top \mathbf{B} = \mathbf{I}, \mathbf{C}^\top \mathbf{C} = \mathbf{I})$ and $\mathcal{G} \in \mathbb{R}^{P \times Q \times R}$ is the core tensor. In the matrix case, the core tensor would correspond to a matrix of singular values and the factor matrices would contain left and right singular vectors.

One can also consider the Kruskal tensor as a special case of the Tucker tensor. This is achieved by forcing the core tensor to be superdiagonal and setting the dimensions of the core tensor appropriately, that is, $P = Q = R$. The factors of a Tucker tensor are typically found by unfolding the tensor to a matrix and using SVD. Tucker tensor is not addressed any further in this chapter. The reader is referred to Sidiropoulos et al. (2017), Kolda and Bader (2009), and Cichocki et al. (2015) for further details.

Unfolding a tensor to a matrix is required in different practical decomposition algorithms and to reveal relevant information contained in the tensor. Moreover, it may be used to visualize high-dimensional tensor data as well as make it more intuitive to interpret. Unfolding facilitates using software and computational methods developed for matrices when manipulating tensors. The unfolding along different modes of a 3-way tensor is described next. A 3-way Kruskal tensor of dimensions: $\mathcal{X} \in \mathbb{R}^{I \times J \times K}$ may be unfolded along each mode as follows:

$$\mathbf{X}_{(1)} = \sum_{r=1}^{R} \gamma_r \mathbf{a}_r \otimes (\mathbf{c}_r \otimes \mathbf{b}_r)^\top = \mathbf{A}\mathrm{diag}(\boldsymbol{\gamma}_r)(\mathbf{C} \odot \mathbf{B})^\top$$

$$\mathbf{X}_{(2)} = \mathbf{B}\boldsymbol{\Gamma}(\mathbf{C} \odot \mathbf{A})^\top$$

$$\mathbf{X}_{(3)} = \mathbf{C}\boldsymbol{\Gamma}(\mathbf{B} \odot \mathbf{A})^\top \quad \text{where } \boldsymbol{\Gamma} \text{ is } \mathrm{diag}(\boldsymbol{\gamma})$$

where $\boldsymbol{\Gamma}$ is $\mathrm{diag}(\boldsymbol{\gamma})$, that is, diagonal matrix of scaling factors. A vectorized version of the tensor is given by

$$\mathrm{vec}(\mathcal{X}) = (\mathbf{C} \odot \mathbf{B} \odot \mathbf{A})\boldsymbol{\gamma}.$$

6.3   Tensor Decompositions

The focus of this section is on statistical robustness of tensor data processing. The key ideas can be conveyed using a widely used tensor decomposition called Canonical Decomposition (CANDECOMP) developed by Carroll and Chang (1970). The same decomposition was developed independently at the same time by Harshman with the name Parallel Factors (PARAFAC) (Harshman, 1970). Consequently, the decomposition is commonly called CANDECOMP/PARAFAC or CP Decomposition for short. It expresses a tensor as a sum of rank-one tensors

$$\mathcal{X} \approx [\![\boldsymbol{\gamma}; \mathbf{A}, \mathbf{B}, \mathbf{C}]\!] = \sum_{r=1}^{R} \gamma_r \mathbf{a}_r \circ \mathbf{b}_r \circ \mathbf{c}_r.$$

As discussed earlier, the number of terms in the preceding sum builds the rank of the tensor. If R is minimal, then it is the rank of the tensor. A related concept is the Kruskal rank of a tensor. It gives sufficient conditions for the decomposition to be unique. It has been shown that CP decomposition is often unique. Assuming that the CP decomposition is exact, the Kruskal rank is given by

$$2R + 2 \leq \kappa_A + \kappa_B + \kappa_C,$$

where $\kappa_A = k$-rank of a matrix $\mathbf{A}$, which is defined as the maximum number k such that any k columns are linearly independent. Tucker decomposition discussed briefly earlier is not unique in contrast to SVD for the case of matrices.

The CP decomposition approximates a tensor $\mathcal{X} \in \mathbb{R}^{I \times J \times K}$ by an estimated tensor $\hat{\mathcal{X}}$ consisting of a sum of rank-1 tensors:

$$\hat{\mathcal{X}} \equiv [\![\boldsymbol{\gamma}; \mathbf{A}, \mathbf{B}, \mathbf{C}]\!] \triangleq \sum_{r=1}^{R} \gamma_r \mathbf{a}_r \circ \mathbf{b}_r \circ \mathbf{c}_r.$$

Thus, $\mathcal{X}$ can be modeled as a sum

$$\mathcal{X} = \sum_{r=1}^{R} \gamma_r \mathbf{a}_r \circ \mathbf{b}_r \circ \mathbf{c}_r + \mathcal{E} \tag{6.1}$$

where $\mathbf{a}_r \in \mathbb{R}^I$, $\mathbf{b}_r \in \mathbb{R}^J$ and $\mathbf{c}_r \in \mathbb{R}^K$, for $r = 1, \ldots, R$, form the unit-norm column vectors of $\mathbf{A} \in \mathbb{R}^{I \times R}$, $\mathbf{B} \in \mathbb{R}^{J \times R}$, and $\mathbf{C} \in \mathbb{R}^{K \times R}$ and the tensor $\mathcal{E} \in \mathbb{R}^{I \times J \times K}$ contains the error terms. The error term describes the approximation error in the absence of noise as well as the error caused by noise if the tensor data are acquired by physical measurements. The goal of approximation or estimating a tensor $\mathcal{X}$ is to minimize the Frobenius norm of the error tensor, that is,

$$\|\mathcal{X} - \hat{\mathcal{X}}\|_F = \|\mathbf{X}_{(k)} - \hat{\mathbf{X}}_{(k)}\|_F.$$

The preceding CP model can be rewritten in matrix form by unfolding the tensor into a matrix along one of the three modes. If the unfolding of $\mathcal{X}$ is done along mode one, we get a $I \times JK$-matrix denoted by $\mathbf{X}_{(1)}$ so that the equivalent representation of (6.1) is

$$\mathbf{X}_{(1)} = \mathbf{A}\boldsymbol{\Gamma}(\mathbf{C} \odot \mathbf{B})^{\top} + E_{(1)}, \tag{6.2}$$

where $\boldsymbol{\Gamma} = \mathsf{diag}(\boldsymbol{\gamma})$ and $E_{(1)}$ denotes the unfolded $I \times JK$ matrix of $\mathcal{E}$. Similarly, unfolding along the other two modes results in the representations:

$$\mathbf{X}_{(2)} = \mathbf{B}\boldsymbol{\Gamma}(\mathbf{C} \odot \mathbf{A})^{\top} + E_{(2)},$$
$$\mathbf{X}_{(3)} = \mathbf{C}\boldsymbol{\Gamma}(\mathbf{B} \odot \mathbf{A})^{\top} + E_{(3)}$$

Commonly, the ALS method is used for CP decomposition (Kolda and Bader, 2009; Cichocki et al., 2015). The ALS method fixes a subset of the unknowns as if they were known and estimates the remaining unknowns using the least square error criterion. Then, a new subset of parameters is fixed and another subset of unknowns is estimated. This is done in an alternating manner for different subsets of parameters until the method converges and the error criterion does not decrease significantly anymore. Alternating estimation is a widely used technique and it can be shown to converge to a local minimum in most estimation problems. Finding the global minimum of the objective function is not guaranteed and using several different initial values may be necessary to find the global optimum. This method works reliably in many practical approximation problems as well as in estimation tasks when the observations are contaminated by Gaussian noise.

Let us consider the ALS method and the case in which $\mathbf{B}$ and $\mathbf{C}$ are fixed and γ_r's are the scales of the columns of $\mathbf{A}$. Hence, $\mathbf{a}_r$'s are not unit vectors anymore, but $\gamma_r = \|\mathbf{a}_r\|$ instead. The goal is to find

$$\min_{\mathbf{A}} \| \mathcal{X} - [\![\boldsymbol{\gamma}; \mathbf{A}, \mathbf{B}, \mathbf{C}]\!] \|^2 = \min_{\mathbf{A}} \| \mathbf{X}_{(1)} - \mathbf{A}(\mathbf{C} \odot \mathbf{B})^\top \|^2. \tag{6.3}$$

The solution minimizing the least square criterion is

$$\hat{\mathbf{A}} = \mathbf{X}_{(1)}(\mathbf{C} \odot \mathbf{B})((\mathbf{C}^\top \mathbf{C}) * (\mathbf{B}^\top \mathbf{B}))^\dagger,$$

where $\dagger$ denotes the Moore-Penrose pseudoinverse. Note that $(\mathbf{C} \odot \mathbf{B})^\top (\mathbf{C} \odot \mathbf{B}) = (\mathbf{B}^\top \mathbf{B}) * (\mathbf{C}^\top \mathbf{C})$ where $*$ denotes entrywise (Hadamard product) multiplication of matrices.

The ALS method solves for each factor matrix of the decomposition in turn, leaving all the others fixed as if they were known. For example, for the case of a 3-way tensor, the ALS will fix $\mathbf{B}$ and $\mathbf{C}$ first and find a least squares solution for $\mathbf{A}$. Then, $\mathbf{A}$ and $\mathbf{C}$ are fixed and a least squares solution for $\mathbf{B}$ is found. Finally, $\mathbf{A}$ and $\mathbf{B}$ are fixed and the least squares solution for $\mathbf{C}$ is determined. This alternating process is iterated until convergence. The complete iterative ALS procedure for a 3-way tensor is:

1. Initialize $\mathbf{B}$ and $\mathbf{C}$ by $\hat{\mathbf{B}}$ and $\hat{\mathbf{C}}$.
2. $\hat{\mathbf{A}} = \mathbf{X}_{(1)}\mathbf{Z}(\mathbf{Z}^\top \mathbf{Z})^{-1}$, where $\mathbf{Z} = \hat{\mathbf{C}} \odot \hat{\mathbf{B}}$
3. $\hat{\mathbf{B}} = \mathbf{X}_{(2)}\mathbf{Z}(\mathbf{Z}^\top \mathbf{Z})^{-1}$, where $\mathbf{Z} = \hat{\mathbf{C}} \odot \hat{\mathbf{A}}$
4. $\hat{\mathbf{C}} = \mathbf{X}_{(3)}\mathbf{Z}(\mathbf{Z}^\top \mathbf{Z})^{-1}$, where $\mathbf{Z} = \hat{\mathbf{B}} \odot \hat{\mathbf{A}}$ $\mathbf{Z}$.
5. Repeat stages 2–4 until the relative change in approximation or estimation error is small.

The performance of tensor decomposition algorithms based on alternating estimation of factor matrices can depend significantly on the choice of good initialization values. For example, random matrices may be used for initialization. One potential alternative is to use the regularized least squares method, in particular because it produces reliable first estimates in practice.

6.4 Robust Tensor Decomposition

Outliers, that is, highly deviating observations, often occur in high-dimensional data. They usually indicate some deviation from the underlying nominal noise model assumptions. For example, the noise present in the observations may come from a heavy-tailed distribution rather than a Gaussian. Moreover, some outlying observations, which differ from the pattern set by the majority of the data, may be present in the observation set. In high-dimensional settings, a data item may not be outlying in any of its components alone, but it may be highly deviating as a multidimensional, vector-valued observation. In the case of high-dimensional and large-scale data to be processed by computers, visual inspection to find such outliers is not feasible. The presence of an outlier may lead to misleading or highly erroneous results in data analysis and may hide relevant information from the data analyst. Tensor decompositions based on a least squares error criterion, such as the ALS method for CP, are highly sensitive to outliers, or heavy-tailed noise, because the influence of such deviating observations is not bounded. Consequently, the results from statistical inference, or parameter estimates, are subject to serious errors, for example, in the form of large bias or accepting incorrect hypotheses based on the observed tensor data.

There is very little research work on robust estimation in the context of tensor estimation and decompositions. The use of robust estimators has been largely neglected in the tensor signal processing community. There is some work in the medical imaging field as well as some tensor factorization with robustness considerations, see Chi and Kolda (2011), Pang et al. (2010), Hubert et al. (2012), and Vorobyov et al. (2005). An obvious way of robustifying a tensor decomposition is to replace the least square error criterion with a robust error criterion where the influence of outliers is bounded. Such an approach can tolerate a sufficiently high proportion of contaminating outliers.

Before robustifying tensor decompositions, let us recall the notation for unfolding. When $\mathcal{X}$ is unfolded along the first mode, one obtains an $I \times JK$-matrix denoted as $\mathbf{X}_{(1)}$ (or briefly as $\mathbf{X}$). Similarly, $\mathbf{X}_{(2)}$ and $\mathbf{X}_{(3)}$ denote the $J \times IK$ and $K \times IJ$ matrices obtained by unfolding the tensor along the 2nd and 3rd modes, respectively. Then, the tensor factorization problem (6.1) can be recast as finding an approximation $\hat{\mathbf{X}}$ to $\mathbf{X}$ of the form

$$\hat{\mathbf{X}} = \mathbf{A}\mathbf{\Gamma}(\mathbf{C} \odot \mathbf{B})^{\top} \tag{6.4}$$

where $\mathbf{\Gamma} = \mathrm{diag}(\boldsymbol{\gamma})$ is a diagonal matrix with γ_r, $r = 1, \ldots, R$, as diagonal elements. For notational convenience, we set $\mathbf{A} \leftarrow \mathbf{A}\mathbf{\Gamma}$, that is, absorb the scaling, so that the weights $\gamma_r = \|\tilde{\mathbf{a}}_r\|$ can be obtained from the norms of the columns of $\mathbf{A}$. Denote $\mathbf{Z} = \mathbf{C} \odot \mathbf{B}$ and let N be the number of columns of the matricized tensor $\mathbf{X}$; in this case, $N = JK$.

A simple way of robustifying a tensor decomposition is to use the least absolute deviation (LAD) criterion. For the regression parameter $\boldsymbol{\beta} = (\beta_1, \ldots, \beta_p)^{\top}$, as discussed in Section 2.4, the criterion minimizing the sum of absolute errors is

$$L_{\mathrm{LAD}}(\boldsymbol{\beta}) = \sum_{j=1}^{n} |r_j(\boldsymbol{\beta})|.$$

When using the LAD criterion, the outliers are less influential than in the case of using the least squares criterion. Let us use the mode-1 unfolding as an example and choose $\mathbf{X} = \mathbf{X}_{(1)}$ in (6.3). For the case of the LAD criterion, the objective function becomes

$$L_{\mathrm{LAD}}(\mathbf{A}, \mathbf{B}, \mathbf{C}) = \|\mathbf{X} - \mathbf{A}\mathbf{Z}^{\top}\|_1 = \sum_{i=1}^{m} \sum_{j=1}^{n} |x_{ij} - \mathbf{z}_j^{\top}\mathbf{a}_i|.$$

It is easy to see that absolute error criterion is not bounded. However, because absolute errors are used instead of squared errors, the method is less sensitive to outliers than the conventional least squares method.

Similarly, popular M-estimators (Maronna et al., 2006; Maronna and Yohai, 2008; Huber and Ronchetti, 2009) could be used instead of the least squares method in ALS. An M-estimation type objective function may be defined by

$$L_\rho(\mathbf{A}, \mathbf{B}, \mathbf{C}) = \sum_{i=1}^{m} \hat{\sigma}_i^2 \sum_{j=1}^{n} \rho \left(\frac{x_{ij} - \mathbf{z}_j^\top \mathbf{a}_i}{\hat{\sigma}_i} \right), \tag{6.5}$$

where $\hat{\sigma}_i$ is a robust scale estimate and a popular choice for $\rho(\cdot)$ is Huber's ρ-function, which is an even and nondecreasing function that was defined in (2.54) as

$$\rho_{\mathrm{H},c}(x) = \frac{1}{2} \times \begin{cases} |x|^2, & \text{for } |x| \le c \\ 2c|x| - c^2, & \text{for } |x| > c, \end{cases} \quad x \in \mathbb{R}.$$

Here, c is a user-defined threshold that influences the degree of robustness. Along with Huber's loss function, one of the most commonly used robust loss functions is Tukey's, which was defined in (2.55) as

$$\rho_{\mathrm{T},c}(x) = \frac{1}{6} \min \left\{ 1, 1 - \left(1 - \frac{|x|^2}{c^2} \right)^3 \right\}, \quad x \in \mathbb{R}.$$

To obtain 85 percent ARE at a nominal Gaussian distribution, c is chosen as $c_{.85} = 3.4437$. Huber's or Tukey's loss function are used to bound the influence of outlying observations by downweighting them. A reliable scale estimate is crucial in identifying and giving lower weights to highly deviating observations as well as for finding the scaling factor γ in the CP decomposition. Numerically, the estimate may be found using, for example, the robust IRWLS method for M-estimation instead of ALS with a conventional least squares criterion.

6.5 Combining Robustness with Sparsity

Analyzing large-scale data, as found in many applications, is a challenge. In particular, optimizing objective functions, which arise in approximation and estimation, is problematic. Taking advantage of some structural properties of matrices and tensors can facilitate the analysis of larger data sets. Exploiting sparsity in tensor decompositions clearly improves the statistical inference of multidimensional data and allows relevant information to be revealed more easily. For example, sparsity can be used for accurate signal recovery (e.g., compressed sensing) or to eliminate unnecessary redundant features (dimensions) of many modern data sets including financial and consumer data, DNA micro-array data, Internet network traffic flows, functional MRIs, and multidimensional radio channel data. Sparsity allows for simpler visualization and exploration of the data. Sparsity in the context of tensors has been considered, for example, in Papalexakis and Sidiropoulos (2011), Kim et al. (2013a, 2014), Sidiropoulos et al. (2017), Croux and Exterkate (2011), and Cichocki et al. (2015).

There are two different forms of sparsity in the context of tensors. The first form of sparsity refers to the case in which a considerable number of data elements in a tensor are zero or close to zero in their magnitude. The second form appears in regularization methods such as the Tikhonov regularization, and the celebrated least absolute shrinkage and selection operator (Lasso) method (Tibshirani, 1996), in which the estimated regression parameters are either shrunk toward zero or driven to zero by increasing the penalty associated with model complexity. These methods optimize criteria with two objectives. As discussed in Chapter 3 for the linear regression model, the first objective favors a high-fidelity fit of the model to the data whilst the second objective promotes a certain type of solution, in this case a sparse solution. The weighting between the two criteria is determined by a regularization parameter λ.

Even though two different forms of sparsity are used in different settings, they are related to a certain degree in the context of tensor data. The underlying sparsity of the tensor data naturally implies that the factor matrices of a tensor decomposition are sparse as well. Consequently, when the tensor data is sparse, or the main features and aspects of a high-dimensional tensor data exhibit some sparse structure, the regularization methods successfully estimate the tensor factors for CP decomposition and identify the sparse components simultaneously. This is not the case with conventional least squares estimation.

In the context of tensor decomposition, combining robustness with identifying sparsity in the model, is appealing. It will allow for dealing with large-scale problems, reduce dimensionality, and identify factor matrices even if there is less data available. We will consider the CP factorization that promotes sparsity. With such a factorization, factor matrices with some nonzero elements can be identified while the processing is statistically robust in the face of outliers and non-Gaussian measurement noise. We will describe methods that utilize a robust bounded loss function for errors in the tensor approximation instead of the least square criterion in ALS. At the same time, the methods promote sparsity with ℓ_1-norm-based regularization for the factor matrices of a data tensor, similarly to Lasso.

As an example of CP decomposition with a more robust sparsity favoring solution, consider the CP Alternating LAD-Lasso (CP-LAD-Lasso) method with the following objective function (Vorobyov et al., 2005):

$$\sum_{i=1}^{m} \left\{ \sum_{j=1}^{n} |x_{ij} - \mathbf{z}_j^\top \mathbf{a}_i| + \lambda_1 \|\mathbf{a}_i\|_1 \right\} + \lambda_2 \|\mathbf{B}\|_1 + \lambda_3 \|\mathbf{C}\|_1.$$

The minimum $\hat{\mathbf{A}} = (\hat{\mathbf{a}}_1 \ \cdots \ \hat{\mathbf{a}}_m)^\top$ can be found by

$$\hat{\mathbf{a}}_i = \min_{\mathbf{a}} \left\{ \sum_{j=1}^{n} |x_{ij} - \mathbf{z}_j^\top \mathbf{a}| + \lambda_1 \|\mathbf{a}\|_1 \right\} \tag{6.6}$$

for $i = 1, \ldots, m$, when $\mathbf{B}$ and $\mathbf{C}$ are fixed. Note that ℓ_1-type loss functions, such as (6.6), are not bounded.

Similarly, M-estimation can be used to ensure that the outliers have a bounded influence. The robust and sparsity promoting objective function may then be written as (Kim et al., 2013b, 2014):

$$\sum_{i=1}^{m} \left\{ \hat{\sigma}_i^2 \sum_{j=1}^{n} \rho \left(\frac{x_{ij} - \mathbf{z}_j^\top \mathbf{a}_i}{\hat{\sigma}_i} \right) + \lambda_1 \|\mathbf{a}_i\|_1 \right\} + \lambda_2 \|\mathbf{B}\|_1 + \lambda_3 \|\mathbf{C}\|_1. \tag{6.7}$$

The decomposition is found by estimating the factors in an alternating manner. The minimum $\hat{\mathbf{A}} = (\hat{\mathbf{a}}_1 \ \cdots \ \hat{\mathbf{a}}_m)^\top$ can be obtained by

$$\hat{\mathbf{a}}_i = \min_{\mathbf{a}} \left\{ \hat{\sigma}_i^2 \sum_{j=1}^{n} \rho \left(\frac{x_{ij} - \mathbf{z}_j^\top \mathbf{a}}{\hat{\sigma}_i} \right) + \lambda_1 \|\mathbf{a}\|_1 \right\} \tag{6.8}$$

for $i = 1, \ldots, m$, while $\mathbf{B}$ and $\mathbf{C}$ are fixed. To keep the results invariant, $\mathbf{x}_i$ and columns of $\mathbf{Z}$ are centered to have median zero. The regularizing parameters λ_i may be selected using, for example, the Bayesian information criterion (BIC) (Stoica and Selen, 2004; Koivunen and Ollila, 2014).

A robust CP decomposition may be obtained by employing an iterative reweighted version of Lasso. In the following we describe an alternating algorithm for minimizing (6.7) with the Tukey objective function defined earlier. The method finds an estimate of a factor matrix while the others are kept fixed as in the ALS procedure described earlier. When $\mathbf{B}$ and $\mathbf{C}$ are fixed, an update of $\mathbf{A} = (\mathbf{a}_1 \ \cdots \ \mathbf{a}_I)^\top$ can be found by solving the I separate optimization problems

$$\hat{\sigma}_i^2 \sum_{j=1}^{n} \rho \left(\frac{x_{ij} - \mathbf{z}_j^\top \mathbf{a}_i}{\hat{\sigma}_i} \right) + \lambda_1 \|\mathbf{a}_i\|_1.$$

Note that an update of $\mathbf{B}$ ($\mathbf{C}$, respectively) can be found in a similar manner, while $\mathbf{A}$ and $\mathbf{C}$ ($\mathbf{B}$, respectively) are fixed. By defining the *weight function* as $w(r) = \rho(r)/r^2$, the preceding objective function is in a form of a *weighted Lasso regression* problem (Kim et al., 2014):

$$\sum_{j=1}^{n} w_{ij}(x_{ij} - \mathbf{z}_j^\top \mathbf{a}_i)^2 + \lambda_1 \|\mathbf{a}_i\|_1,$$

where, for $i = 1, \ldots, I$,

$$w_{ij} = w(r_{ij}) = \left(\frac{x_{ij} - \mathbf{z}_j^\top \mathbf{a}_i}{\hat{\sigma}_i} \right)^{-2} \rho \left(\frac{x_{ij} - \mathbf{z}_j^\top \mathbf{a}_i}{\hat{\sigma}_i} \right)$$

denote the weights of the residuals r_{ij}. Then, finding the minimum $\hat{\mathbf{A}} = (\hat{\mathbf{a}}_1 \ \cdots \ \hat{\mathbf{a}}_I)^\top$ is equivalent to computing I separate weighted Lasso solutions (i.e., a Lasso regression with response variable $x_{ij}^* = \sqrt{w_{ij}} x_{ij}$ and predictor $\mathbf{z}_j^* = \sqrt{w_{ij}} \mathbf{z}_j$):

$$\hat{\mathbf{a}}_i = \arg\min_{\mathbf{a}_i} \left\{ \sum_{j=1}^{n} \left(x_{ij}^* - (\mathbf{z}_j^*)^\top \mathbf{a}_i \right)^2 + \lambda_1 \|\mathbf{a}_i\|_1 \right\}. \tag{6.9}$$

An alternating algorithm, called the CP alternating iteratively reweighted Lasso (CP-IRW-Lasso) is used in Section 6.6 (Stoica and Selen, 2004).

An estimate of the regularization parameter $\lambda = (\lambda_1, \lambda_2, \lambda_3)^\top$ can be found by minimizing the BIC defined as (Kim et al., 2013b) and (Stoica and Selen, 2004)

$$\mathrm{BIC}(\lambda) = 2M \ln \hat{\sigma} + w \cdot \mathrm{df}(\lambda) \cdot \ln M \qquad (6.10)$$

where $M = I \cdot J \cdot K$, $\hat{\sigma} = \mathrm{mad}_{i,j}(r_{ij})$ is a scale estimate of the residuals $r_{ij} = (\mathbf{X}_{(1)} - \hat{\mathbf{A}} \hat{\Gamma} \mathbf{Z}^\top)_{ij}$ with $\mathbf{Z} = \hat{\mathbf{C}} \odot \hat{\mathbf{B}}$ and $\mathrm{df}(\lambda)$ denotes the number of degrees of freedom of the model, defined as the sum of the number of nonzero elements in the factor matrices ($\hat{\mathbf{A}}$, $\hat{\mathbf{B}}$, and $\hat{\mathbf{C}}$). Furthermore, $w \geq 1$ is a weight that can be assigned by the user. The default is $w = 1$ and $w > 1$ can be used to favor sparsity. The BIC is computed over a grid of penalty parameter values, $\lambda_1 \times \lambda_2 \times \lambda_3$, and the triplet that produces the minimum BIC is selected.

6.6   Simulation Examples

In the following we will present simulation results demonstrating how to perform tensor factorizations in a statistically robust manner while promoting sparse solutions. More detailed results can be found in Kim et al. (2013a, 2013b, 2014). We generate the observed 3-way tensor data as follows: $\mathbf{X} = \mathbf{X}_0 + \mathbf{\mathcal{E}}$, where $\mathbf{X}_0 = \sum_{r=1}^{R} \gamma_r \mathbf{a}_r \circ \mathbf{b}_r \circ \mathbf{c}_r$ is the Kruskal tensor, $\mathbf{\mathcal{E}}$ is the noise tensor, and the rank R is assumed to be known. The factor matrices and the true 3-way tensor $\mathbf{X}_0$ are sparse in the absence of noise. The following performance criteria are used in demonstrating statistical robustness and capability to identify sparsity in the data. The quality of the obtained estimate $\hat{\mathbf{X}}$ is measured using the normalized mean square error (NMSE) criterion given by:

$$\mathrm{NMSE}(\hat{\mathbf{X}}) = \frac{\|\mathbf{X}_0 - \hat{\mathbf{X}}\|_2^2}{\|\mathbf{X}_0\|_2^2}.$$

A simple way to quantify if the sparse components are identified correctly is to use contingency tables. A 2×2 contingency table is simply

		Estimate of $\mathbf{A}$		
		0	$\neq 0$	sum
True	0	n_{1C}	n_{1M}	n_1
$\mathbf{A}$	$\neq 0$	n_{2M}	n_{2C}	n_2
	sum	n_1'	n_2'	$I \cdot R$

In this table n_{1C} (n_{2C}, respectively) denotes the number of entries in the estimate $\hat{\mathbf{A}}$ "correctly classified" as being zero (nonzero, sparse components, respectively) and n_{1M} (respectively, n_{2M}) is the number of entries in $\hat{\mathbf{A}}$ "misclassified" as being nonzero (zero, respectively). A compact way to describe the performance is the *classification error rate*:

$$\mathrm{CER}(\hat{\mathbf{A}}) = (n_{1M} + n_{2M})/(I \cdot R)$$

and *recovery rate*: $\mathrm{RER}(\hat{\mathbf{A}}) = 1 - \mathrm{CER}(\hat{\mathbf{A}})$.

Table 6.1 Simulation results for Gaussian noise (upper table) and heavy-tailed Cauchy noise (lower table). The robust methods provide reliable estimates of the factor matrices and identify sparse components even in the face of Cauchy noise.

CP Alternating-	Classifying zeros		RER(%)	Average
method	Correct(%)	Incorrect(%)		NMSE (std)
CP-ALS	0	0	50	0.0685 (0.1152)
CP-Lasso	77.3	3.7	86.8	0.0057 (0.0015)
CP-LAD-Lasso	83.6	8.0	87.8	0.0153 (0.0398)
CP-IRW-Lasso	73.5	3.9	84.8	0.0087 (0.0234)
CP-ALS	0	0	50	$1.0{\cdot}10^7$ $(7{\cdot}10^7)$
CP-Lasso	61.2	53.2	54.0	$1.0 \cdot 10^7$ $(7 {\cdot}10^7)$
CP-LAD-Lasso	93.40	11.7	90.9	0.0232 (0.0440)
CP-IRW-Lasso	48.47	1.7	73.4	0.0021 (0.0005)

The dimensionalities of the 3-way tensor in the simulation are $I = 1000, J = 20$, and $K = 20$. The results are averaged over $M = 50$ tensor realizations. The Kruskal tensor $\mathcal{X}_0$ has rank $R = 3$, and only the factor matrix $\mathbf{A} \in \mathbb{R}^{1000 \times 3}$ is sparse where A_{ij} is either equal to a zero or a number drawn independently from a normalized Gaussian distribution $\mathcal{N}(0, 1)$ with equal probability 0.5.

The simulation results for statistical robustness and simultaneous identification of sparse components are provided in Table 6.1. In the simulation example, the noise is drawn from a Cauchy distribution, which has heavy tails. The heavy-tailed noise tensor $\mathcal{E} \in \mathbb{R}^{1000 \times 20 \times 20}$ is based on data from a Cauchy distribution with symmetry center 0 and a scale parameter $1/2$. The λ-values used for favoring a sparse solution are selected using the BIC criterion. Both the obtained NMSE and the capability to identify sparse components are considered. In the case of heavy-tailed Cauchy noise, the CP-ALS and CP alternating Lasso (CP-Lasso) methods provided highly deteriorated estimates of the factor matrices leading to meaningless NMSE values and useless factorization. Both of the robust sparse methods (CP-IRW-Lasso and CP-LAD-Lasso) provided highly reliable results despite the heavy-tailed noise. The CP-IRW-Lasso method gives superior performance in terms of NMSE magnitude in comparison to its robust competitor, the CP-LAD-Lasso method. In terms of the RER criterion, the CP-LAD-Lasso method gives the best performance. For the Gaussian noise case, all the methods promoting a sparse solution (CP-Lasso, CP-LAD-Lasso, CP-IRW-Lasso) outperform the conventional CP-ALS method. In terms of NMSE values, the CP-IRW-Lasso has significantly better performance than the CP-ALS. The CP-Lasso gives the lowest NMSE for Gaussian noise, as expected. For the RER, the sparse methods give good results, as expected, all achieving the rates that are higher than 80 percent for the case of Gaussian noise.

The simulation setup for identifying sparsity is described next. The entries of the factor matrix $\mathbf{A}$ are generated as detailed in the preceding text. The entries of $\mathbf{B} \in \mathbb{R}^{20 \times 3}$ and $\mathbf{C} \in \mathbb{R}^{20 \times 3}$ are independent and drawn from a standard Gaussian distribution $\mathcal{N}(0, 1)$. Moreover, the columns of $\mathbf{A}$, $\mathbf{B}$, and $\mathbf{C}$ are normalized to have unit norm. The noise tensor $\mathcal{E} \in \mathbb{R}^{1000 \times 20 \times 20}$ elements are independent and are consistent with a standard Gaussian distribution. The scale factors in CP are $\gamma_1 = 1000, \gamma_2 = 500$, and $\gamma_3 = 500$. In the simulation, the sparsity factor (SF), defining the average number of zero elements in $\mathcal{X}_0$, is 12.6 percent. The SNR is defined as the average value of $\|\mathcal{X}_0\|^2/\|\tilde{\mathcal{E}}\|^2$, which, using a linear scale, is SNR $= 4.2894$.

The CP-ALS method gives the average NMSE of 0.0652. The method is obviously not suitable for identifying sparsity because the criterion favors only a tight fit, but there is no term promoting sparsity. Hence, none of the elements are set to zero. The average of confusion matrices for identifying sparse factor matrix $\mathbf{A} \in \mathbb{R}^{1000 \times 3}$ is:

$$\begin{pmatrix} 0 & 1507 \\ 0 & 1493 \end{pmatrix}.$$

Both the CER and RER are about 50 percent, indicating that the method basically produces random guesses about the sparsity. The following contingency table for the CP Alternating Lasso in which sparsity is promoted is obtained:

$$\begin{pmatrix} 1290 & 216 \\ 83 & 1411 \end{pmatrix}.$$

The rate of correctly identifying zero features is $1290/1507 \approx 85.6$ percent and the overall recovery rate is $\text{RER}(\hat{\mathbf{A}}) = (1290 + 1411)/3000 = 90.3$ percent. The achieved average NMSE is 0.0088.

6.7   Concluding Remarks

In this section, we considered robustness in high-dimensional data analysis problems where tensor models are used for the data. Multilinear techniques using tensor decompositions provide a unifying framework and simplified algebra for high-dimensional data analysis. Tensor decompositions, such as, CP use the ALS method extensively. The tensor decomposition was robustified by replacing the least square criterion with a robust estimator such as the LAD or the M-estimator. Furthermore, we considered a method that favors a sparse solution for the decomposition. This is achieved by adding a regularizing term, which promotes sparsity, in the objective function as in the highly popular Lasso method. Statistically robust and ℓ_1-regularized tensor decompositions clearly improve the reliability of data analysis and inference in the face of heavy-tailed noise, or outliers, while identifying the sparse components reliably.

7 Robust Filtering

Robust filtering constitutes an important field of research with a long and rich history, dating back to before the mathematical formalization of robust statistics by Huber in 1964. The literature on robust filtering is large and a complete survey is beyond the scope of the book. Based on our experience, we introduce in this chapter the basic principles that are of importance. Only the real-valued case is treated; for complex valued filtering, the reader is referred, for example, to Mandic and Goh (2009) and Schreier and Scharf (2010). We begin by introducing robust Wiener filtering and then consider nonparametric nonlinear robust filters and smoothers, such as the ones using the weighted median and the weighted myriad. To illustrate the application of nonparametric nonlinear robust filtering, we consider the postprocessing of a signal that indicates heart-rate variability and that is derived from the electrocardiogram (ECG). The major focus of this chapter lies on robust (extended) Kalman filtering. As the Kalman filter relies on a state transition model and a measurement model, two types of outliers may occur: (state) innovation outliers and additive/replacement (measurement) outliers. For both outlier cases, robust filters of different computational complexity, ranging from easy to use ad-hoc modifications to rigorously derived filters, are presented. We also show how robust extended Kalman filtering can be applied to track mobile user equipment given TOA measurements from fixed terminals (anchors/base stations) in harsh conditions. Robust particle filtering is beyond the scope of this chapter and Calvet et al. (2015) is a useful reference. For further reading on robust filtering the excellent surveys by Kassam and Poor (1985), Schick and Mitter (1994), Ruckdeschel et al. (2014), and Calvet et al. (2015) are recommended.

7.1 Robust Wiener Filtering

Historically, filtering was pioneered by Kolmogorov and Wiener who developed the theory for a linear time-invariant (LTI) system, which is designed to produce, at its output, the MMSE estimate of a wide-sense stationary (WSS) random process that is of interest. To understand (robust) Wiener filters, some definitions are required, and these are given for the real case:

DEFINITION 17 *Second-order moment function (SOMF)*
The SOMF of a real-valued random process $x_t(\zeta)$, $t \in \mathbb{Z}$, as illustrated in Figure 9.1, is defined as

$$r_{xx}(t_1, t_2) = \mathsf{E}\left[x_{t_1}(\zeta)x_{t_2}(\zeta)\right], \quad t_1, t_2 \in \mathbb{Z}. \tag{7.1}$$

DEFINITION 18   *Wide-sense stationarity (WSS)*
A real-valued random process $x_t(\zeta)$, $t \in \mathbb{Z}$, is said to be WSS if

1. $\mathsf{E}[x_t(\zeta)]$ *is a constant.*
2. $r_{xx}(t_1, t_2) = r_{xx}(m)$, *where* $m = |t_2 - t_1|$.

A process that is stationary to order 2 or greater is certainly WSS.

DEFINITION 19   *Cross-second-order moment function (cross-SOMF)*
The cross-SOMF of the real-valued random processes $x_t(\zeta)$, $y_t(\zeta)$, $t \in \mathbb{Z}$, is defined as

$$r_{xy}(t_1, t_2) = \mathsf{E}\left[x_{t_1}(\zeta)y_{t_2}(\zeta)\right], \quad t_1, t_2 \in \mathbb{Z}. \tag{7.2}$$

DEFINITION 20   *Joint wide-sense stationarity (Joint WSS)*
Two real-valued random processes $x_t(\zeta)$, $y_t(\zeta)$, $t \in \mathbb{Z}$, are said to be jointly WSS if they are WSS according to Definition 18 and

$$r_{xy}(t_1, t_2) = r_{xy}(m), \quad \text{where} \quad m = |t_2 - t_1|. \tag{7.3}$$

DEFINITION 21   *Central cross-second-order moment function (central cross-SOMF)*
The central cross-SOMF of the real-valued random processes $x_t(\zeta)$, $y_t(\zeta)$, $t \in \mathbb{Z}$, is defined as

$$c_{xy}(t_1, t_2) = \mathsf{E}\left[(x_{t_1}(\zeta) - \mathsf{E}\left[x_{t_1}(\zeta)\right])(y_{t_2}(\zeta) - \mathsf{E}\left[y_{t_2}(\zeta)\right])\right], \quad t_1, t_2 \in \mathbb{Z}. \tag{7.4}$$

or equivalently

$$c_{xy}(t_1, t_2) = r_{xy}(t_1, t_2) - \mathsf{E}\left[x_{t_1}(\zeta)\right]\mathsf{E}\left[y_{t_2}(\zeta)\right]. \tag{7.5}$$

7.1.1   Wiener Filtering

Consider a signal of interest, which is defined by x_t, $t = 1, \ldots, N$, and measurements/observations of this signal modeled according to

$$y_t = x_t + v_t, \quad t = 1, \ldots, N,$$

where v_t, $t = 1, \ldots, N$, defines one signal from a WSS random process. The objective of Wiener filtering is to estimate the signal x_t from y_t, $t = 1, \ldots, N$, by finding the unit sample response h_k, $k \in \mathbb{Z}$, of the LTI system that minimizes the following objective function

$$L_{\text{WF}} = \mathsf{E}\left[\left(x_t - \sum_{k=-\infty}^{\infty} h_k y_{t-k}\right)^2\right] = \mathsf{E}\left[e_t^2\right], \tag{7.6}$$

where $e_t = x_t - \sum_{k=-\infty}^{\infty} h_k y_{t-k}$. The solution to this problem is found by solving

$$\frac{\partial L_{\text{WF}}}{\partial h} = -2 \sum_{m=-\infty}^{\infty} \mathsf{E}\left[e_t y_{t-m}\right] = 0.$$

This equation implies, for the optimal filter, that the error is orthogonal to the output of the filter, that is,

$$\mathsf{E}\left[e_t y_{t-m}\right] = 0, \quad \forall m \in \mathbb{Z}. \tag{7.7}$$

Let x_t and e_t be two jointly WSS zero-mean random processes. Thus, according to Definition 19, (7.7) is equivalent to the cross-SOMF, that is,

$$\mathsf{E}\left[e_t y_{t-m}\right] = r_{ey}(m) = 0, \quad \forall m \in \mathbb{Z}. \tag{7.8}$$

Under these assumptions, (7.7) is also equivalent to

$$\mathsf{E}\left[(e_t - \underbrace{\mathsf{E}\left[e_t\right]}_{=0})(y_{t-m} - \underbrace{\mathsf{E}\left[y_{t-m}\right]}_{=0}) \right] = 0. \tag{7.9}$$

The coefficients of the unit sample response h_k, $k \in \mathbb{Z}$, of the Wiener filter can then be obtained from the following system of equations

$$\sum_{k=-\infty}^{\infty} h_k r_{yy}(m-k) = r_{xy}(m).$$

If we assume that the signal and noise are uncorrelated, that is, $c_{xv}(m) = c_{vx}(m) = 0$, $\forall m$, which for zero-mean processes is equivalent to $r_{xv}(m) = r_{vx}(m) = 0$, $\forall m$, it follows that

$$r_{yy}(m) = r_{xx}(m) + r_{vv}(m)$$

and

$$r_{xy}(m) = r_{xx}(m).$$

Under these conditions, the frequency response, $H(e^{J\omega})$, which is the Fourier transform of h_k, $k \in \mathbb{Z}$,

$$H(e^{J\omega}) = \sum_{k=-\infty}^{\infty} h_k e^{-J\omega k}$$

with $\omega = 2\pi f / f_s$, where f_s denotes the sampling frequency, becomes

$$H(e^{J\omega}) = \frac{S_{xx}(e^{J\omega})}{S_{xx}(e^{J\omega}) + S_{vv}(e^{J\omega})}. \tag{7.10}$$

Here, $S_{xx}(e^{J\omega})$ and $S_{vv}(e^{J\omega})$ are the power spectral densities (PSDs) of the signal $x_t(\zeta)$ and the noise $v_t(\zeta)$, which, according to the Wiener–Khintchine theorem, are the Fourier transforms of $r_{xx}(m)$ and $r_{vv}(m)$, respectively; see Definition 25. Therefore, the minimization problem of (7.6) depends entirely on the signal and noise PSD, that is, $S_{xx}(e^{J\omega})$ and $S_{vv}(e^{J\omega})$.

7.1.2   Robust Wiener Filtering

Robust Wiener filtering (Vastola and Poor, 1984; Kassam and Poor, 1985) acknowledges the fact that, in practice, the PSDs that are chosen for designing $H(e^{J\omega})$ may differ from the true ones. The uncertainty can be modeled in different ways, that is, by use of the ε-contamination model, the total variation model, or the band-model. Modeling the uncertainty in the signal PSD with the ε-contamination model dictates that $S_{xx}(e^{J\omega}) \in \mathcal{S}_{xx}(e^{J\omega})$, with

$$S_{xx}(e^{J\omega}) = (1-\varepsilon)S_{xx}^{o}(e^{J\omega}) + \varepsilon S_{xx}^{c}(e^{J\omega}), \quad \forall \omega \in [-\pi,\pi)$$

$$\int_{-\pi}^{\pi} S_{xx}^{c}(e^{J\omega})\lambda(d\omega) = \int_{-\pi}^{\pi} S_{xx}^{o}(e^{J\omega})\lambda(d\omega),$$

where λ is a finite measure, for example, the Lebesgue measure on $[-\pi,\pi)$ and $S_{xx}^{o}(e^{J\omega})$ and $S_{xx}^{c}(e^{J\omega})$ denote the nominal and contaminating PSD, respectively. Of course, the uncertainty in the noise PSD, $S_{vv}(e^{J\omega})$, can be modeled equivalently. The ε-contamination model is the most commonly used uncertainty class, as it models the idea that we have a fraction ε of completely general uncertainty about the PSD. Other uncertainty models are detailed in Vastola and Poor (1984).

A robust Wiener filter has a frequency response $H_{\text{rob}}(e^{J\omega})$, which minimizes the MSE for worst-case choices of $S_{xx}(e^{J\omega})$ and $S_{vv}(e^{J\omega})$ within the considered uncertainty classes of the signal and noise PSDs. Because of this min-max procedure, $H_{\text{rob}}(e^{J\omega})$ is also called the *most robust* Wiener filter for these uncertainty classes. Designing a robust Wiener filter, thus, requires deriving the least favorable PSDs for the given uncertainty model and then finding the optimal (Wiener) solution for this case. The interested reader is referred, for example, to Vastola and Poor (1984) for the derivation of some specific most robust Wiener filters.

7.2   Nonparametric Nonlinear Robust Filters

In this section, we consider the task of estimating a univariate signal, that is, a time-varying trend, given observations that are corrupted by impulsive noise, that is,

$$y_t = x_t + v_t, \quad t = 1, \dots, N. \tag{7.11}$$

The signal $\mathbf{x}_t = (x_1, \dots, x_N)^\top$ is assumed to be smooth, but may exhibit changing trends and level shifts. The distribution of $\mathbf{v}_t = (v_1, \dots, v_N)^\top$ is consistent with impulsive noise, which appears as outliers in the observations $\mathbf{y}_t = (y_1, \dots, y_N)^\top$. This fundamental problem has been addressed by many researchers by applying a robust location estimator to a running/moving windowed version of the observations.

Running location estimators are obtained by running a window function, for example, the rectangular window through the data and, for each window position, computing the location estimate. Let the length of the observation window be $N_w = 2n_c + 1$, where n_c denotes the index of the central element of the window function. For example, the running median filter (Tukey, 1977) applies the median location estimator, as defined in Eq. (1.12), by evaluating

$$\hat{x}_t = \text{med}(y_{t-N_w+1}, \ldots, y_t), \tag{7.12}$$

while the running median smoother also takes future observations into account

$$\hat{x}_t = \text{med}(y_{t-n_c}, \ldots, y_t, \ldots, y_{t+n_c}). \tag{7.13}$$

Many other running robust location estimation-based filters and smoothers have been proposed, for example, those based on M-estimation, trimming, or order statistics (Lee and Kassam, 1985; Gonzalez and Arce, 2001; Aysal and Barner, 2007; Pitas and Venetsanopoulos, 2013). For example, a running M-type filter/smoother applies the M-estimator of the location parameter, as defined in Chapter 1, to the windowed data. The M-estimates can be computed, for example, with the MATLAB$^{©}$ implementations of Huber's and Tukey's M-estimates of location with previously computed scale, that are provided in the RobustSP toolbox. Meridian filters (Aysal and Barner, 2007), by contrast, use the meridian location estimate that is the MLE under the Meridian distribution. This distribution is a member of the generalized Cauchy distribution.

It is clear from (7.12) and (7.13) that the median filters and smoothers are temporally blind, that is, they do not take the temporal correlation of the signal $\mathbf{x}_t$ into account. This means that all observations within the window are treated equally. However, for correlated signals, it is often the case that the estimate $\hat{x}_t$ should rely more on y_t than on the other samples. It is usual for y_{t-1} and y_{t+1} to carry more information on x_t than y_{t-2} and y_{t+2} and so on. To utilize this correlated information, weighted median filters

$$\hat{x}_t = \text{med}(w_1 \diamond y_{t-N_w+1}, \ldots, w_{N_w} \diamond y_t), \tag{7.14}$$

and smoothers

$$\hat{x}_t = \text{med}(w_1 \diamond y_{t-n_c}, \ldots, w_{n_c+1} \diamond y_t, \ldots, w_{N_w} \diamond y_{i+n_c}) \tag{7.15}$$

have been proposed. Here, $\mathbf{w}_t = (w_1, \ldots, w_{N_w})^\top$ is a positive integer-valued weighting function and $\diamond$ denotes the replica operator. If, for example, $w_t = 3$

$$w_t \diamond y_t = 3 \diamond y_t = \{y_t, y_t, y_t\}.$$

By choice of $\mathbf{w}_t$, weighted median smoothers can emphasize or deemphasize specific observations. For illustration purposes, consider the center weighted median smoother, whose output is

$$\hat{x}_t = \text{med}(y_{t-n_c}, \ldots, w_{n_c+1} \diamond y_t, \ldots, y_{t+n_c}), \tag{7.16}$$

which is obtained from (7.15) by letting $\mathbf{w}_t = (1, \ldots, 1, w_{n_c+1}, 1, \ldots, 1)^\top$, where $w_{n_c+1} \in \mathbb{Z}^+$ is an odd number (assuming, without loss of generality, that the index of

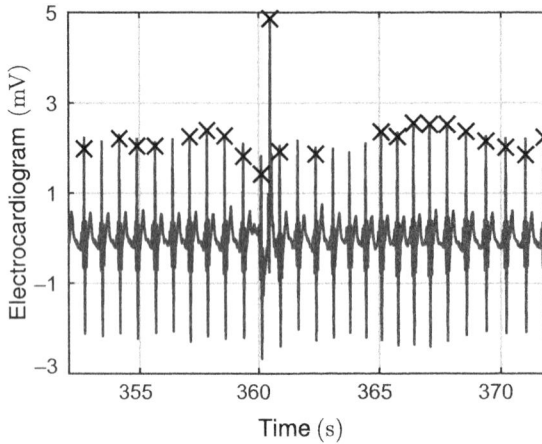

Figure 7.1 Excerpt of an ECG. The detected R-peak positions are marked by black crosses.

the window begins with 1). For $w_{n_c+1} = 1$, (7.16) reduces to the median smoother given in (7.13), while for $w_{n_c+1} > N_w$, (7.16) reduces to an identity operation.

Example 14   Running Median Filtering for ECG Postprocessing

To illustrate the effect and practical applicability of nonparametric nonlinear filtering, consider the postprocessing of an R-R interval series that has been obtained from an ECG. The R-R interval series carries information on the heart rate and the heart-rate variability. Figure 7.1 shows an excerpt of an ECG recording, which was taken during a psychophysiological study (Kelava et al., 2014). It was recorded at Technische Universität Darmstadt by the Department of Psychology using the Biopac MP 150 System and the AcqKnowledge 4.2 Software. The data was sampled with a sampling frequency of 250 Hz and the R-R intervals were extracted by using the R-peak detector proposed by Pan and Tompkins (1985).

The resulting R-R series, which describes the durations of successive cardiac cycles is shown in Figure 7.2. The vast majority of R-peaks has been correctly identified by using the detector proposed by Pan and Tompkins (1985), however, occasionally, as shown in the excerpt of Figure 7.1, some peaks are missed (e.g., around $t = 354$ s, 356 s, and 362–364 s). Furthermore, at approximately $t = 361$ s, an additional peak is detected, which actually has been produced by a motion artifact. These errors are displayed in Figure 7.2, where a missed detection results in an increase of the R-R interval and a false alarm causes a decrease. For this example, where the subject had a very regular heart beat, a simple running median filter with $N_w = 10$ is able to extract an accurate estimate of the R-R series from the outlier contaminated original series.

Of course, such a simple nonlinear robust filter is only applicable if the observations follow the time-varying trend model of (7.11), which is not the case, for example, for atrial fibrillation (Schäck et al., 2017), or for measurements taken during extensive physical activity (Schäck et al., 2015). Furthermore, it is important to notice that the

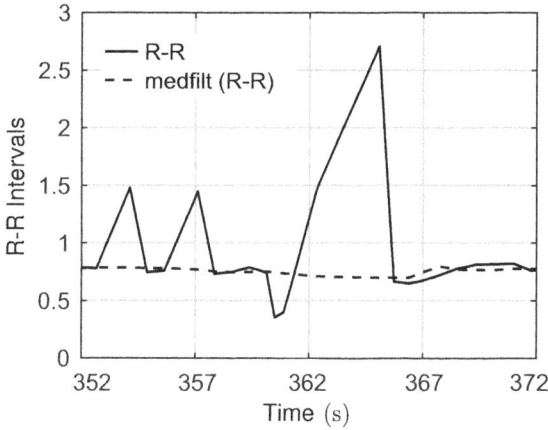

Figure 7.2 The estimated R-R intervals corresponding to the ECG from Figure 7.1 and the median filtered R-R intervals.

window length (N_w) is a user-defined design parameter. A long window produces a smooth output whereas a short window more closely follows the input of the median filter. This trade-off must be taken into consideration when applying a running median filter.

The running myriad filter (Kalluri and Arce, 2000; Gonzalez and Arce, 2001) is similar to the running median filter. Myriad filters have a solid theoretical basis, for example, they are optimal for α-stable noise. The myriad filter is defined by

$$\hat{x}_t = \arg\min_{\beta} \sum_{k=t-N_w+1}^{t} \log(c^2 + (y_k - \beta)), \qquad (7.17)$$

where c is a tuning parameter that trades off robustness and efficiency under a Gaussian model. For $c \to \infty$, the myriad converges to the sample average (Gonzalez and Arce, 2001). The weighted running myriad filter is defined by

$$\hat{x}_t = \arg\min_{\beta} \sum_{k=t-N_w+1}^{t} \log(c^2 + w_k(y_k - \beta)), \qquad (7.18)$$

where $\mathbf{w}_t = (w_{t-Nw+1}, \ldots, w_t)^\top$ is a positive weight vector. (Weighted) running myriad smoothers are obtained by taking into account future values analogously to the (weighted) running median smoothers. Interestingly, myriad filters are very general in the sense that they subsume traditional linear finite impulse response (FIR) filters. As described in Barner and Arce (2003, chapter 5), for $c \to \infty$, the weighted myriad filter, as defined in (7.18), reduces to a constrained linear FIR filter. This suggests that a linear FIR filter can be provided with resistance to impulsive noise by simply reducing c. Such a transformation is referred to as *myriadization* of linear FIR filters. Myriadization, thus, allows for a linear FIR filter for Gaussian or noiseless environments to be first designed and modified so that the filter has resistance against impulsive noise. An example of robust FIR bandpass filter design is given in Barner and Arce (2003, chapter 5).

7.3   Robust Kalman Filtering

A fundamental drawback of the robust Wiener filtering approach of Section 7.1 is the assumption of stationarity, which in many signal processing applications does not hold. The nonparametric nonlinear robust filters that have been discussed in Section 7.2 assume a univariate signal with a time-varying trend. This section discusses a class of filters that can be applied to a much broader range of problems.

Consider a discrete time linear dynamic system where an unobservable vector valued state $\mathbf{x}_t \in \mathbb{R}^{p \times 1}$ evolves according to

$$\mathbf{x}_t = \mathbf{F}_t \mathbf{x}_{t-1} + \mathbf{v}_t. \tag{7.19}$$

Here, $\mathbf{F}_t \in \mathbb{R}^{p \times p}$ is the state transition matrix, and $\mathbf{v}_t \in \mathbb{R}^{p \times 1}$ is additive white noise. A measurement (observation) is modeled by

$$\mathbf{y}_t = \mathbf{H}_t \mathbf{x}_t + \mathbf{w}_t. \tag{7.20}$$

Thus, the observable data $\mathbf{y}_t \in \mathbb{R}^{q \times 1}$ is represented as a linear transform $\mathbf{H}_t \in \mathbb{R}^{q \times p}$ of the unobservable state $\mathbf{x}_t \in \mathbb{R}^{p \times 1}$, with an additive white noise component $\mathbf{w}_t \in \mathbb{R}^{q \times 1}$ that is independent of $\mathbf{v}_t$. Both $\mathbf{v}_t$ and $\mathbf{w}_t$ are independent of $\mathbf{x}_t$.

Under the zero-mean Gaussian assumption, that is,

$$\mathbf{v}_t \sim \mathcal{N}(\mathbf{0}, \mathbf{Q}_t) \tag{7.21}$$

$$\mathbf{w}_t \sim \mathcal{N}(\mathbf{0}, \mathbf{R}_t), \tag{7.22}$$

where

$$\mathbf{Q}_t = \mathsf{E}\left[\mathbf{v}_t \mathbf{v}_t^\top\right] \in \mathbb{R}^{p \times p} \tag{7.23}$$

is the *state error covariance matrix*, and

$$\mathbf{R}_t = \mathsf{E}\left[\mathbf{w}_t \mathbf{w}_t^\top\right] \in \mathbb{R}^{q \times q} \tag{7.24}$$

is the *measurement error covariance matrix*, the Kalman filter is the optimal recursive estimator of the state $\mathbf{x}_t$ given observations $\mathcal{Y}_t = \{\mathbf{y}_0, \mathbf{y}_1, \ldots, \mathbf{y}_t\}$ (Kalman, 1960).

As described in Algorithm 10, after an initialization, the Kalman filter performs a model-based prediction and a data-driven correction step. The prediction $\mathbf{x}_{t|t-1}$ of the state, as given in (7.25), is corrected in (7.27) by taking into account the new observation $\mathbf{y}_t$. The Kalman gain $\mathbf{K}_t$, as defined in (7.29), determines the weight given to the prediction, or the data-driven correction, considering the *state-prediction error covariance* $\boldsymbol{\Sigma}_{t|t-1}$ and the *measurement error covariance* $\mathbf{R}_t$.

Kalman filtering has been applied to a large range of problems, not only for state estimation, but also for simultaneously estimating model parameters, choosing among

---

Algorithm 10: KalmanFilter: computes the standard Kalman filter algorithm.

---

input : $\mathbf{F}_t \in \mathbb{R}^{p \times p}$, $\mathbf{H}_t \in \mathbb{R}^{q \times p}$, $\mathbf{y}_t \in \mathbb{R}^{q \times 1}$, $\mathbf{Q}_t \in \mathbb{R}^{p \times p}$, $\mathbf{R}_t \in \mathbb{R}^{q \times q}$

output : $\mathbf{x}_{t|t}$

initialize: $\mathbf{x}_{0|0}$, $\boldsymbol{\Sigma}_{0|0}$

1 **prediction**:

$$\mathbf{x}_{t|t-1} = \mathbf{F}_t \mathbf{x}_{t-1|t-1} \tag{7.25}$$

$$\boldsymbol{\Sigma}_{t|t-1} = \mathbf{F}_t \boldsymbol{\Sigma}_{t-1|t-1} \mathbf{F}_t^\top + \mathbf{Q}_t \tag{7.26}$$

2 **correction**:

$$\mathbf{x}_{t|t} = \mathbf{x}_{t|t-1} + \mathbf{K}_t (\mathbf{y}_t - \mathbf{H}_t \mathbf{x}_{t|t-1}) \tag{7.27}$$

$$\boldsymbol{\Sigma}_{t|t} = (\mathbf{I}_p - \mathbf{K}_t \mathbf{H}_t) \boldsymbol{\Sigma}_{t|t-1} \tag{7.28}$$

where

$$\mathbf{K}_t = \boldsymbol{\Sigma}_{t|t-1} \mathbf{H}_t^\top (\mathbf{H}_t \boldsymbol{\Sigma}_{t|t-1} \mathbf{H}_t^\top + \mathbf{R}_t)^{-1} \tag{7.29}$$

is the Kalman gain and $\boldsymbol{\Sigma}_{t|t-1}$, $\boldsymbol{\Sigma}_{t|t}$ are the state prediction error and correction error covariance matrices, respectively.

---

several competing models and detecting abrupt changes in the states, the parameters, or the form of the model.

However, in spite of its many advantages, which have lead to its great popularity, the Kalman filter is very sensitive to outliers. Even rare occurrences of unusually large observations $\mathbf{y}_t$ severely degrade the performance of the Kalman filter, resulting in poor state estimates, nonwhite residuals, and invalid inference.

Engineering practitioners and statisticians have developed a variety of heuristic or rigorously derived approaches to robust Kalman filtering. The minimum requirements (Schick and Mitter, 1994) that robust Kalman filters should fulfill are:

1. The state estimation error must remain bounded, even if a single outlying observation is arbitrarily large.
2. The effect of a single outlying observation must not be spread out over time by the filter dynamics, that is, the propagation of a single outlier onto multiple residuals must be prevented.
3. The residual sequence should remain nearly white when the observation noise is Gaussian, except for some isolated outliers.

In general, in the context of state estimation, it is possible to distinguish two general types of outliers: *additive outliers*, that is, outliers in the observations noise sequence $\mathbf{w}_t$, or *innovation outliers*, that is, outliers in the state transition noise sequence $\mathbf{v}_t$. For example, outliers in $\mathbf{v}_t$ can model maneuvers in a target tracking problem, while outliers in $\mathbf{w}_t$ may be induced by NLOS propagation paths in mixed LOS/NLOS environments (Hammes and Zoubir, 2011).

In the following sections, examples of robust Kalman filtering, which can deal with additive and innovation outliers, are provided. We begin with the simplest ad-hoc robustification, the 3σ-rejection Kalman filter.

7.3.1 3σ-Rejection and Score Function Type Kalman Filter

A simple (in the sense of maintaining the linear filter) approach to deal with outliers in $\mathbf{w}_t$ is to discard them and then run a Kalman filter based on the cleaned observations. This simple ad-hoc robustification is equivalent to introducing a score function in (7.27) that is linear within some interval and zero outside it, that is,

$$\psi_{3\sigma}(x) = \begin{cases} x & \text{if } |x| \le c \\ 0 & \text{if } |x| > c. \end{cases} \tag{7.30}$$

A Kalman filter that rejects large observations is such that (7.27) is replaced by

$$\mathbf{x}_{t|t} = \mathbf{x}_{t|t-1} + \mathbf{K}_t \psi_{3\sigma}(\mathbf{y}_t - \mathbf{H}_t \mathbf{x}_{t|t-1}) \tag{7.31}$$

and this reduces to the standard Kalman filter for $c = \infty$. Often, a three-sigma rejection rule is used, which means that $c = 3\hat{\sigma}$, where

$$\hat{\sigma} = \hat{\sigma}(\mathbf{y}_t - \mathbf{H}_t \mathbf{x}_{t|t-1})$$

is a robust scale estimate, that is, the median absolute deviation, see (1.25). While such approaches are attractive because they maintain the computational complexity of the (linear) Kalman filter, they also have some drawbacks. Firstly, ignoring outlying observations completely in some cases leads to a large loss of information, and therewith a decreased efficiency. Secondly, a discontinuous redescending score function as in (7.30) causes a discontinuity of the estimator as a function of the data. Further ad-hoc robustifications can be obtained by choosing $\psi(\cdot)$ from the monotone or redescending class as defined in Chapter 1. This leads to the score function type Kalman filter, where (7.27) is replaced by

$$\mathbf{x}_{t|t} = \mathbf{x}_{t|t-1} + \mathbf{K}_t \psi(\mathbf{y}_t - \mathbf{H}_t \mathbf{x}_{t|t-1}). \tag{7.32}$$

Algorithm 11 details the implementation of such a Kalman filter.

7.3.2   The Masreliez Approximate Conditional Mean Filter for Additive Outliers

Masreliez (1975) proposed two robust Kalman filters, the first one deals with outliers in the observations noise sequence $\mathbf{w}_t$ in (7.20), while the second is designed to be robust against outliers in the state transition noise sequence $\mathbf{v}_t$ in (7.19). The first approximate conditional mean (ACM) filter proposed by Masreliez is based on the first-order approximation of the conditional observation density prior to updating

$$f(\mathbf{y}_t | \mathcal{Y}_{t-1}) \tag{7.38}$$

Algorithm 11: KalmanFilterScore: computes the score function type Kalman filter algorithm. The 3σ-rejection Kalman filter algorithm is obtained by using $\psi(x) = \psi_{3\sigma}(x)$.

---

input   : $\mathbf{F}_t \in \mathbb{R}^{p \times p}$, $\mathbf{H}_t \in \mathbb{R}^{q \times p}$, $\mathbf{y}_t \in \mathbb{R}^{q \times 1}$, $\mathbf{Q}_t \in \mathbb{R}^{p \times p}$, $\mathbf{R}_t \in \mathbb{R}^{q \times q}$
output : $\mathbf{x}_{t|t}$
initialize: $\mathbf{x}_{0|0}$, $\mathbf{\Sigma}_{0|0}$

1 **prediction**:

$$\mathbf{x}_{t|t-1} = \mathbf{F}_t \mathbf{x}_{t-1|t-1} \tag{7.33}$$

$$\mathbf{\Sigma}_{t|t-1} = \mathbf{F}_t \mathbf{\Sigma}_{t-1|t-1} \mathbf{F}_t^\top + \mathbf{Q}_t \tag{7.34}$$

2 **correction**:

$$\mathbf{x}_{t|t} = \mathbf{x}_{t|t-1} + \mathbf{K}_t \psi (\mathbf{y}_t - \mathbf{H}_t \mathbf{x}_{t|t-1}) \tag{7.35}$$

$$\mathbf{\Sigma}_{t|t} = (\mathbf{I}_p - \mathbf{K}_t \mathbf{H}_t) \mathbf{\Sigma}_{t|t-1} \tag{7.36}$$

where

$$\mathbf{K}_t = \mathbf{\Sigma}_{t|t-1} \mathbf{H}_t^\top (\mathbf{H}_t \mathbf{\Sigma}_{t|t-1} \mathbf{H}_t^\top + \mathbf{R}_t)^{-1} \tag{7.37}$$

is the Kalman gain and $\mathbf{\Sigma}_{t|t-1}$, $\mathbf{\Sigma}_{t|t}$ are the state prediction error and correction error covariance matrices, respectively.

---

and assumes that the state transition uncertainty $\mathbf{v}_t$ is Gaussian and the observation noise $\mathbf{w}_t$ is heavy tailed. The predicted state density conditioned on past observations

$$f(\mathbf{x}_t | \mathcal{Y}_{t-1}) \tag{7.39}$$

is assumed to be Gaussian with mean $\mathbf{x}_{t|t-1}$ and covariance matrix $\mathbf{Q}_t$. Masreliez proposed a Kalman filter as detailed in Algorithm 12. Interestingly, if the conditional observation density prior to updating in (7.38) is Gaussian, that is,

$$\mathbf{g}_t(\mathbf{y}_t) = (\mathbf{H}_t \mathbf{\Sigma}_{t|t-1} \mathbf{H}_t^\top + \mathbf{R}_t)^{-1} (\mathbf{y}_t - \mathbf{H}_t \mathbf{x}_{t|t-1}) \tag{7.40}$$

and

$$\mathbf{G}_t(\mathbf{y}_t) = \mathbf{G}_t = (\mathbf{H}_t \mathbf{\Sigma}_{t|t-1} \mathbf{H}_t^\top + \mathbf{R}_t)^{-1}, \tag{7.41}$$

then the ACM filter coincides with the Kalman filter and Algorithm 12 reduces to Algorithm 10.

A further important remark is that $\mathbf{g}_t(\mathbf{y}_t)$ takes the role of a score function for $f(\mathbf{y}_t | \mathcal{Y}_{t-1})$. For example, when $f(\mathbf{y}_t | \mathcal{Y}_{t-1})$ is Gaussian, $\mathbf{g}_t(\mathbf{y}_t)$ is linear in $(\mathbf{y}_t - \mathbf{H}_t \mathbf{x}_{t|t-1})$. It then follows that a reasonable ad-hoc strategy for developing robust Kalman type filters is to insert, for $\mathbf{g}_t(\mathbf{y}_t)$, one of the well-known ψ-functions, for example, Huber's as given in (1.21).

Algorithm 12: ACMFilterAO: computes the approximate conditional mean filter algorithm that is robust against outliers in the observations (additive outliers).

---

input   : $\mathbf{F}_t \in \mathbb{R}^{p \times p}$, $\mathbf{H}_t \in \mathbb{R}^{q \times p}$, $\mathbf{y}_t \in \mathbb{R}^{q \times 1}$, $\mathbf{Q}_t \in \mathbb{R}^{p \times p}$, $\mathbf{R}_t \in \mathbb{R}^{q \times q}$

output : $\mathbf{x}_{t|t}$

initialize: $\mathbf{x}_{0|0}$, $\boldsymbol{\Sigma}_{0|0}$

1 **prediction**:

$$\mathbf{x}_{t|t-1} = \mathbf{F}_t \mathbf{x}_{t-1|t-1} \tag{7.42}$$

$$\boldsymbol{\Sigma}_{t|t-1} = \mathbf{F}_t \boldsymbol{\Sigma}_{t-1|t-1} \mathbf{F}_t^\top + \mathbf{Q}_t \tag{7.43}$$

2 **correction**:

$$\mathbf{x}_{t|t} = \mathbf{x}_{t|t-1} + \boldsymbol{\Sigma}_{t|t-1} \mathbf{H}_t^\top \mathbf{g}_t(\mathbf{y}_t) \tag{7.44}$$

$$\boldsymbol{\Sigma}_{t|t} = \boldsymbol{\Sigma}_{t|t-1} - \boldsymbol{\Sigma}_{t|t-1} \mathbf{H}_t^\top \mathbf{G}_t(\mathbf{y}_t) \mathbf{H}_t \boldsymbol{\Sigma}_{t|t-1} \tag{7.45}$$

where the components $i = 1, \ldots q$ of the column vector $\mathbf{g}_t(\mathbf{y}_t) \in \mathbb{R}^{q \times 1}$ are given by

$$- \frac{\partial f(\mathbf{y}_t | \mathcal{Y}_{t-1})}{\partial (\mathbf{y}_t)_i} \cdot (\mathbf{y}_t | \mathcal{Y}_{t-1}))^{-1} \tag{7.46}$$

and $\mathbf{G}_t(\mathbf{y}_t)$ is a matrix with elements i, j equal to

$$\frac{\partial \mathbf{g}_t(\mathbf{y}_t)_i}{\partial (\mathbf{y}_t)_j}, \tag{7.47}$$

that is, $\mathbf{G}_t(\mathbf{y}_t)$ is the derivative of $\mathbf{g}_t(\mathbf{y}_t)$.

7.3.3   The Masreliez Approximate Conditional Mean Filter for Innovation Outliers

The second ACM filter proposed by Masreliez (1975) deals with the innovation outlier case, that is, the case in which the predicted state density conditioned on past observations $f(\mathbf{x}_t | \mathcal{Y}_{t-1})$ is heavy tailed, such that $\mathbf{v}_t$ contains outliers. The conditional observation density prior to updating $f(\mathbf{y}_t | \mathcal{Y}_{t-1})$, by contrast, is assumed to be Gaussian, such that $\mathbf{w}_t$ is zero mean Gaussian with covariance matrix $\mathbf{R}_t$. For such cases, Algorithm 13 provides robust state estimates. The choice of $\mathbf{g}_t(\cdot)$ relies on knowledge of $f(\mathbf{y}_t | \mathcal{Y}_{t-1})$, and robust filters can be designed such that $\mathbf{g}_t$ provides robustness over an uncertainty class of distributions including $f(\mathbf{y}_t | \mathcal{Y}_{t-1})$.

7.3.4   The Schick and Mitter Approximate Conditional Mean Filter for Additive Outliers

Schick and Mitter (1994) proposed a first-order approximation for the conditional prior distribution of the state (given all past observations) for the case in which the observation noise belongs to the ε-contaminated Gaussian neighborhood. The approximation is based on using an asymptotic expansion around a small parameter involving a fraction of contamination ε.

Algorithm 13: ACMFilterIO: computes the approximate conditional mean filter algorithm, which is robust against outliers in the state transition equation (innovation outliers).

---

input   : $\mathbf{F}_t \in \mathbb{R}^{p \times p}$, $\mathbf{H}_t \in \mathbb{R}^{q \times p}$, $\mathbf{y}_t \in \mathbb{R}^{q \times 1}$, $\mathbf{Q}_t \in \mathbb{R}^{p \times p}$, $\mathbf{R}_t \in \mathbb{R}^{q \times q}$

output : $\mathbf{x}_{t|t}$

initialize: $\mathbf{x}_{0|0}$, $\Sigma_{0|0}$

1 **prediction**:

$$\mathbf{x}_{t|t-1} = \mathbf{F}_t \mathbf{x}_{t-1|t-1} \qquad (7.48)$$

$$\Sigma_{t|t-1} = \mathbf{F}_t \Sigma_{t-1|t-1} \mathbf{F}_t^\top + \mathbf{Q}_t \qquad (7.49)$$

2 **correction**:

$$\mathbf{x}_{t|t} = \underbrace{(\mathbf{H}_t^\top \Sigma_{t|t-1} \mathbf{H}_t)^{-1}}_{\triangleq \mathbf{T}_t} \mathbf{H}_t^\top (\mathbf{R}_t^{-1} \mathbf{y}_t - \mathbf{g}_t(\mathbf{y}_t)) \qquad (7.50)$$

$$\Sigma_{t|t} = \mathbf{T}_t - \mathbf{T}_t \mathbf{H}_t^\top \mathbf{G}_t(\mathbf{y}_t) \mathbf{H}_t^\top \mathbf{T}_t \qquad (7.51)$$

where the components $i = 1, \ldots q$ of the column vector $\mathbf{g}_t(\mathbf{y}_t) \in \mathbb{R}^{q \times 1}$ are given by

$$-\frac{\partial f(\mathbf{y}_t|\mathcal{Y}_{t-1})}{\partial(\mathbf{y}_t)_i} \cdot (\mathbf{y}_t|\mathcal{Y}_{t-1}))^{-1} \qquad (7.52)$$

and $\mathbf{G}_t(\mathbf{y}_t)$ is a matrix with elements i, j equal to

$$\frac{\partial \mathbf{g}_t(\mathbf{y}_t)_i}{\partial(\mathbf{y}_t)_j}, \qquad (7.53)$$

i.e., $\mathbf{G}_t(\mathbf{y}_t)$ is the derivative of $\mathbf{g}_t(\mathbf{y}_t)$.

This approximation makes use of the exponential stability of the Kalman filter, which ensures that the effects of past outliers attenuates quickly. The first-order approximation to the conditional prior distribution is then used, based on a generalization of a result due to Masreliez (1975), to derive a first-order approximation to the conditional mean of the state (given all past observations and the current one).

Schick and Mitter (1994) combine robust location M-estimation with the stochastic approximation method of Robbins and Monro to develop a robust recursive estimator of the state of a linear dynamic system. Both point estimation and filtering estimate parameters are based on observations contaminated by noise, but while the parameters to be estimated are fixed in the former case, they vary according to some (possibly stochastic) model in the latter. The Robbins-Monro type stochastic approximation algorithms for location estimation, assuming a symmetric distribution, are of the form

$$\hat{\mu}_t = \hat{\mu}_{t-1} + \frac{k}{t} \psi(y_t - \hat{\mu}_{t-1}), \qquad (7.54)$$

where $k > 0$ is an appropriate constant and $\psi(\cdot)$ is chosen consistent with one of the functions discussed in Chapter 1. The resulting non-Gaussian estimator has the form of banks of Kalman filters and optimal smoothers that are weighted by the posterior probabilities that each observation was an outlier. Clearly, this filter has a high computational cost, which, depending on the application may not always be justified by the performance gain it provides.

7.3.5   Robust Regression-Based Kalman Filter

A further general approach to robust Kalman filtering is to recognize the equivalence of the state space formulation based on (7.19) and (7.20) to the linear regression problem (2.8); see, for example, Duncan and Horn (1972), Durovic and Kovacevic (1999), and Gandhi and Mili (2010). Under this framework, outliers in $\mathbf{w}_t$ correspond to vertical outliers, while outliers in $\mathbf{v}_t$ correspond to leverage points as displayed in Figures 2.7 and 2.8.

Consistent with Gandhi and Mili (2010), consider the linear dynamic system defined by (7.19) and (7.20) as a batch-mode linear regression:

$$\begin{pmatrix} \mathbf{y}_t \\ \mathbf{x}_{t|t-1} \end{pmatrix} = \begin{pmatrix} \mathbf{H}_t \\ \mathbf{I} \end{pmatrix} \mathbf{x}_t + \begin{pmatrix} \mathbf{w}_t \\ \mathbf{x}_{t|t-1} - \mathbf{x}_t \end{pmatrix}. \tag{7.55}$$

By defining

$$\mathbf{e}_t = \begin{pmatrix} \mathbf{w}_t \\ \mathbf{x}_{t|t-1} - \mathbf{x}_t \end{pmatrix}, \tag{7.56}$$

the measurement error covariance matrix becomes

$$\mathsf{E}\left[\mathbf{e}_t\mathbf{e}_t^\top\right] = \begin{pmatrix} \mathbf{R}_t & \mathbf{0} \\ \mathbf{0} & \mathbf{\Sigma}_{t|t-1} \end{pmatrix} = \mathbf{L}_t\mathbf{L}_t^\top \tag{7.57}$$

with $\mathbf{R}_t$ and $\mathbf{\Sigma}_{t|t-1}$ given in (7.24) and (7.26), respectively. Next, $\mathbf{L}_t^{-1}$, the inverse of the Choleski decomposition given in (7.57), is multiplied through (7.55) from the left-hand side. This pre-whitening operation yields a linear regression model

$$\tilde{\mathbf{y}}_t = \tilde{\mathbf{X}}_t\boldsymbol{\beta}_t + \tilde{\mathbf{v}}_t, \tag{7.58}$$

where

$$\tilde{\mathbf{y}}_t = \mathbf{L}_t^{-1}\begin{pmatrix} \mathbf{y}_t \\ \mathbf{x}_{t|t-1} \end{pmatrix},$$

$$\tilde{\mathbf{X}}_t = \mathbf{L}_t^{-1}\begin{pmatrix} \mathbf{H}_t \\ \mathbf{I} \end{pmatrix}, \tag{7.59}$$

$$\boldsymbol{\beta}_t = \mathbf{x}_t,$$

and

$$\tilde{\mathbf{w}}_t = \mathbf{L}_t^{-1}\mathbf{e}_t.$$

Because of the equivalence of (7.58) to (2.8), in principle, any robust linear regression estimator, for example, one of the ones that were discussed in Chapter 2, can be used to determine $\mathbf{x}_t$.

Example 15   Generalized M-Estimation-Based Robust Kalman Filtering

Gandhi and Mili (2010) proposed a robust Kalman filter based on a generalized M-estimator, which incorporates a prewhitening step that is based on a robust multivariate estimator of location and covariance (see Chapter 4). The filter can simultaneously deal with measurement and innovation outliers, that is, outliers in $\mathbf{w}_t$ and $\mathbf{v}_t$.

The generalized M-estimator minimizes the following objective function

$$L_{\mathrm{GM}}(\boldsymbol{\beta}_t) = \sum_{t=1}^{N} \varpi_t^2 \rho \left(\frac{\tilde{v}_t}{\hat{\sigma}\, \varpi_t} \right), \tag{7.60}$$

where $\rho(\cdot)$ is, for example, Huber's function as given in (1.20) and $\hat{\sigma}$ is the normalized median absolute deviation (1.25) of the residuals, that is,

$$\hat{\sigma} \equiv \hat{\sigma}(\tilde{\mathbf{v}}_t) = \hat{\sigma}(\tilde{\mathbf{y}}_t - \tilde{\mathbf{X}}_t \boldsymbol{\beta}_t)$$

with $\tilde{\mathbf{v}}_t = (\tilde{v}_1, \dots, \tilde{v}_N)^\top$. Vertical outliers are taken care of by choice of $\rho(x)$, while leverage points are down-weighted by the generalized M-estimator proposed by Gandhi and Mili (2010) using

$$\varpi_t = \min \left(1, \frac{c^2}{\mathrm{PS}_t^2} \right), \tag{7.61}$$

where $c > 0$ is a tuning constant and PS_t is the largest distance of the one-dimensional projection of the point $\tilde{v}_t$ to the cloud of all points that are contained in $\tilde{\mathbf{v}}_t$. An algorithm to compute PS_t is given in Gandhi and Mili (2010).

The generalized M-estimator is obtained from (7.60) by solving the following non-linear system of equations with an IRWLS algorithm

$$\frac{\partial L_{\mathrm{GM}}(\boldsymbol{\beta}_t)}{\partial \boldsymbol{\beta}_t} = -\sum_{t=1}^{N} \frac{\varpi_t \tilde{\mathbf{x}}_t^\top}{\hat{\sigma}} \psi(\tilde{v}_t) = \mathbf{0}. \tag{7.62}$$

In (7.62), $\psi(x) = \frac{d\rho(x)}{dx}$ and $\tilde{\mathbf{x}}_i$ is the tth column of $\tilde{\mathbf{X}}_t$, as defined in (7.59).

Algorithm 14 summarizes the implementation of the generalized M-estimator-based robust Kalman filter.

7.4   Robust Extended Kalman Filtering

In all of the Kalman filtering approaches, considered in previous sections, it was assumed that the relationship between the states $\mathbf{x}_t$ and the observations $\mathbf{y}_t$ can be appropriately represented by a linear model, that is,

---

Algorithm 14:   GMKalman: algorithm that implements the generalized M-estimator-based robust Kalman filter.

---

input   : $\mathbf{F}_t \in \mathbb{R}^{p \times p}$, $\mathbf{H}_t \in \mathbb{R}^{q \times p}$, $\mathbf{y}_t \in \mathbb{R}^{q \times 1}$, $\mathbf{Q}_t \in \mathbb{R}^{p \times p}$, $\mathbf{R}_t \in \mathbb{R}^{q \times q}$
output   : $\mathbf{x}_{t|t}$
initialize: $N_{\text{iter}} \in \mathbb{N}$, $\mathbf{x}_{0|0}$, $\mathbf{\Sigma}_{0|0}$

1 **prediction**:

$$\mathbf{x}_{t|t-1} = \mathbf{F}_t \mathbf{x}_{t-1|t-1}$$

$$\mathbf{\Sigma}_{t|t-1} = \mathbf{F}_t \mathbf{\Sigma}_{t-1|t-1} \mathbf{F}_t^\top + \mathbf{Q}_t$$

2 **correction**:
3 **for** $n = 0, 1, \ldots, N_{\text{iter}}$ **do**
4

$$\mathbf{x}_{t|t}^{(n+1)} = (\tilde{\mathbf{X}}_t \mathbf{D}^{(n)} \tilde{\mathbf{X}}_t)^{-1} \tilde{\mathbf{X}}_t^\top \mathbf{D}^{(n)} \tilde{\mathbf{y}}_t$$

with $\tilde{\mathbf{X}}_t$ given in (7.59) and

$$\mathbf{D}^{(n)} = \text{diag}\left(\frac{\psi\left(\tilde{r}_1^{(n)}\right)}{\tilde{r}_1^{(n)}}, \ldots, \frac{\psi\left(\tilde{r}_N^{(n)}\right)}{\tilde{r}_N^{(n)}} \right)$$

where $\tilde{r}_t^{(n)} = \frac{\tilde{v}_t^{(n)}}{\hat{\sigma} \varpi_t}$.

5

$$\mathbf{\Sigma}_{t|t} = \frac{\mathsf{E}\left[\psi^2(\tilde{n}_i)\right]}{\mathsf{E}\left[\psi'(\tilde{n}_i)\right]} (\tilde{\mathbf{X}}_t^\top \tilde{\mathbf{X}}_t)^{-1} (\tilde{\mathbf{X}}_t^\top \text{diag}\left(\varpi_i^2\right) \tilde{\mathbf{X}}_t)(\tilde{\mathbf{X}}_t^\top \tilde{\mathbf{X}}_t)^{-1}$$

$$\mathbf{y}_t = \mathbf{H}_t \mathbf{x}_t + \mathbf{w}_t$$

as previously given in (7.20). However, for some applications, such as the localization of moving user equipment (agent), this assumption does not hold.

In the more general case, the nonlinear state and measurement equations are given by

$$\mathbf{x}_t = \mathbf{f}_t(\mathbf{x}_{t-1}) + \mathbf{v}_t \tag{7.63}$$

and

$$\mathbf{y}_t = \mathbf{h}_t(\mathbf{x}_t) + \mathbf{w}_t, \tag{7.64}$$

respectively. Here, $\mathbf{f}_t(\cdot) \in \mathbb{R}^{p \times 1}$ is a nonlinear vector function that describes the relationship between successive states, while the nonlinear vector function $\mathbf{h}_t(\cdot) \in \mathbb{R}^{q \times 1}$ maps the observations to the state vector. The uncertainties in the state equation, as for the linear case are modeled by the i.i.d. random process $\mathbf{v}_t \in \mathbb{R}^{p \times 1}$, while the uncertainties in the observations are captured by $\mathbf{w}_t \in \mathbb{R}^{q \times 1}$.

From a Bayesian perspective, the aim of nonlinear filtering is to estimate the unknown pdf $f(\mathbf{x}_t|\mathcal{Y}_t)$, based on the model formulated in (7.63) and (7.64), given the observations $\mathcal{Y}_t = \{\mathbf{y}_1,\ldots,\mathbf{y}_t\}$. For known pdfs of $\mathbf{v}_t$ and $\mathbf{w}_t$, the unknown state $\mathbf{x}_t$ can be estimated, for example, by maximizing the posterior pdf $f(\mathbf{x}_t|\mathcal{Y}_t)$, which is known as the maximum a posteriori estimator (MAP), or by minimizing the mean-squared error (MMSE) based on calculating the conditional expectation

$$\mathsf{E}[\mathbf{x}_t|\mathcal{Y}_t] = \int \mathbf{x}_t f(\mathbf{x}_t|\mathcal{Y}_t)d\mathbf{x}_t. \tag{7.65}$$

For linear functions $\mathbf{f}_t(\cdot)$ and $\mathbf{h}_t(\cdot)$ in (7.63) and (7.64), and assuming white zero-mean mutually independent Gaussian random processes $\mathbf{v}_t$ and $\mathbf{w}_t$, the MAP estimator and the MMSE estimator coincide and are known as the Kalman filter. See Algorithm 10.

However, for the case of nonlinear functions $\mathbf{f}_t(\cdot)$ and/or $\mathbf{h}_t(\cdot)$, and non-Gaussian pdfs for $\mathbf{v}_t$ and/or $\mathbf{w}_t$, finding an analytical solution for $f(\mathbf{x}_t|\mathcal{Y}_t)$ often becomes intractable and suboptimal approaches are the preferred option.

In this section, we focus on the robust extended Kalman filter (REKF), which linearizes $\mathbf{h}_t(\mathbf{x}) = (h_1(\mathbf{x}),\ldots,h_q(\mathbf{x}))^\top$ around the predicted state estimate:

$$\mathbf{h}_t(\mathbf{x}_t) \approx \mathbf{h}_t(\mathbf{x}_{t|t-1}) + \mathbf{J}_{\mathbf{h}}(\mathbf{x}_{t|t-1})(\mathbf{x}_t - \mathbf{x}_{t|t-1}), \tag{7.66}$$

where

$$\mathbf{J}_{\mathbf{h}}(\mathbf{x}_{t|t-1}) = \left. \frac{\partial \mathbf{h}_t(\mathbf{x}_t)}{\partial \mathbf{x}_t} \right|_{\mathbf{x}_t = \mathbf{x}_{t|t-1}} \tag{7.67}$$

is the Jacobian matrix of partial derivatives

$$\mathbf{J}_{\mathbf{h}}(\mathbf{x}) = \begin{pmatrix} \frac{\partial h_1}{\partial x_1} & \frac{\partial h_1}{\partial x_2} & \cdots & \frac{\partial h_1}{\partial x_p} \\ \vdots & \vdots & \ddots & \vdots \\ \frac{\partial h_q}{\partial x_1} & \frac{\partial h_q}{\partial x_2} & \cdots & \frac{\partial h_q}{\partial x_p} \end{pmatrix} \in \mathbb{R}^{q \times p} \tag{7.68}$$

evaluated at the predicted state estimate $\mathbf{x}_{t|t-1}$.

Similarly, to linearize $\mathbf{f}_t(\mathbf{x}) = (f_1(\mathbf{x}),\ldots,f_p(\mathbf{x}))^\top$, the EKF uses a first-order Taylor series expansion around $\mathbf{x}_{t-1|t-1}$

$$\mathbf{f}_t(\mathbf{x}_{t-1}) \approx \mathbf{f}_t(\mathbf{x}_{t-1|t-1}) + \mathbf{J}_{\mathbf{f}}(\mathbf{x}_{t-1|t-1}), \tag{7.69}$$

where $\mathbf{J}_{\mathbf{f}}(\mathbf{x})$ is the Jacobian matrix of partial derivatives

$$\mathbf{J}_{\mathbf{f}}(\mathbf{x}) = \begin{pmatrix} \frac{\partial f_1}{\partial x_1} & \frac{\partial f_1}{\partial x_2} & \cdots & \frac{\partial f_1}{\partial x_p} \\ \vdots & \vdots & \ddots & \vdots \\ \frac{\partial f_p}{\partial x_1} & \frac{\partial f_p}{\partial x_2} & \cdots & \frac{\partial f_p}{\partial x_p} \end{pmatrix} \in \mathbb{R}^{p \times p}. \tag{7.70}$$

Assuming white zero-mean mutually independent Gaussian random processes $\mathbf{v}_t$ and $\mathbf{w}_t$, the EKF algorithm is summarized in Algorithm 15.

Algorithm 15: ExtKalmanFilter: computes the standard EKF algorithm.

---

input : $\mathbf{f}_t(\cdot) \in \mathbb{R}^{p \times p}$, $\mathbf{h}_t(\cdot) \in \mathbb{R}^{q \times p}$, $\mathbf{y}_t \in \mathbb{R}^{q \times 1}$, $\mathbf{Q}_t \in \mathbb{R}^{p \times p}$, $\mathbf{R}_t \in \mathbb{R}^{q \times q}$

1 **initialization**: $\mathbf{x}_{0|0}$, $\Sigma_{0|0}$

2 **prediction**:

$$\mathbf{x}_{t|t-1} \approx \mathbf{f}_t(\mathbf{x}_{t-1|t-1}) \tag{7.71}$$

$$\Sigma_{t|t-1} = \mathbf{J_f}(\mathbf{x}_{t-1|t-1})\Sigma_{t-1|t-1}\mathbf{J_f}^{\top}(\mathbf{x}_{t-1|t-1}) + \mathbf{Q}_t \tag{7.72}$$

3 **correction**:

$$\mathbf{x}_{t|t} \approx \mathbf{x}_{t|t-1} + \mathbf{K}_t(\mathbf{y}_t - \mathbf{h}(\mathbf{x}_{t|t-1})) \tag{7.73}$$

$$\Sigma_{t|t} = \left(\mathbf{I}_p - \mathbf{K}_t\mathbf{J_h}(\mathbf{x}_{t|t-1})\right)\Sigma_{t|t-1} \tag{7.74}$$

where

$$\mathbf{K}_t = \Sigma_{t|t-1}\mathbf{J_h}^{\top}(\mathbf{x}_{t|t-1})\left(\mathbf{J_h}(\mathbf{x}_{t|t-1})\Sigma_{t|t-1}\mathbf{J_h}^{\top}(\mathbf{x}_{t|t-1}) + \mathbf{R}_t\right)^{-1} \tag{7.75}$$

is the EKF gain and $\Sigma_{t|t-1}$, $\Sigma_{t|t}$ are the state prediction error and correction error covariance matrices, respectively.

7.4.1   Robust Extended Kalman Filter for the Tracking of Mobile User Equipment

Let

$$\mathbf{x}_t = (x_t, y_t, dx_t/dt, dy_t/dt)^{\top} \tag{7.76}$$

be the unknown state vector that describes the positions and velocities of an agent in a two-dimensional plane; the agent being consistent with mobile user equipment. Furthermore, let the movement of the agent be described by the following nonlinear discrete-time statistical state equation

$$\mathbf{x}_t = \mathbf{F}\mathbf{x}_{t-1} + \mathbf{G}\mathbf{v}_{t-1} \tag{7.77}$$

and measurement equation

$$\mathbf{y}_t = \mathbf{h}(\mathbf{x}_t) + \mathbf{w}_t. \tag{7.78}$$

Here,

$$\mathbf{F} = \begin{pmatrix} \mathbf{I}_2 & \Delta t \mathbf{I}_2 \\ \mathbf{0}_2 & \mathbf{I}_2 \end{pmatrix}$$

models the (linear) state transitions, which in this example are constant over time, namely $\mathbf{F}_t \equiv \mathbf{F}$. The matrix

$$\mathbf{G} = \begin{pmatrix} \Delta t^2/2 \cdot \mathbf{I}_2 \\ \Delta t \cdot \mathbf{I}_2 \end{pmatrix},$$

with Δt denoting the sampling period, maps the accelerations to the position changes of the agent. The uncertainty in the motion model defined by (7.77) is modeled by the zero-mean white Gaussian random process $\mathbf{v}_t = (v_1(t), v_2(t))^\top \sim \mathcal{N}(\mathbf{0}_2, \mathbf{Q}_t)$.

Eq. (7.78) relates the observations at the fixed terminals (base stations, anchors) to the state vector $\mathbf{x}_t$ describing the positions and velocities of the agent through the nonlinear function $\mathbf{h}(\cdot) \in \mathbb{R}^{q \times 1}$, where q in this case is the number of anchors. Herein, $\mathbf{h}(\cdot)$ depends on the measured signal parameter. For example, analogously to (2.95), for TOA based tracking, at each anchor $m = 1, \ldots, q$

$$h_m(\boldsymbol{\beta}_t) = \sqrt{(x_{\text{UE},t} - x_{\text{BS},m})^2 + (y_{\text{UE},t} - y_{\text{BS},m})^2} \qquad (7.79)$$

with $\boldsymbol{\beta}_t = (x_{\text{UE},t}, y_{\text{UE},t})^\top$ denoting the unknown positions of the agent at time instant t and $(x_{\text{BS},m}, y_{\text{BS},m})^\top$ being the fixed (and known) position of the mth anchor. Similarly, for angle-of-arrival (AOA) based tracking, at each anchor $m = 1, \ldots, q$

$$h_m(\boldsymbol{\beta}_t) = \tan^{-1}\left(\frac{y_{\text{UE},t} - y_{\text{BS},m}}{x_{\text{UE},t} - x_{\text{BS},m}}\right). \qquad (7.80)$$

In this example, we assume that all measurements are TOA measurements that are consistent with (7.79) so that $\mathbf{h}(\cdot) \equiv \mathbf{h}_m(\cdot)$.

When all anchors are in LOS contact to the agent, it is justified (e.g., by the central limit theorem) to assume that the uncertainty in the measurements can be described by a zero-mean white Gaussian random process $\mathbf{w}_t$. For example, an EKF, as given in Algorithm 15, can be used to obtain excellent localization results.

However, in mixed LOS/NLOS conditions, the zero-mean additive white Gaussian noise assumption is not valid and algorithms that are based upon $\mathbf{w}_t \sim \mathcal{N}(\mathbf{0}_q, \mathbf{R}_t)$ drastically degrade in accuracy. For example, for TOA-based tracking, the NLOS measurements take a longer path and are positively biased (Hammes and Zoubir, 2011). Such a phenomenon can be described, for example, by an ε-contamination model. In this case, the pdf of the elements of the vector $\mathbf{w}_t$ is given by

$$f_w = (1 - \varepsilon)\mathcal{N}(0, \sigma_{\text{LOS}}^2) + \varepsilon\mathcal{H}(w), \qquad (7.81)$$

where $\mathcal{H}(w)$ is an unspecified contaminating distribution with positive mean and variance $\sigma_{\text{NLOS}}^2 > \sigma_{\text{LOS}}^2$.

Next, an example of a REKF (Hammes and Zoubir, 2011) for mixed LOS/NLOS conditions is given. Analogously to the case described for the Kalman filter in Section 7.3.5, a batch-mode linear regression estimation technique can be applied. Starting from the linear state equation (7.77), that is,

$$\mathbf{x}_t = \mathbf{F}\mathbf{x}_{t-1} + \mathbf{G}\mathbf{v}_{t-1}$$

and linearizing the nonlinear vector function $\mathbf{h}(\mathbf{x}_t)$ in the measurement equation (7.78) around the predicted state estimate $\mathbf{x}_{t|t-1}$ based on (7.66), yields

$$\mathbf{y}_t \approx \mathbf{h}(\mathbf{x}_{t|t-1}) + \mathbf{J}_\mathbf{h}(\mathbf{x}_{t|t-1})(\mathbf{x}_t - \mathbf{x}_{t|t-1}) + \mathbf{w}_t, \qquad (7.82)$$

where $\mathbf{J_h}(\mathbf{x})$ is the Jacobian matrix, which, from (7.68) and (7.76), takes the form

$$
\mathbf{J_h}(\mathbf{x}_{t|t-1}) = \begin{pmatrix} \left.\frac{dh_1(\mathbf{x}_t)}{dx_t}\right|_{\mathbf{x}_t=\mathbf{x}_{t|t-1}} & \left.\frac{dh_1(\mathbf{x}_t)}{dy_t}\right|_{\mathbf{x}_t=\mathbf{x}_{t|t-1}} & 0 & 0 \\ \vdots & \vdots & \vdots & \vdots \\ \left.\frac{dh_q(\mathbf{x}_t)}{dx_t}\right|_{\mathbf{x}_t=\mathbf{x}_{t|t-1}} & \left.\frac{dh_q(\mathbf{x}_t)}{dy_t}\right|_{\mathbf{x}_t=\mathbf{x}_{t|t-1}} & 0 & 0 \end{pmatrix}.
$$

The linear state-space model defined by (7.77) and (7.78) can be rewritten as a linear regression as follows

$$
\begin{pmatrix} \mathbf{F}\mathbf{x}_{t-1|t-1} \\ \mathbf{y}_t - \mathbf{h}(\mathbf{x}_{t|t-1}) + \mathbf{J_h}(\mathbf{x}_{t|t-1})\mathbf{x}_{t|t-1} \end{pmatrix} = \begin{pmatrix} \mathbf{I}_p \\ \mathbf{J_h}(\mathbf{x}_{t|t-1}) \end{pmatrix}\mathbf{x}_t + \mathbf{e_t} \tag{7.83}
$$

where

$$
\mathbf{e_t} = \begin{pmatrix} -\mathbf{F}(\mathbf{x}_{t-1} - \mathbf{x}_{t-1|t-1}) + \mathbf{G}\mathbf{v}_{t-1} \\ \mathbf{w}_t \end{pmatrix}
$$

with

$$
\mathsf{E}\left[\mathbf{e}_t\mathbf{e}_t^\top\right] = \begin{pmatrix} \mathbf{\Sigma}_{t|t-1} & \mathbf{0} \\ \mathbf{0} & \mathbf{R}_t \end{pmatrix} = \mathbf{L}_t\mathbf{L}_t^\top
$$

and

$$
\mathbf{\Sigma}_{t|t-1} = \mathbf{F}\mathbf{\Sigma}_{t-1|t-1}\mathbf{F}^\top + \mathbf{G}\mathbf{Q}_t\mathbf{G}^\top.
$$

Multiplying the inverse of the Choleski decomposition $\mathbf{L}_t$ to the left of (7.83) results in a linear regression system

$$
\tilde{\mathbf{y}}_t = \tilde{\mathbf{X}}_t\boldsymbol{\beta}_t + \tilde{\mathbf{v}}_t, \tag{7.84}
$$

where

$$
\tilde{\mathbf{y}}_t = \mathbf{L}_t^{-1}\begin{pmatrix} \mathbf{x}_{t|t-1} \\ \mathbf{y}_t - \mathbf{h}(\mathbf{x}_{t|t-1}) + \mathbf{J_h}(\mathbf{x}_{t|t-1})\mathbf{x}_{t|t-1} \end{pmatrix},
$$

$$
\tilde{\mathbf{X}}_t = \mathbf{L}_t^{-1}\begin{pmatrix} \mathbf{I}_p \\ \mathbf{J_h}(\mathbf{x}_{t|t-1}) \end{pmatrix}, \tag{7.85}
$$

$$
\boldsymbol{\beta}_t = \mathbf{x}_t,
$$

and

$$
\tilde{\mathbf{w}}_t = \mathbf{L}_t^{-1}\mathbf{e}_t.
$$

By recognizing the equivalence of (7.84) and (2.8), any robust regression estimator, including those discussed in Chapter 2, can be used to estimate $\mathbf{x}_t$, that is, to track the agent given the TOA measurements.

Algorithm 16: RobExtKalmanFilter: computes the M-estimation-based EKF algorithm for the tracking of a mobile agent.

---

input : $\mathbf{F} \in \mathbb{R}^{p \times p}$, $\mathbf{G} \in \mathbb{R}^{p \times 2}$, $\mathbf{h}_t(\cdot) \in \mathbb{R}^{q \times p}$, $\mathbf{y}_t \in \mathbb{R}^{q \times 1}$, $\mathbf{Q}_t \in \mathbb{R}^{p \times p}$, $\mathbf{R}_t \in \mathbb{R}^{q \times q}$

output : $\mathbf{x}_{t|t}$

initialize: $\mathbf{x}_{0|0}$, $\boldsymbol{\Sigma}_{0|0}$

1 **prediction**:

$$\mathbf{x}_{t|t-1} \approx \mathbf{F}(\mathbf{x}_{t-1|t-1}) \tag{7.86}$$

$$\boldsymbol{\Sigma}_{t|t-1} = \mathbf{F}\boldsymbol{\Sigma}_{t-1|t-1}\mathbf{F}^\top(\mathbf{x}_{t-1|t-1}) + \mathbf{GQ}_t\mathbf{G}^\top \tag{7.87}$$

2 **correction**:

3 Use the Algorithm 4 to compute the M-estimate based on the equivalent linear regression model (7.84) to obtain $\mathbf{x}_{t|t}$.

4 Approximate the correction error covariance matrix by

$$\boldsymbol{\Sigma}_{t|t} \approx (\tilde{\mathbf{X}}_t^\top \tilde{\mathbf{X}}_t)^{-1}, \tag{7.88}$$

where $\tilde{\mathbf{X}}_t$ is given in (7.85).

Example 16 TOA-Based Tracking of a Mobile Agent in a Mixed LOS/NLOS Environment.

This example compares the performance of the M-estimation-based extended Kalman filter (REKF), as implemented in Algorithm 16, to that of the standard EKF, as implemented in Algorithm 15, for a tracking problem in a mixed LOS/NLOS environment.

The arrangement of the $q = 5$ anchors is displayed in Figure 7.3. The anchors are located at the following (known) positions: $(x_{\text{BS},1} = 2000\,\text{m}, y_{\text{BS},1} = 7000\,\text{m})^\top$, $(x_{\text{BS},2} = 12,000\,\text{m}, y_{\text{BS},2} = 7000\,\text{m})^\top$, $(x_{\text{BS},3} = 7000\,\text{m}, y_{\text{BS},3} = 12,000\,\text{m})^\top$, $(x_{\text{BS},4} = 7000\,\text{m}, y_{\text{BS},4} = 2000\,\text{m})^\top$, and $(x_{\text{BS},5} = 7000\,\text{m}, y_{\text{BS},5} = 7000\,\text{m})^\top$. The trajectory of the agent, which is shown in Figure 7.3, was randomly generated based on (7.77) with $\mathbf{x}_{0|0} = (4300, 4300, 2, 2)^\top$, $\Delta t = 0.2$, $N = 2000$, and $\boldsymbol{\Sigma}_{0|0} = \text{diag}(50^2\,\text{m}^2, 50^2\,\text{m}^2, 6^2\,\text{m/s}^2, 6^2\,\text{m/s}^2$.

The uncertainty about the state transition is described by a white Gaussian random process $\mathbf{v}_t \sim \mathcal{N}(\mathbf{0}_p, \mathbf{I}_p)$. This means that no innovations outliers are present. By contrast, the mixed LOS/NLOS conditions are modeled as a Markovian process that switches between the LOS condition where $w_t \sim \mathcal{N}(0, \sigma_{\text{LOS}}^2)$, with $\sigma_{\text{LOS}} = 150$, and the NLOS condition, where the NLOS effects on the TOA measurements are modeled by a pdf with positive mean and larger variance, namely

$$w_t \sim \mathcal{N}(\mu_{\text{NLOS}}, \sigma_{\text{NLOS}}^2).$$

In this simulation example, $\mu_{\text{NLOS}} = 1400\,\text{m}$ and $\sigma_{\text{NLOS}} = 400\,\text{m}$. The probability of being in NLOS condition is set to $\varepsilon = 0.4$.

Algorithm 16 with Huber's score function, as defined in (1.20) with $c = 1.5$, is used to compute the REKF, while Algorithm 15 computes the EKF. Figure 7.4 displays the

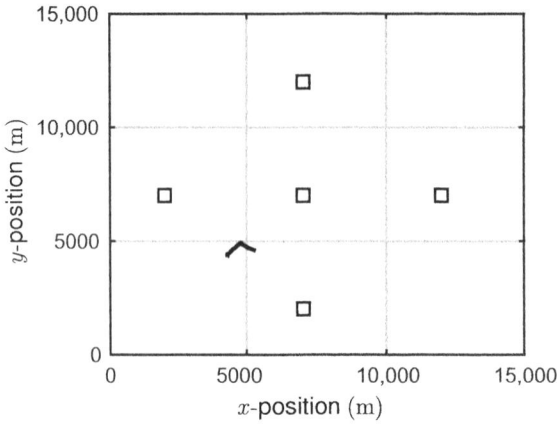

Figure 7.3 True trajectory of an agent surrounded by $q = 5$ anchors.

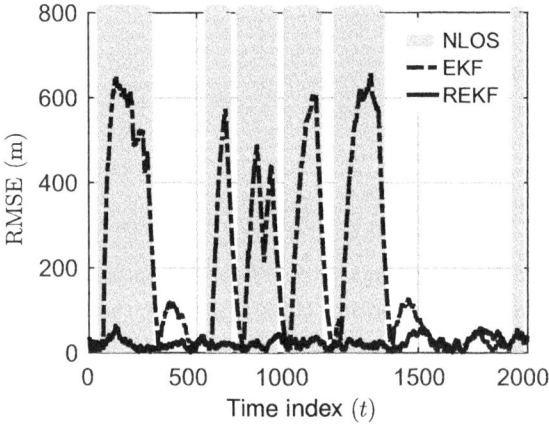

Figure 7.4 RMSE of the EKF and REKF trackers. The mixed LOS/NLOS conditions are modeled as a Markovian process.

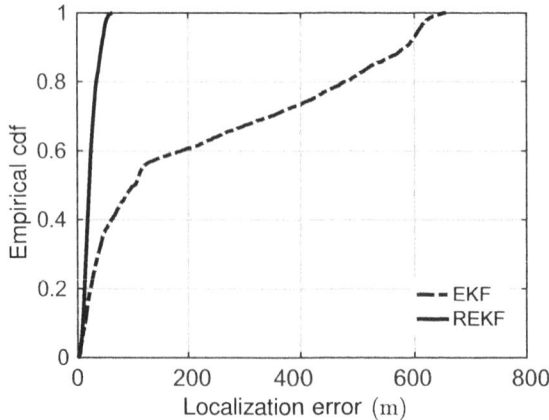

Figure 7.5 Empirical cdf of the localization error for the EKF and REKF trackers.

root-mean-squared-errors of the two trackers, while Figure 7.5 shows the empirical cdf. The bounding effect of NLOS outliers in the REKF becomes clear, when looking at the NLOS time-instances (shaded gray) in Figure 7.4. The EKF is significantly influenced, while the REKF maintains nearly the same performance throughout the tracking task. This is also reflected in the empirical cdfs that are shown in Figure 7.5, where the localization error distribution of the EKF is much heavier tailed compared to that of the REKF.

7.5 Concluding Remarks

In this chapter, the important robust filtering frameworks, such as robust Wiener filtering, nonlinear nonparametric robust filtering, and robust Kalman filtering, have been detailed. In addition to these, there exist many other frameworks, such as H_∞-filtering (Dong et al., 2010), robust particle filtering (Calvet et al., 2015), or important contributions to the Kalman filtering framework, such as optimally robust Kalman filters (Ruckdeschel et al., 2014). Other related areas of importance include multivariate robust nonparametric filtering, such as vector median filters (Astola et al., 1990), nonlinear filtering of multichannel images under robust error criteria (Koivunen, 1996), multichannel Wiener filters, robust filtering for complex data, and distributed adaptive robust filtering (Al-Sayed et al., 2016). Such research underpins the practical importance and relevance that robust filters have and will continue to have.

8   Robust Methods for Dependent Data

Correlated data streams are commonly measured, for example, in engineering, data analytics, economics, biomedicine, radar, and speech signal processing. Although there exists a considerable body of literature on robust estimation methods under the assumption of i.i.d. observations, for a long time, estimators for dependent data have been limited. Although the first contributions were made in the mid-seventies and eighties (Kassam and Thomas, 1975; Martin and Thomson, 1982; Kassam and Poor, 1985; Bustos and Yohai, 1986), there has not been much progress on methods for dependent data for a long time. This is mainly due to the fact that existing robust estimators and measures of robustness for i.i.d. data are not easily extendable to the dependent data case. Outliers in time series are more complex (Tsay, 1988) and the measures of robustness depend on the temporal structure of the contamination process. For example, the IF of an estimator is different when the outliers appear isolated compared to when they occur in patches.

To illustrate the nature of occurrence of outliers, the example of monitoring of intracranial pressure (ICP) signals is considered. Measuring such signals is common practice for patients who have suffered a traumatic brain injury. A danger for these patients is that the primary brain damage, caused by the accident, can lead to a secondary pathophysiological damage, which usually occurs together with a significantly high or low ICP value. Currently, manual observation and assessment of the ICP signals is done by nurses who predict whether ICP levels are likely to rise or drop significantly and, when appropriate, call a doctor who administers medication (Han et al., 2013). Such a procedure is subject to human errors and is relatively ineffective. Accurate ICP forecasting enables active and early intervention for more effective control of ICP levels. Figure 8.1 plots an example of an ICP measurement, where some artifacts are highlighted by crosses.

8.1   Signal and Outlier Models

Although nonstationary data is frequently encountered in engineering applications, in many cases, it is valid to assume that the data is (locally) wide-sense stationary after taking appropriate measures, for example, differentiation or data transformation. Autoregressive moving-average (ARMA) models are among the most popular models for characterizing dependent data and they have long been used in numerous real-world

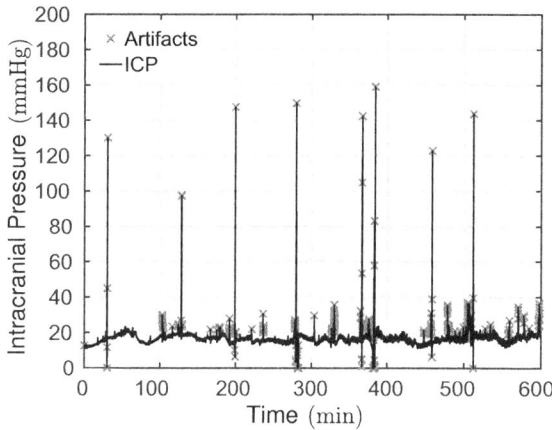

Figure 8.1 A 10-hour excerpt of a typical ICP measurement; artifacts are highlighted by crosses.

applications. This chapter therefore focuses on robust parameter estimation for ARMA models associated with random processes for which the majority of the samples are appropriately modeled by a stationary and invertible ARMA model and a minority consists of outliers with respect to the ARMA model. For such cases and, in general, classical estimators are unreliable and may break down completely (Maronna et al., 2006; Zoubir et al., 2012; Muma and Zoubir, 2017).

The nature of the outliers depends on the application. For example, motion artifacts are often evident in biomedical signals such as ICP (Han et al., 2013; Muma, 2014), ECG (Spangl and Dutter, 2007; Strasser et al., 2012), and photoplethysmographic (PPG) signals (Schäck et al., 2015) while in electricity consumption forecasting outliers are associated with holidays, major sporting events, and strikes (Chakhchoukh et al., 2009; Zoubir et al., 2012). Also in many other applications, there is a clear need for robust methods that can, to some extent, resist outliers.

8.1.1   Autoregressive Moving-Average Models

Let

$$\boldsymbol{y}_t = (\ldots, y_{t-2}, y_{t-1}, y_t) \tag{8.1}$$

denote a semi-infinite sequence of observations, generated by a stationary and invertible ARMA(p, q) process up to time t according to

$$y_t = \mu + \sum_{i=1}^{p} \phi_i(y_{t-i} - \mu) + a_t(\boldsymbol{\beta}) - \sum_{i=1}^{q} \theta_i a_{t-i}(\boldsymbol{\beta}) \tag{8.2}$$

where the parameter vector $\boldsymbol{\beta} = (\boldsymbol{\phi}, \boldsymbol{\theta}, \mu)$, $\boldsymbol{\phi} = (\phi_1, \ldots, \phi_p)$ and $\boldsymbol{\theta} = (\theta_1, \ldots, \theta_q)$. Here, a_t are i.i.d. random variables with a symmetric distribution and it is also assumed that $\mathsf{E}\left[\log^+(|a_t|)\right] < \infty$, where $\log^+(x) = \max\{\log(x), 0\}$.

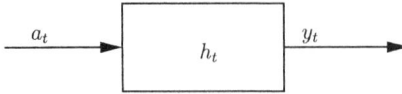

Figure 8.2 The ARMA process from a linear filtering perspective.

It is also possible to define the parameter vector $\boldsymbol{\beta} = (\boldsymbol{\phi}, \boldsymbol{\theta}, \mu)$ by the polynomial operators

$$\phi(B) = 1 - \sum_{i=1}^{p} \phi_i B^i \tag{8.3}$$

and

$$\theta(B) = 1 - \sum_{i=1}^{q} \theta_i B^i, \tag{8.4}$$

respectively. To restrict the parameter space in a manner that is consistent with a stationary and invertible ARMA model, the polynomials $\phi(B)$ and $\theta(B)$ must have all their roots outside the unit circle.

Then, by defining

$$a_t^e(\boldsymbol{\beta}) = \theta^{-1}(B)\phi(B)(y_t - \mu), \tag{8.5}$$

the following recursion follows:

$$a_t^e(\boldsymbol{\beta}) = y_t - \mu - \sum_{i=1}^{p} \phi_i(y_{t-i} - \mu) + \sum_{i=1}^{q} \theta_i a_{t-i}^e(\boldsymbol{\beta}) \tag{8.6}$$

and $a_t^e(\boldsymbol{\beta}) = a_t$, such that the *innovations sequence* $\boldsymbol{a}_t = (\ldots, a_{t-2}, a_{t-1}, a_t)$ can be reconstructed. To be able to uniquely identify an ARMA model, it is further required that $\phi(B)$ and $\theta(B)$ do not have common roots.

Figure 8.2 displays an ARMA process from a linear filtering perspective. Here, the random i.i.d. noise sequence $\boldsymbol{a}_t = (\ldots, a_{t-2}, a_{t-1}, a_t)$ is filtered by a stable, linear, and time-invariant system with unit sample response h_t, $t \in \mathbb{Z}$ and transfer function

$$H(e^{J\omega}) = \frac{1 - \sum_{i=1}^{q} \theta_i e^{-J\omega i}}{1 - \sum_{i=1}^{p} \phi_i e^{-J\omega i}}. \tag{8.7}$$

In real-world applications, the observations y_t never exactly follow (8.2). The robust parameter estimation methods that will be described in this chapter rely on much less restrictive assumptions about the data. It is only required that the ARMA models are associated with random processes for which the majority of the samples are appropriately modeled by a stationary and invertible ARMA model and a minority consists of outliers with respect to the ARMA model.

8.1.2   Outlier Models

There exist several statistical models for outliers in dependent data; see for example, Tsay (1988), Deutsch et al. (1990), Ljung (1993), Louni (2008), Maronna et al. (2006), and Zoubir et al. (2012). The following provides a brief review of important models.

The *additive outlier (AO) model* defines contaminated observations y_t^ε according to

$$y_t^\varepsilon = y_t + \xi_t^\varepsilon w_t, \tag{8.8}$$

where y_t follows an ARMA model, as given in (8.2), w_t defines the contaminating process that is independent of y_t, and ξ_t^ε is a stationary random process for which

$$\xi_t^\varepsilon = \begin{cases} 1 & \text{with probability } \varepsilon \\ 0 & \text{with probability } (1-\varepsilon). \end{cases} \tag{8.9}$$

For the *replacement outlier (RO) model*, outliers are drawn from a second so-called replacement process that is independent from the nominal process, that is,

$$y_t^\varepsilon = (1 - \xi_t^\varepsilon)y_t + \xi_t^\varepsilon w_t, \tag{8.10}$$

where w_t is independent of y_t and ξ_t is defined by (8.9).

A different type of outlier is represented by the *innovation outlier (IO) model*, which models outliers that occur in the innovation sequence a_t. These outliers can be additive when a_t in the ARMA model is contaminated through

$$a_t^\varepsilon = a_t + \xi_t^\varepsilon w_t, \tag{8.11}$$

where w_t is independent of a_t and ξ_t is defined by (8.9). Or individual samples in a_t can be replaced via

$$a_t^\varepsilon = (1 - \xi_t^\varepsilon)a_t + \xi_t^\varepsilon w_t, \tag{8.12}$$

where w_t defines the contaminating process that is independent of a_t.

The AOs or ROs are more frequent in practice than IOs, and are much more difficult to deal with in the estimation procedure. Figures 8.3 and 8.4 display the outlier models from a linear filtering perspective. It becomes clear from Figure 8.3 that IOs carry information about the model, whereas AOs and ROs inject information about the external errors. In summary, AOs and ROs are pure measurement errors, while IOs are exceptional shocks that satisfy the feed-through mechanism of the nominal ARMA process. As will be shown in Section 8.2, a single AO or RO can have a devastating effect on parameter estimation due to the so-called propagation of outliers.

Outliers may also differ in their temporal structure. For *isolated outliers*, ξ_t^ε takes on the value 1, such that at least one nonoutlying observation occurs between two outliers (e.g., ξ_t^ε follows an independent Bernoulli distribution). For *patchy outliers*, $\xi_t^\varepsilon, t \in 1, \ldots, N$, takes on the value 1 for $N_{\text{patch}} \le N/2$ subsequent samples. Some other types of outlier models exist in the literature, such as level shift, variance change, and seasonal outliers (Tsay, 1988), to mention a few.

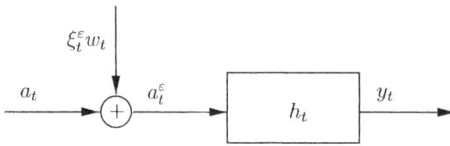

Figure 8.3 The IO process from a linear filtering perspective.

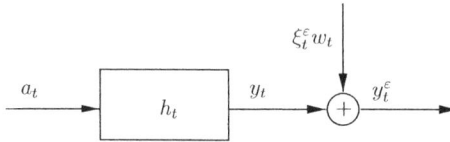

Figure 8.4 The AO process from a linear filtering perspective. RO processes share the same structure by letting $w_t = v_t - y_t$, where v_t is a contamination process that is independent of y_t.

8.2   Propagation of Outliers

ARMA parameter estimation, that is, determining $\hat{\boldsymbol{\beta}}$, is often based on minimizing some function of the reconstructed innovation sequence. However, as can be seen from (8.5), one AO or RO in y_t^ε can propagate onto multiple innovations $a_t^e(\boldsymbol{\beta})$. In the extreme case, all entries of the innovations sequence are disturbed by a single outlier.

The phenomenon of outlier propagation is illustrated in Figure 8.5 for four different ARMA models:

   Figure 8.5a: AR(1) with $\phi_1 = 0.9$,
   Figure 8.5b: AR(6) with $\boldsymbol{\phi} = (-.4, 0.05, 0.1, 0.2, 0.3, -.4)^\top$,
   Figure 8.5c: MA(1) with $\theta_1 = 0.9$,
   Figure 8.5d: ARMA(1,1) with $\boldsymbol{\beta} = (\phi_1, \theta_1)^\top$, where $\phi_1 = -0.8$ and $\theta_1 = 0.9$.

All processes are created from the same zero-mean unit variance innovation sequence a_t, $t = 1, \ldots, N$, with $N = 50$. In all of the subplots, the true innovation sequence a_t, $t = 1, \ldots, N$, is marked by "+." Next, a single sample of y_t, at position $t = 25$, is replaced by the value $y_{25} = 50$. This corresponds to the RO model of (8.10), which yields y_t^ε, where $\varepsilon = 2$ percent, the contaminating pdf $f(w_t)$ is only nonzero at $w_t = 50$, and $\xi_t^\varepsilon = 0$, $\forall t$, except for $\xi_{25}^\varepsilon = 1$. Here, $\delta(\cdot)$ is the Kronecker delta function. Figure 8.5 plots $a_t^e(\boldsymbol{\beta})$, that is, the results of applying the recursion (8.6) to the four different contaminated ARMA models y_t^ε.

Figure 8.5a and 8.5b illustrates that in an AR(p) model, a single RO creates $p + 1$ outliers in the reconstructed innovations $a_t^e(\boldsymbol{\beta})$, $t = 1, \ldots, N$. Likewise, when representing the AR(p) model as a regression, $p + 1$ estimating equations are affected. This phenomenon, which also occurs for AO, is called propagation of outliers. Because of the propagation of outliers, the classical strategies of robust estimation, for example,

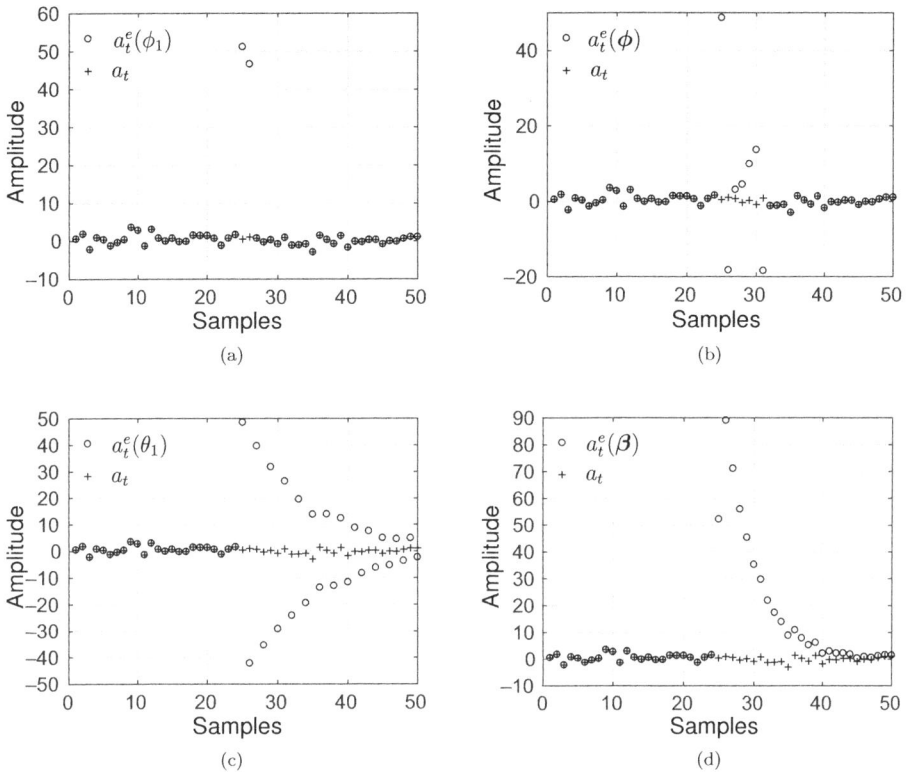

Figure 8.5 Examples of the propagation of a single RO of amplitude 50 at $t = 25$ for different ARMA models. (a) AR(1) with $\phi_1 = 0.9$. (b) AR(6) with $\boldsymbol{\phi} = (-.4, 0.05, 0.1, 0.2, 0.3, -.4)^\top$. (c) MA(1) with $\theta_1 = 0.9$. (d) ARMA(1,1) with $\boldsymbol{\beta} = (\phi, \theta)^\top$, where $\phi_1 = -0.8$ and $\theta_1 = 0.9$.

minimizing a robust function of the reconstructed innovations $a_t^e(\boldsymbol{\beta})$, $t = 1, \ldots, N$, do not succeed.

Unfortunately, for AR(p) models, because of the propagation of outliers, regression estimators such as the LAD, Rank-LAD, M, LTS, S, MM, or τ, which have been discussed in Chapter 2, have a BP that is inversely proportional to $p+1$. For ARMA models with $q > 0$, such as the MA(1) model or the ARMA(1,1) model that are illustrated in Figure 8.5c and 8.5d, the situation is even worse. Here, a single RO or AO can effect all subsequent reconstructed innovations $a_t^e(\boldsymbol{\beta})$. Thus, for ARMA models with $q > 0$, BP = 0.

Two approaches to deal with the propagation of outliers have been proposed in the literature:

1.   The use of robust filters
2.   Use of the bounded influence propagation model

These approaches are discussed in the following sections.

8.2.1   Robust Approximate Conditional Mean Type Filters

The first approach to prevent the propagation of outliers is to combine robust estimators with *ACM type filters*; see Masreliez (1975), Martin and Thomson (1982), Spangl and Dutter (2007), and Maronna et al. (2006). For ARMA(p,q) models, the state equation relates the unobservable r-dimensional state vector $\mathbf{x}_t$ of dimension $r = \max(p, q+1) \times 1$ to the previous state $\mathbf{x}_{t-1}$ via the state transition matrix $\boldsymbol{\Phi}$:

$$\mathbf{x}_t = \boldsymbol{\Phi}\mathbf{x}_{t-1} + \boldsymbol{\theta}a_t, \tag{8.13}$$

where

$$\boldsymbol{\Phi} = \begin{bmatrix} \boldsymbol{\phi}_{r-1} & \mathbf{I}_{r-1} \\ \phi_r & \mathbf{0}_{r-1} \end{bmatrix} \tag{8.14}$$

with $\boldsymbol{\phi}_{r-1} = (\phi_1, \ldots, \phi_i, \ldots, \phi_{r-1})^\top$, $\mathbf{I}_{r-1}$ denoting the $(r-1)$-dimensional identity matrix, $\mathbf{0}_{r-1}$ being the $(r-1)$-dimensional all zeros row-vector, and $\phi_i = 0$ for $i > p$. Furthermore, $\boldsymbol{\theta} = (1, -\theta_1, -\theta_2, \ldots, -\theta_{r-1})^\top$ with $\theta_r = 0$ for $r > q$ in case $p > q$.

The measurement equation represents observations, which, depending on the application, may follow, for example, the AO model (8.8)

$$y_t^\varepsilon = y_t + \xi_t^\varepsilon w_t,$$

or the RO model (8.10)

$$y_t^\varepsilon = (1 - \xi_t^\varepsilon)y_t + \xi_t^\varepsilon w_t$$

with $t = 1, \ldots, N$. To avoid mathematical complexity, it is assumed in this section that $\mu = 0$. In practice, a robust location estimate, as discussed in Section 1.2.1 can be subtracted. The case when $\mu \neq 0$ is discussed, for example, in Maronna et al. (2006).

The state recursion of the ACM filter is given by

$$\hat{\mathbf{x}}_{t|t} = \boldsymbol{\Phi}\hat{\mathbf{x}}_{t-1|t-1} + \frac{\boldsymbol{\sigma}_{1,t}}{\sigma_t} \, \psi\left(\frac{y_t^\varepsilon - \hat{y}_{t|t-1}}{\sigma_t}\right), \tag{8.15}$$

where $\hat{y}_{t|t-1}$ is the first element of the predicted state vector

$$\hat{\mathbf{x}}_{t|t-1} = \boldsymbol{\Phi}\hat{\mathbf{x}}_{t-1|t-1},$$

and $\boldsymbol{\sigma}_{1,t}$ denotes the first column of the prediction error covariance matrix

$$\boldsymbol{\Sigma}_t = \mathsf{E}\left[(\hat{\mathbf{x}}_{t|t-1} - \mathbf{x}_t)(\hat{\mathbf{x}}_{t|t-1} - \mathbf{x}_t)^\top\right], \tag{8.16}$$

and $\psi\,(\cdot)$ is often chosen as a redescending function (Martin and Thomson, 1982; Spangl and Dutter, 2007), for example (1.23). The one-step prediction error variance σ_t^2 is the first element of $\boldsymbol{\sigma}_1$. Furthermore, $\hat{y}_{t|t-1}$ is the robust one-step ahead prediction of y_t.

The robustly filter-cleaned process $y_t^f(\boldsymbol{\beta}, \sigma_t)$ at time t is the first element of $\hat{\mathbf{x}}_{t|t}$. Filtered robust estimates are obtained by using the filtered innovations, which are recursively obtained for $t \geq p + 1$ by

$$d_t^f(\boldsymbol{\beta}, \sigma_t) = y_t - \sum_{i=1}^{p} \phi_i y_{t-i}^f(\boldsymbol{\beta}, \sigma_t) + \sum_{i=1}^{q} \theta_i \sigma_t \psi\left(\frac{d_{t-i}^f(\boldsymbol{\beta}, \sigma_t)}{\sigma_t}\right), \tag{8.17}$$

instead of the innovations $a_t^e(\boldsymbol{\beta})$ from (8.6), in the estimating equation (Maronna et al., 2006).

The ACM filter state prediction error covariance recursions is computed via

$$\boldsymbol{\Sigma}_t = \boldsymbol{\Phi}\boldsymbol{P}_{t-1}\boldsymbol{\Phi} + \sigma_a^2\boldsymbol{\theta}\boldsymbol{\theta}^\top \tag{8.18}$$

where the state filtering error covariance recursion $\boldsymbol{P}_{t-1}$ is recursively obtained from

$$\boldsymbol{P}_{t-1} = \boldsymbol{\Sigma}_{t-1} - \frac{1}{\sigma_{t-1}^2}W\left(\frac{a_{t-1}^f(\boldsymbol{\beta},\sigma_{t-1})}{\sigma_{t-1}}\right)\boldsymbol{\sigma}_{1,t-1}\boldsymbol{\sigma}_{1,t-1}^\top, \tag{8.19}$$

and σ_a^2 is the variance of the innovations sequence, which is unknown, and must be estimated (Maronna et al., 2006). The weights $W(\cdot)$ are defined in the same manner as in the location estimation problem (1.24) and $\boldsymbol{\sigma}_{1,t-1}$ denotes the first column of $\boldsymbol{\Sigma}_{t-1}$. The first element of this vector is σ_{t-1}. The initial value for the state recursion of the ACM filter can be chosen as

$$\hat{\mathbf{x}}_{0|0} = (0,\ldots,0)^\top.$$

For the initial value $\boldsymbol{P}_0$ of the state filtering error covariance, a robust covariance estimate (see Chapter 4) of

$$\begin{pmatrix} y_1^\varepsilon & y_2^\varepsilon & \cdots & y_p^\varepsilon \\ y_2^\varepsilon & y_3^\varepsilon & \cdots & y_{p+1}^\varepsilon \\ \vdots & \vdots & & \vdots \\ y_{N-p+1}^\varepsilon & y_{N-p+2}^\varepsilon & \cdots & y_N^\varepsilon \end{pmatrix}$$

can be used.

Figure 8.6 gives an example of the filtered innovations that have been computed, for $t \geq p+1$, via (8.17) for the previously discussed AR(6) with $\phi = (-.4, 0.05, 0.1, 0.2, 0.3, -.4)^\top$. Due to the ACM filter, the single RO of amplitude 50 at $t = 25$ affects only one innovation sample, that is, $a_{25}^f(\phi,\sigma_t)$. This is very different from the case when the innovations are reconstructed from (8.6), where the single RO affected $p+1$ innovations samples, $a_{25}^e(\phi),\ldots,a_{31}^e(\phi)$, as shown in Figure 8.5b. This example demonstrates that robust filters suppress the propagation of outliers.

Robust filters can be coupled with any robust estimator that is based on the innovations by replacing $a_t^e(\boldsymbol{\beta})$ with the filtered innovations $a_t^f(\boldsymbol{\beta},\sigma_t)$ in the estimating equation. A disadvantage of the filtered estimators is that they are intractable in terms of robustness and asymptotic statistical analysis.

8.2.2   Bounded Influence Propagation Model

Muler et al. (2009) introduced an alternative approach for preventing the propagation of outliers in ARMA parameter estimation. They proposed the use of an auxiliary model

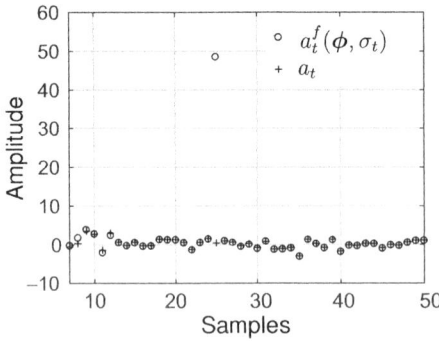

Figure 8.6 Example of the filtered innovations computed, for $t \geq p + 1$, via (8.17) for the previously discussed AR(6) with $\phi = (-.4, 0.05, 0.1, 0.2, 0.3, -.4)^\top$. Due to the ACM filter, the single RO of amplitude 50 at $t = 25$ affects only one innovation sample.

that inherently prevents the propagation of outliers. It is called the bounded influence propagation (BIP)-ARMA model (Muler et al., 2009) and is defined as

$$y_t = \mu + \sum_{i=1}^{p} \phi_i(y_{t-i} - \mu) + a_t - \sum_{i=1}^{r} \left(\phi_i a_{t-i} + (\theta_i - \phi_i)\sigma \eta \left(\frac{a_{t-i}}{\sigma} \right) \right). \quad (8.20)$$

Here, $r = \max(p, q)$, if $r > p$, $\phi_{p+1} = \ldots = \phi_r = 0$, and if $r > q$, $\theta_{q+1} = \ldots = \theta_r = 0$. ARMA models are included by setting $\eta(x) = x$, while the propagation of outliers is prevented by choosing $\eta(x)$ to be one of the well-known robust score functions, for example, Tukey's (1.23). For such a choice, all innovations that lie within some region around μ are left untouched and, the effect of a single AO or RO is bounded to a single corrupted innovation.

In (8.20), σ is a robust M-scale of a_t (see Section 1.2.1), that is, it solves

$$\mathsf{E} \left[\psi \left(\frac{a_t}{\sigma} \right) \cdot \left(\frac{a_t}{\sigma} \right) \right] = b, \quad (8.21)$$

where, as for (1.26), the positive constant b must satisfy $0 < b < \rho(\infty)$ and is chosen as $\mathsf{E}[\rho(u)]$, where u is a standard normal random variable to achieve consistency with the Gaussian distribution. Examples for $\rho(x)$ are given in (1.20) and (1.22).

From (8.20), the innovations sequence can be recursively obtained, for $t \geq p + 1$, according to

$$a_t^b(\boldsymbol{\beta}, \sigma) = y_t - \mu - \sum_{i=1}^{p} \phi_i(y_{t-i} - \mu) + \sum_{i=1}^{r} \left(\phi_i a_{t-i}^b(\boldsymbol{\beta}, \sigma) + (\theta_i - \phi_i)\sigma \eta \left(\frac{a_{t-i}^b(\boldsymbol{\beta}, \sigma)}{\sigma} \right) \right). \quad (8.22)$$

Figure 8.7 gives an example of the innovations that have been computed, for $t \geq p + 1$, via (8.22) for the previously discussed AR(6) with $\phi = (-.4, 0.05, 0.1, 0.2, 0.3, -.4)^\top$. Due to the BIP-AR model, the single RO of amplitude 50 at $t = 25$ affects only one innovation sample, that is, $a_{25}^b(\phi, \sigma)$. This is very different from the case when the innovations are reconstructed from (8.6), as shown in Figure 8.5b, and it is similar to the filtered innovations (cf. Figure 8.6).

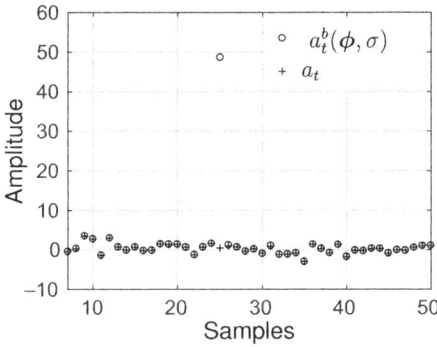

Figure 8.7 Example of the innovations computed, for $t \geq p + 1$, via (8.22) for the previously discussed AR(6) with $\boldsymbol{\phi} = (-.4, 0.05, 0.1, 0.2, 0.3, -.4)^\top$. Due to the BIP-AR model, the single RO of amplitude 50 at $t = 25$ affects only one innovation sample.

Similarly to robust filters, the BIP-ARMA model can be combined with any robust estimator, which is based on the innovations, by replacing $a_t^e(\boldsymbol{\beta})$ by the BIP-ARMA innovations $a_t^b(\boldsymbol{\beta}, \sigma_t)$ in the estimating equation. To compute $\sigma \equiv \sigma(\boldsymbol{\beta})$ in (8.22), the MA-infinity representation of the BIP-ARMA model is used:

$$
y_t = \mu - a_t + \sum_{i=1}^{\infty} \lambda_i(\boldsymbol{\beta}) \sigma \eta \left(\frac{a_{t-i}}{\sigma} \right),
\tag{8.23}
$$

where $\lambda_i(\boldsymbol{\beta})$ are the coefficients of $\phi^{-1}(B)\theta(B)$ as defined in (8.3) and (8.4). It then follows that

$$
\sigma^2(\boldsymbol{\beta}) = \frac{\sigma_y^2}{1 + \kappa^2 \sum_{i=1}^{\infty} \lambda_i^2(\boldsymbol{\beta})},
\tag{8.24}
$$

where σ_y is the standard deviation of y_t and

$$
\kappa^2 = \mathsf{var}\left[\eta \left(\frac{a_t}{\sigma} \right) \right] = \mathsf{E}\left[\left(\eta \left(\frac{a_t}{\sigma} \right) - \mathsf{E}\left[\eta \left(\frac{a_t}{\sigma} \right) \right] \right)^2 \right].
\tag{8.25}
$$

The estimate of σ in (8.24) can then be computed according to

$$
\hat{\sigma}^2(\boldsymbol{\beta}) = \frac{\hat{\sigma}_y^2}{1 + \kappa^2 \sum_{i=1}^{q_{\text{long}}} \lambda_i^2(\boldsymbol{\beta})},
\tag{8.26}
$$

where q_{long} is chosen sufficiently large to approximate the MA-infinity representation. The scale estimate $\hat{\sigma}_y$ is a robust scale estimate of $\mathbf{y}_N = (y_1, \ldots, y_N)^\top$, for example, an M-scale as defined in Section 1.2.1.

Statistical and robustness analysis is possible for BIP-ARMA-based estimators. For example, the consistency and asymptotic normality have been established for BIP-M, BIP-S, and BIP-MM estimators in Muler et al. (2009). For the BIP-τ estimator the consistency and asymptotic normality have been established, and the IF has been computed for AR(1) models with AOs, in Muma and Zoubir (2017).

8.3   An Overview of Robust Autoregressive Moving-Average Parameter Estimators

Research on robust ARMA parameter estimation may be loosely grouped into two categories and based on the approach taken: the diagnostic approach, for example, Tsay (1988), Deutsch et al. (1990), Ljung (1993), McQuarrie and Tsai (2003), Louni (2008), and Dehling et al. (2015), and the statistically robust approach, for example, de Luna and Genton (2001), Maronna et al. (2006), Muler et al. (2009), Zoubir et al. (2012), Chakhchoukh et al. (2009, 2010), Chakhchoukh (2010a), Spangl and Dutter (2007), Andrews (2008), Shariati et al. (2014), Dürre et al. (2015), and Muma and Zoubir (2017). Diagnostic approaches enhance robustness via detection and hard rejection of outliers, followed by a classical parameter estimation method that handles missing values. Statistically robust methods utilize the entire data set and accommodate the outliers by bounding their influence on the ARMA parameter estimates. Robust statistical theory also provides measures, such as the IF, the BP, and the maximum bias curve, which have been discussed for i.i.d. data in previous chapters. These measures characterize quantitative and qualitative robustness and allow for an analytical comparison of different estimators. However, as mentioned earlier, quantifying robustness in the dependent data setting is challenging when compared to the i.i.d. case. This section focuses on robust ARMA parameter estimators that use robust ACM type filters or BIP models, and briefly discusses robust autocorrelation-based estimators.

8.3.1   *M*-Estimation

In the case of ARMA models, an *M*-estimate is obtained from

$$\hat{\boldsymbol{\beta}}_{\mathrm{M}} = \arg\min_{\boldsymbol{\beta}} \sum_{t=p+1}^{N} \rho\left(\frac{a_t^e(\boldsymbol{\beta})}{\hat{\sigma}(\boldsymbol{a}_N(\boldsymbol{\beta}))}\right). \tag{8.27}$$

Here, $\hat{\sigma}(\boldsymbol{a}_N(\boldsymbol{\beta}))$ is a robust *M*-scale estimate of $\boldsymbol{a}_N(\boldsymbol{\beta}) = (a_{p+1}^e(\boldsymbol{\beta}), \ldots, a_N^e(\boldsymbol{\beta}))^{\top}$, as defined by (1.19). It is assumed that $\rho(x)$ is a real-valued function with the following properties: $\rho(0) = 0, \rho(x) = \rho(-x)$, and $\rho(x)$ is continuous, nonconstant, and nondecreasing with respect to $|x|$. If the derivative $\psi(x) = \frac{d\rho(x)}{dx}$ exists, the *M*-estimating equation is given by

$$\sum_{t=p+1}^{N} \psi\left(\frac{a_t^e(\boldsymbol{\beta})}{\hat{\sigma}(\boldsymbol{a}_N(\boldsymbol{\beta}))}\right) \nabla a_t^e(\boldsymbol{\beta}) = \mathbf{0} \tag{8.28}$$

with

$$\nabla a_t^e(\boldsymbol{\beta}) \triangleq \left(\frac{\partial a_t^e(\boldsymbol{\beta})}{\partial \phi_i}, \frac{\partial a_t^e(\boldsymbol{\beta})}{\partial \theta_i}, \frac{\partial a_t^e(\boldsymbol{\beta})}{\partial \mu}\right)^{\top} \tag{8.29}$$

where

$$\frac{\partial a_t^e(\boldsymbol{\beta})}{\partial \phi_i} = -\theta^{-1}(B)(y_{t-i} - \mu) = -\phi^{-1}(B)a_{t-i}, \quad 1 \le i \le p, \tag{8.30}$$

$$\frac{\partial a_t^e(\boldsymbol{\beta})}{\partial \theta_i} = -\theta^{-2}(B)\phi(B)(y_{t-i} - \mu) = \theta^{-1}(B)a_{t-i}, \quad 1 \le i \le q, \tag{8.31}$$

and

$$\frac{\partial a_t^e(\boldsymbol{\beta})}{\partial \mu} = -\frac{1 - \sum_{i=1}^{p} \phi_i}{1 - \sum_{i=1}^{q} \theta_i}. \tag{8.32}$$

For the AR(p) model, the M-estimator is defined according to (8.27) by

$$\hat{\boldsymbol{\phi}}_{\mathrm{M}} = \arg\min_{\boldsymbol{\phi}} \sum_{t=p+1}^{N} \rho\left(\frac{a_t^e(\boldsymbol{\phi})}{\hat{\sigma}(\boldsymbol{a}_N(\boldsymbol{\phi}))}\right), \tag{8.33}$$

where, from (8.6), with $\mu = 0$ and $q = 0$, it follows, for $t = p + 1, \ldots, N$, that

$$a_t^e(\boldsymbol{\phi}) = y_t - \sum_{i=1}^{p} \phi_i y_{t-i} \tag{8.34}$$

and $\hat{\sigma}(\boldsymbol{a}_N(\boldsymbol{\phi}))$ is a robust M-scale estimate of $\boldsymbol{a}_N(\boldsymbol{\phi}) = (a_{p+1}^e(\boldsymbol{\phi}), \ldots, a_N^e(\boldsymbol{\phi}))^\top$, as defined by (1.19).

As discussed in Section 8.2, due to the propagation of outliers, the BP of an M-estimator is not larger than $1/(p + 1)$ for an AR(p) process and is equal to zero for ARMA(p,q) models with $q > 0$. So-called filtered M-estimators (Martin et al., 1982) incorporate an ACM-type filter based on initial estimates to clean the observations via

$$\hat{y}_t = y_t^{\varepsilon} - d_t^f(\boldsymbol{\beta}, \sigma_t) + \sigma_t \psi\left(\frac{d_t^f(\boldsymbol{\beta}, \sigma_t)}{\sigma_t}\right) \tag{8.35}$$

before applying a nonlinear LS solver to iteratively find a solution to an M-estimating equation. However, no robustness and asymptotic statistical analysis exists for filtered M-estimators. Similarly, BIP M-estimators are defined by evaluating (8.27) where $a_t^e(\boldsymbol{\beta})$ is replaced by $a_t^b(\boldsymbol{\beta}, \sigma)$, which have been evaluated from (8.22). BIP M-estimators are consistent and asymptotically normal (Muler et al., 2009).

8.3.2   *S*-Estimation

In the case of ARMA models, an S-estimate is obtained from

$$\hat{\boldsymbol{\beta}}_{\mathrm{s}} = \arg\min_{\boldsymbol{\beta}} \hat{\sigma}(\boldsymbol{a}_N(\boldsymbol{\beta})), \tag{8.36}$$

where $\boldsymbol{a}_N(\boldsymbol{\beta}) = (a_{p+1}^e(\boldsymbol{\beta}), \ldots, a_N^e(\boldsymbol{\beta}))^\top$ and $\hat{\sigma}(\cdot)$ is an M-scale as defined in (1.19).

Filtered S-estimators are computed by evaluating (8.36), but with $a_t^e(\boldsymbol{\beta})$ replaced by $d_t^f(\boldsymbol{\beta}, \sigma_t)$, which have been computed from (8.17). Similarly, BIP S-estimators are defined by evaluating (8.36) where $a_t^e(\boldsymbol{\beta})$ is replaced by $a_t^b(\boldsymbol{\beta}, \sigma)$, which have been evaluated from (8.22). BIP S-estimators are consistent and asymptotically normal (Muler et al., 2009). As for the case that was discussed in Section 1.3, maximizing $\mathsf{ARE}\left(\hat{\boldsymbol{\beta}}_{\mathrm{s}}\right)$ conflicts with maximizing $\mathsf{GES}\left(\hat{\boldsymbol{\beta}}_{\mathrm{s}}\right)$. S-estimators trade-off robustness against outliers, and efficiency under a Gaussian ARMA model, by tuning the ρ function that is used in the M-scale.

8.3.3   MM-Estimation

In the case of ARMA models, an MM-estimate is obtained from

$$\hat{\boldsymbol{\beta}}_{MM} = \arg\min_{\boldsymbol{\beta}} \sum_{t=p+1}^{N} \rho\left(\frac{a_t^e(\boldsymbol{\beta})}{\hat{\sigma}(\boldsymbol{a}_N(\boldsymbol{\beta}_s))}\right). \tag{8.37}$$

The MM-estimate is therefore an M-estimate where the previously computed M-scale $\hat{\sigma}(\boldsymbol{a}_N(\boldsymbol{\beta}_s))$ is based on the innovations $\boldsymbol{a}_N(\boldsymbol{\beta}_s)$ of an S-estimator. Filtered-MM and BIP MM-estimators are obtained by replacing a_t^e by d_t^f and a_t^b, respectively. BIP MM-estimators are consistent and asymptotically normal (Muler et al., 2009) and can be simultaneously efficient and highly robust. As for the regression case, MM-estimators require a two-step procedure, and improve upon the highly robust initial estimate $\hat{\boldsymbol{\beta}}_s$ in terms of efficiency. This is done by using two different ρ functions. In the first step, to compute $\hat{\boldsymbol{\beta}}_s$, the ρ function in (8.36) is tuned for robustness. The ρ function in (8.37) that is used for the second (M-estimation) step is tuned for efficiency. For example,

$$\rho_2(x) = \begin{cases} 0.5x^2 & \text{if } |x| \le 2 \\ 0.002x^8 - 0.052x^6 + 0.432x^4 - 0.972x^2 + 1.792 & \text{if } 2 < |x| \le 3 \\ 3.25 & |x| > 3, \end{cases} \tag{8.38}$$

and

$$\rho_1(x) = \rho_2(x/c_1), \tag{8.39}$$

with $c_1 = 0.4050$.

8.3.4 τ-Estimation

In the case of ARMA models, a τ-estimate is obtained from

$$\hat{\boldsymbol{\beta}}_\tau = \arg\min_{\boldsymbol{\beta}} \hat{\sigma}_\tau(\boldsymbol{a}_N(\boldsymbol{\beta})), \tag{8.40}$$

where $\hat{\sigma}_\tau(\boldsymbol{a}_N(\boldsymbol{\beta}))$ is the robust and efficient τ-estimate (Yohai and Zamar, 1988) of the scale of the innovations $\boldsymbol{a}_N(\boldsymbol{\beta}) = (a_{p+1}^e(\boldsymbol{\beta}), \dots, a_N^e(\boldsymbol{\beta}))^\top$ and is defined as

$$\hat{\sigma}_\tau(\boldsymbol{a}_N(\boldsymbol{\beta})) = \hat{\sigma}(\boldsymbol{a}_N(\boldsymbol{\beta}))\sqrt{\frac{1}{N-p} \sum_{t=p+1}^{N} \rho_2\left(\frac{a_t^e(\boldsymbol{\beta})}{\hat{\sigma}(\boldsymbol{a}_N(\boldsymbol{\beta}))}\right)}. \tag{8.41}$$

Here, $\hat{\sigma}(\boldsymbol{a}_N(\boldsymbol{\beta}))$ is an M-estimate of the scale of $\boldsymbol{a}_N(\boldsymbol{\beta})$, which is based on $\rho_1(x)$, where $\rho_1(x)$ is tuned for robustness. The function $\rho_2(x)$ is tuned for efficiency and must, additionally, satisfy the constraint of $2\rho_2(x) - \psi_2(x)x \ge 0$, where $\psi_2(x) = \frac{d\rho_2(x)}{dx}$. For example, ρ_1 and ρ_2 can be chosen consistent with (8.39) and (8.38). Similar to the

MM-estimator, for an appropriate choice of ρ_1 and ρ_2, the τ-estimator becomes both highly robust and efficient.

As shown in Muma and Zoubir (2017), the τ-estimator of the ARMA parameters satisfies an M-estimating equation. Differentiating (8.40) yields the following system of equations:

$$
\nabla \hat{\sigma}_\tau (a_N(\beta))^2 = 2\hat{\sigma}(a_N(\beta))\nabla\hat{\sigma}(a_N(\beta))\frac{1}{N-p}\sum_{t=p+1}^{N}\rho_2\left(\frac{a_t^e(\beta)}{\hat{\sigma}(a_N(\beta))}\right)
$$

$$
+\frac{1}{N-p}\sum_{t=p+1}^{N}\psi_2\left(\frac{a_t^e(\beta)}{\hat{\sigma}(a_N(\beta))}\right)
$$

$$
\times\left(\nabla a_t^e(\beta)\hat{\sigma}(a_N(\beta)) - a_t^e(\beta)\nabla\hat{\sigma}(a_N(\beta))\right) = \mathbf{0}.
$$

$$(8.42)$$

Here,

$$
\nabla\hat{\sigma}(a_N(\beta)) = -\hat{\sigma}(a_N(\beta))\frac{\sum_{t=p+1}^{N}\psi_1\left(\frac{a_t^e(\beta)}{\hat{\sigma}(a_N(\beta))}\right)\nabla a_t^e(\beta)}{\sum_{t=p+1}^{N}\psi_1\left(\frac{a_t^e(\beta)}{\hat{\sigma}(a_N(\beta))}\right)a_t^e(\beta)}
$$

$$(8.43)$$

with $\nabla a_t^e(\beta)$ defined in (8.29). Substituting (8.43) in (8.42) and defining

$$
W(\beta) = \frac{\sum_{t=p+1}^{N}2\rho_2\left(\frac{a_t^e(\beta)}{\hat{\sigma}(a_N(\beta))}\right) - \psi_2\left(\frac{a_t^e(\beta)}{\hat{\sigma}(a_N(\beta))}\right)\frac{a_t^e(\beta)}{\hat{\sigma}(a_N(\beta))}}{\sum_{t=p+1}^{N}\psi_1\left(\frac{a_t^e(\beta)}{\hat{\sigma}(a_N(\beta))}\right)\frac{a_t^e(\beta)}{\hat{\sigma}(a_N(\beta))}},
$$

$$(8.44)$$

the τ-estimate satisfies an M-estimating equation

$$
\sum_{t=p+1}^{N}\psi_\tau\left(\frac{a_t^e(\beta)}{\hat{\sigma}(a_N(\beta))}\right)\nabla a_t^e(\beta) = \mathbf{0},
$$

$$(8.45)$$

with data-adaptive ψ_τ given by

$$
\psi_\tau(x) = W(\beta)\psi_1\left(\frac{a_t^e(\beta)}{\hat{\sigma}(a_N(\beta))}\right) + \psi_2\left(\frac{a_t^e(\beta)}{\hat{\sigma}(a_N(\beta))}\right),
$$

$$(8.46)$$

as long as $\rho_2(x)$ satisfies the inequality $2\rho_2(x) - \psi_2(x)x \geq 0$. Special cases are (i) $\rho_2(x) = 1/2\ x^2$, which results in $W(\beta) = 0$ and the τ-estimator being equivalent to an LSE, and (ii) $\rho_1(x) = \rho_2(x)$, which results in the τ-estimator being equivalent to an S-estimator.

8.3.5 Robust Autocorrelation-Based Estimators

The so-called ratio-of-medians (RM) and the median-of-ratios (MR) estimators (Chakhchoukh, 2010b; Zoubir et al., 2012) use robust autocorrelation estimates, based on sample medians, to estimate the parameters of an AR model.

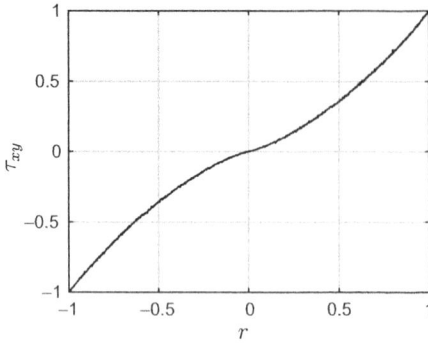

Figure 8.8 An explicit relation between τ_{xy} and r does not seem to exist; it is found numerically. A MATLAB© code is provided in the RobustSP toolbox.

Consider a zero-mean Gaussian random vector $(x, y) \in \mathbb{R}^{1 \times 2}$ that is defined by its pdf

$$f(x, y) = \frac{1}{2\pi\sigma^2\sqrt{1-r^2}} e^{\frac{-(x^2+y^2-2rxy)}{2\sigma^2(1-r^2)}}, \tag{8.47}$$

where $\sigma^2 = \sigma_x^2 = \sigma_y^2$ is the variance, and r is the (Pearson) correlation between x and y. Let $(\mathbf{x}_N, \mathbf{y}_N) \in \mathbb{R}^{N \times 2}$ be a sample of observations that were generated consistent with (8.47). Then, the RM is defined as

$$\tau_{xy} = \frac{\text{med}(\mathbf{x}_N \circ \mathbf{y}_N)}{\text{med}(\mathbf{x}_N \circ \mathbf{x}_N)}, \tag{8.48}$$

where $\text{med}(\cdot)$ is the sample median as defined in (1.12) and $\circ$ is the Hadamard product (also known as the Schur product or the entrywise product). Although an explicit relation between τ_{xy} and r does not seem to exist, it is easily found numerically. The MATLAB© code to evaluate the relation between τ_{xy} and r is provided in the RobustSP toolbox and yields the graph that is displayed in Figure 8.8.

For an AR(p) process, consider the following RM

$$\tau_{yy}(k) = \frac{\text{med}(\mathbf{y}_N \circ \mathbf{y}_{N-k})}{\text{med}(\mathbf{y}_N \circ \mathbf{y}_N)}, \quad k = 1, \ldots, p, \tag{8.49}$$

where, $\mathbf{y}_N \triangleq (y_{k+1}, y_{k+2}, \ldots, y_N)^\top$, $\mathbf{y}_{N-k} \triangleq (y_1, y_2, \ldots, y_{N-k})^\top$ follow an AR(p) model. A robust autocorrelation estimate $\hat{r}_{yy}(k), k = 1, \ldots, p$, can then be computed from $\tau_{yy}(k)$ by using the relation shown in Figure 8.8.

The *RM estimate* $\hat{\boldsymbol{\beta}}_{\text{RM}}$ is obtained from $(\hat{r}_{yy}(0), \hat{r}_{yy}(1), \ldots, \hat{r}_{yy}(p))^\top$ by solving the Yule–Walker equations

$$\begin{pmatrix} \hat{r}_{yy}(0) & \hat{r}_{yy}(-1) & \cdots & \hat{r}_{yy}(-p+1) \\ \hat{r}_{yy}(1) & \hat{r}_{yy}(0) & \cdots & \hat{r}_{yy}(p-2) \\ \vdots & \vdots & \ddots & \vdots \\ \hat{r}_{yy}(p-1) & \hat{r}_{yy}(p-2) & \cdots & \hat{r}_{yy}(0) \end{pmatrix} \begin{pmatrix} \phi_1 \\ \phi_2 \\ \vdots \\ \phi_p \end{pmatrix} = \begin{pmatrix} \hat{r}_{yy}(1) \\ \hat{r}_{yy}(2) \\ \vdots \\ \hat{r}_{yy}(p) \end{pmatrix} \tag{8.50}$$

with $\hat{r}_{yy}(0) = 1$. These equations can be efficiently solved via the well-known Levinson–Durbin recursion.

Let $\mathbf{Y}_{N-k} \in \mathbb{R}^{(N-k)\times(N-k)}$ be a diagonal matrix with entries $1/y_1, 1/y_2, \dots, 1/y_{N-k}$. Then the MR provides a robust autocorrelation estimate using

$$\hat{r}_{yy}(k) = \mathsf{med}(\mathbf{Y}_{N-k} \cdot \mathbf{y}_N), \quad k = 1, \dots, p \tag{8.51}$$

and the *MR estimate* $\hat{\boldsymbol{\beta}}_{\mathrm{MR}}$ is obtained from $(1, \hat{r}_{yy}(1), \dots, \hat{r}_{yy}(p))^{\top}$ by solving (8.50).

MATLAB$^{©}$ code to compute RM and MR estimators is provided in the RobustSP toolbox. Strong consistency and asymptotic normality have been established for a Gaussian AR process. Measures of robustness, such as the maximum bias, BP, and IF have been derived for an AR(1) model (Chakhchoukh, 2010b). In Chakhchoukh et al. (2009), these estimators were successfully applied for short-term load forecasting by coupling the estimator with a robust ACM-type filter cleaner that rejects outlying observations. Subsequent processing allows the estimation of the parameters of a seasonal integrated ARMA (SARIMA) model.

8.3.6   Other Estimators

In Chakhchoukh et al. (2010), estimators have been proposed that minimize a robust efficient *minimum Hellinger distance estimator* of scale of the innovations sequence. To mitigate the propagation of outliers, robustly filtered innovations, based on (8.17), were considered. Alternatively, BIP-ARMA innovations (8.22) could be used.

Bustos and Yohai (1986) proposed estimators, called *robust autocovariance (RA) estimators*, which are based on a RA estimate of the innovations sequence. RA estimators are consistent and asymptotically normal. However, as the innovations sequence is derived from the ARMA model, it suffers from the propagation of outliers. RA estimators based on (8.17) or (8.22) could provide increased robustness, especially for large values of p or $q > 0$, for which the BP of the RA estimators becomes zero.

8.4   Robust Model Order Selection

In practical applications, the orders of the ARMA model are unknown. Robust model order selection requires finding a suitable statistical model to describe the majority of the data while preventing outliers, or other contaminants, from having overriding influence on the final conclusions. The task of model order selection for ARMA processes is to choose appropriate values for p and q given a finite set of observations. This can be done by use of the information criteria (IC) (Stoica and Selen, 2004), which provides a trade-off between data fit and model complexity according to

$$\mathrm{IC}(p,q) = \underbrace{-2\,L\!\left(p,q,\hat{\boldsymbol{\beta}}\,\big|\mathbf{y}\right)}_{\text{data fit}} + \underbrace{\alpha(p,q)}_{\text{model complexity penalty}} \tag{8.52}$$

Table 8.1 The penalty terms of some well-known IC.

Criterion	Penalty
AIC	$\frac{2(p+q)}{N}$
AICc	$\frac{2(p+q)}{N-(p+q)-2}$
BIC	$\frac{p+q}{N} \ln(N)$
HQ	$2(p+q)\frac{\ln(\ln(N))}{N}$

for a set of candidate model orders $p = 0, 1, \ldots, p_{max}$ and $q = 0, 1, \ldots, q_{max}$. Here, $L\left(p, q, \hat{\beta} \middle| \mathbf{y}\right)$ is the log-likelihood function, given estimates $\hat{\beta} = (\hat{\phi}, \hat{\theta})$ of the ARMA parameters based on $\mathbf{y} = (y_1, \ldots, y_N)^{\top}$. Different penalty terms of some well-known IC are listed in Table 8.1. Here, AIC stands for Akaike's criterion, AICc includes a small sample correction factor, BIC is the Bayesian information criterion, and HQ is the criterion by Hannan and Quinn (1979).

Robust ARMA model order selection, in the presence of AOs, has not received significant attention in the robust statistics community with the notable exceptions of Le et al. (1996), Ronchetti (1997), Shi and Tsai (1998), and Maronna et al. (2006) for the case of AR processes and Agostinelli (2004) and Muma (2014) for general ARMA processes. Under the Gaussian assumption the IC (8.52) becomes

$$\mathrm{IC}_{\hat{\sigma}}(p, q) = \underbrace{N \ln(\hat{\sigma}^2(a_N(\hat{\beta}(p, q))))}_{\text{data fit}} + \underbrace{\alpha(p, q)}_{\text{model complexity penalty}} \tag{8.53}$$

and a straightforward approach to robust model order selection consists of replacing the non-robust standard deviation and parameter estimates by their robust counterparts. To prevent the propagation of outliers, innovations based on (8.17) or (8.22) are required. For example, in Muma (2014), model order selection based on BIP-τ estimation was proposed according to

$$\mathrm{IC}_{\hat{\sigma}_\tau}(p, q) = N\ln(\hat{\sigma}_\tau^2(a_N^b(\hat{\beta}(p, q), \hat{\sigma}(p, q)))) + \alpha(p, q), \tag{8.54}$$

where, $\hat{\sigma}_\tau(a_N^b(\beta(p, q), \sigma(p, q)))$ is the τ-scale estimate (8.41) of the innovations series obtained by minimizing the τ-estimator's objective function (8.40) based on (8.22) for candidate model orders $p = 0, 1, \ldots, p_{max}$ and $q = 0, 1, \ldots, q_{max}$. The scale parameter $\sigma(p, q)$ of the BIP model can be estimated using (8.26). Possible choices for $\alpha(p, q)$ are provided in Table 8.1.

A second possibility is to derive IC based on a robust likelihood $L_{\mathrm{rob}}\left(p, q, \hat{\beta} \middle| \mathbf{y}\right)$ (Le et al., 1996; Agostinelli, 2004; Muma, 2014):

$$\mathrm{IC}_{L_{\mathrm{rob}}}(p, q) = \underbrace{-2\, L_{\mathrm{rob}}\left(p, q, \hat{\beta} \middle| \mathbf{y}\right)}_{\text{robust data fit}} + \underbrace{\alpha(p, q)}_{\text{model complexity penalty}} \tag{8.55}$$

For BIP-τ estimation, $L_{\text{rob}}\left(p, q, \hat{\boldsymbol{\beta}} \middle| \mathbf{y}\right)$ becomes

$$-\frac{N}{2}\ln(2\pi) - \frac{N}{2}\ln(\hat{\sigma}_\tau^2(\mathbf{a}_N^b(\hat{\boldsymbol{\beta}}(p,q), \hat{\sigma}(p,q)))) - \frac{1}{2}\sum_{t=1}^{N}\rho_\tau\left(\frac{a_t^b(\hat{\boldsymbol{\beta}}(p,q), \hat{\sigma}(p,q))}{\hat{\sigma}_\tau^2(\mathbf{a}_N^b(\hat{\boldsymbol{\beta}}(p,q), \hat{\sigma}(p,q)))}\right),$$
(8.56)

where

$$\rho_\tau(x) = W(\boldsymbol{\beta})\rho_1(x) + \rho_2(x)$$

with $W(\boldsymbol{\beta})$ given in (8.44) and $\rho_1(x)$, $\rho_2(x)$, for example, chosen consistent with (8.39), (8.38). For $\rho_2(x)$, it must hold that $2\rho_2(x) - \psi_2(x)x \geq 0$, where $\psi_2(x) = \frac{d\rho_2(x)}{dx}$.

A slightly different approach was proposed by Agostinelli (2004) who defined a robust Akaike IC for ARMA models via a weighted likelihood. Details on possible choices and calculation of the weights are specified in the paper. The key idea is to compute weights that down-weight the influence of outlying observations via Pearson residuals $\hat{r}_t(p, q)$ and a residual adjustment function $f_{\text{RAF}}(\hat{r}_t(p, q))$ that operates on the Pearson residuals. This allows the creation of different measures of data fit, such as, for example, the squared Hellinger or the Kullback–Leibler measures. For example, the Hellinger residual adjustment function, which is used to measure the "outlyingness" of the innovations samples, is given by

$$f_{\text{RAF}}(\hat{r}_t(p, q)) = 2(\sqrt{(\hat{r}_t(p, q) - 1)} - 1),$$

where $\hat{r}_t(p, q)$ are the Pearson residuals that are defined as the (normalized) difference between a nonparametric kernel density estimate and one that uses a Gaussian model; see Agostinelli (2004). The weights are then obtained via

$$w_t = \min\left(1, \frac{\left[f_{\text{RAF}}(\hat{r}_t(p, q)) + 1\right]^+}{\hat{r}_t(p, q) + 1}\right),$$

where $[\cdot]^+$ indicates the positive part. An extension to BIP-τ estimation is given in Muma (2014).

Robust model order selection based on BIP-τ estimation has been successfully applied to real data applications in biomedicine. In particular, the case of ICP signals, which are contaminated by motion artifacts, see Figure 8.1, is treated in Muma (2014), and robust model order selection for heart-rate variability estimates, which contain outliers due to errors in the R-peak detection of the ECG signal, is detailed in Muma and Zoubir (2017). Section 11.5 discusses an example of robust data cleaning for PPG-based pulse-rate variability analysis, where the model order is determined from (8.54).

8.5   Measures of Robustness

8.5.1   Influence Function for Dependent Data

The IF defines the bias impact of an infinitesimal contamination at an arbitrary point on the estimator, when normalized by the fraction of contamination. There exist two definitions of the IF for the dependent data case (Martin and Yohai, 1986; Künsch, 1984), which differ, but are mathematically related; see Maronna et al. (2006) and Martin and Yohai (1984) for a comparative discussion. The IFs of some robust estimators for the case of an AR(1) or MA(1) have been evaluated theoretically in Martin and Yohai (1986), Chakhchoukh et al. (2009), and Muma and Zoubir (2017); however, no result exists for higher-order AR or MA processes, or for an ARMA(p, q) process with $p \neq 0$ and $q \neq 0$. This is due to the complexity introduced by the correlation, where for an AR(p) process, for example, one must consider the joint distribution of $(y_t, y_{t-1}, \ldots, y_{t-p})$.

For dependent data, the influence functions also change depending on the outlier model, which makes the definition more general than in the i.i.d. case because a contamination process does not have to be represented by a Dirac distribution. It can, for example, be a Gaussian process with a different variance and correlation (Martin and Yohai, 1986; Chakhchoukh et al., 2009). Assume that the observations follow an ARMA model that is contaminated by AOs or ROs as in (8.8) or (8.10). The temporal structure of the outliers may be patchy or i.i.d., depending on the choice of the outlier indicator process ξ_t^ε.

DEFINITION 22   *IF for Dependent Data*
The dependent data IF is defined (Martin and Yohai, 1986) as the directional derivative of the functional representation of an estimator $\mathbf{T} = \hat{\boldsymbol{\beta}}$ *at $F(y)$, that is,*

$$\mathsf{IF}\big(F(w); \mathbf{T}, \{F(y, \xi^\varepsilon, w)\}\big) = \lim_{\downarrow \varepsilon} \frac{\mathbf{T}(F(y^\varepsilon)) - \mathbf{T}(F(y))}{\varepsilon} = \left[\frac{\partial}{\partial \varepsilon} \mathbf{T}(F(y^\varepsilon)) \right]_{\varepsilon=0} \quad (8.57)$$

provided that the limit exists. Here, $F(y)$, $F(w)$, $F(\xi^\varepsilon)$, and $F(y^\varepsilon)$ are the cdfs of y_t, w_t, ξ^ε, and y_t^ε, respectively. Furthermore, $F(y, \xi^\varepsilon, w)$ is the joint distribution of y_t, w_t, ξ^ε.

In general, the IF defined by (8.57) is a functional on a specified distribution space. This is in contrast to the IF for i.i.d. data, for example, (2.81), which is a function on a finite-dimensional space. For dependent data, the IF not only depends upon the estimator $\mathbf{T}$, the nominal distribution $F(y)$, and the contamination process distribution $F(w)$, but also upon the particular trajectory of contamination distributions $\{F(y, \xi^\varepsilon, w)\}$. In fact, the derivative in (8.57) is taken along particular trajectories $\{F(y, \xi^\varepsilon, w)\}$ to $F(y)$ as ε tends to zero.

$\mathsf{IF}(F(w); \mathbf{T}, \{F(y, \xi^\varepsilon, w)\})$ is defined for quite general functionals; however, all subsequent results are for a special class of functionals, so-called $\boldsymbol{\Psi}$ *functionals*, which may be computed as a solution of the estimating equation

$$\int \boldsymbol{\Psi}(\mathbf{y}_t, \mathbf{T}) dF(\mathbf{y}_t) = 0, \quad (8.58)$$

where $F(\mathbf{y}_t)$ is the distribution of the stationary and ergodic random process $\mathbf{y}_t$, as given in $\mathbf{y}_t$. It is assumed that (8.58) either has a unit root, or that a well-defined solution exists in the case of multiple roots. For a detailed discussion of Ψ functionals and the associated influence functionals, the interested reader is referred to Martin and Yohai (1984). The class of estimators that can be represented in a form that is consistent with (8.58) is quite large and contains both classical and robust parameter estimators, for example, the M-estimators, the generalized M-estimator, and estimators based on residual autocovariances (RA-estimators) (Martin and Yohai, 1986).

It was shown in Muma and Zoubir (2017) that the τ-estimators of the ARMA parameters are also of the Ψ-type, that is, they satisfy (8.58). In the following, an example for the IF of the τ-estimator for the particular case of AR(1) models with AOs, is given. Let y_t^ε follow (8.8), with y_t satisfying (8.2) for the case of $p = 1$, $q = 0$, and $\mu = 0$. Furthermore, let ξ_t^ε be an independently distributed $0 - 1$ sequence that is independent of y_t and w_t. If $P(w_t = c_w) = 1$ for a constant c_w, the IF has the appealing heuristic interpretation of displaying the influence of a contamination value c_w on the estimator, that is, $\mathsf{IF}(F(w); \mathbf{T}, \{F(y, \xi^\varepsilon, w)\}) = \mathsf{IF}(F(w), \mathbf{T}, \phi_1)$, in a similar manner to that of Hampel's definition (Hampel, 1974) for i.i.d. data. Under some assumptions as detailed in Muma and Zoubir (2017), the IF of the τ-estimator $\hat{\boldsymbol{\beta}}_\tau$, with associated functional representation $\mathbf{T}_\tau$, is

$$\mathsf{IF}(F(w), \mathbf{T}_\tau, \phi_1) = \frac{(1 - \phi_1^2)^{1/2}}{I_1} \mathsf{E}\left[(y_{t-1} + w_{t-1})(1 - \phi_1^2)^{1/2} \psi_\tau (a_t - \phi_1 w_{t-1})\right],$$
(8.59)

Computing this IF requires the evaluation of the following integrals:

$$I_1 = \int_{-\infty}^{\infty} v^2 \left. \frac{\partial (\psi_\tau (x))}{\partial x} \right|_{x=u} \frac{1}{2\pi} e^{-\frac{u^2+v^2}{2}} du dv$$

$$I_2 = \int_{-\infty}^{\infty} \int_{-\infty}^{\infty} \int_{-\infty}^{\infty} (y_{t-1} + w_{t-1})(1 - \phi_1^2)^{1/2} \psi_\tau (a_t - \phi_1 w_{t-1})$$
$$f(y_t, y_{t-1}; \phi_1) f(w_{t-1}) dy_t dy_{t-1} dw_{t-1}$$

where

$$f(y_t, y_{t-1}; \phi_1) = f(y_t | y_{t-1}; \phi_1) f(y_{t-1}; \phi_1)$$

with

$$f(y_t | y_{t-1}; \phi_1) = \frac{1}{\sqrt{2\pi\sigma^2}} e^{-\frac{1}{2} \frac{(y_t - \phi_1 y_{t-1})^2}{\sigma^2}}$$

$$f(y_{t-1}; \phi_1) = \frac{\sqrt{1 - \phi_1^2}}{\sqrt{2\pi\sigma^2}} e^{-\frac{1}{2} \frac{y_{t-1}^2 (1 - \phi_1^2)}{\sigma^2}}$$

and v and u are independent standard normal random variables.

Figure 8.9 displays the IF of the τ-estimator, and that of the LSE, for the preceding example of an AR(1), with $\phi_1 = -0.5$, for independent AOs of magnitude c_w, with ρ_1 and ρ_2 as defined in (8.39) and (8.38), respectively, and for which $\eta(x) = d\rho_2(x)/dx$.

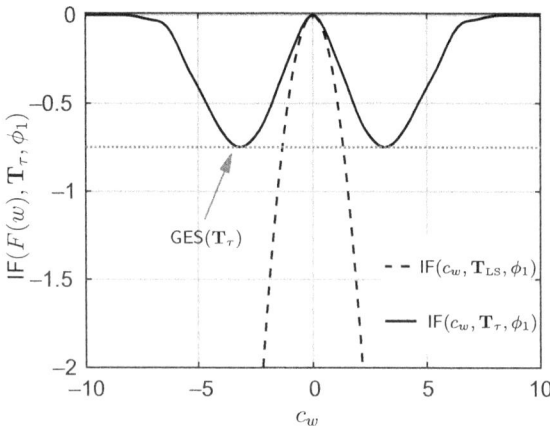

Figure 8.9 The IF of the τ-estimator, and that of the LSE, for the AR(1) model with $\phi_1 = -0.5$ and for the case of independent AOs of magnitude c_w. The supremum of the IF is the GES.

IFs for this model have also been computed for M-estimators and RA estimators in Martin and Yohai (1986). Likewise, the IFs of the MR estimator and RM estimator have been published in Chakhchoukh (2010a, 2010b). Because the computation of IFs is quite involved for the dependent data case, the only models for which IF expressions have explicitly been computed are the AR(1) and MA(1) models. However, expressions for patchy and i.i.d. outlier models exist (Martin and Yohai, 1986).

An alternative to the IF is the sensitivity curve (SC), which was already described for the i.i.d. data case, for location and scale models in Chapter 1. It is easily computed for the preceding example by evaluating

$$\mathrm{SC}_N\left(c_w, \hat{\phi}_1\right) \triangleq N \cdot \left(\hat{\phi}_1(y_1, y_2, \ldots, y_{N-1}, c_w) - \hat{\phi}_1(y_1, y_2, \ldots, y_{N-1})\right). \qquad (8.60)$$

The SC of an estimator $\hat{\phi}_1$ displays the bias of an AR(1) parameter estimator when an additional observation, which takes on the value c_w, is added to a sample $\mathbf{y}_{N-1} = (y_1, y_2, \ldots, y_{N-1})^\top$. Figure 8.10 shows the SC based on $N = 50$ samples of the τ-estimator, and that of the LSE, for the AR(1) model with $\phi_1 = -0.5$ and for the case of independent AOs of magnitude c_w. The SC confirms the findings that were obtained for the IFs, in particular, the empirical GES of the τ-estimator is similar to the GES that is obtained from its IF. Furthermore, the LS is extremely sensitive to extreme outliers, while the τ-estimator ignores them completely due to the redescending ψ_τ function; see (8.46).

8.5.2 Maximum Bias Curve for Dependent Data

The MBC provides information on the maximum asymptotic bias of an estimator with respect to a given fraction of contamination ε. For dependent data, the MBC is defined as for the i.i.d. case, but it also depends on the outlier model. In practice, in the dependent

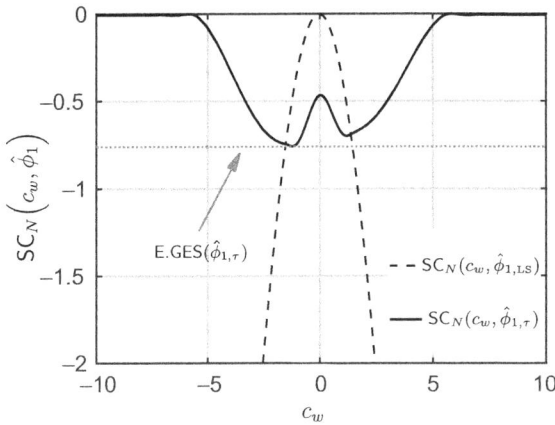

Figure 8.10 The SC of the τ-estimator, and that of the LSE, for the AR(1) model with $\phi_1 = -0.5$ and for the case of independent AOs of magnitude c_w. The supremum of the SC is the empirical gross-error sensitivity (E.GES).

data setting, the MBC is usually approximated by using MC simulations (Maronna et al., 2006; Chakhchoukh et al., 2009, 2010; Muma and Zoubir, 2017) according to

$$\text{MBC}(\varepsilon) = \sup_{c_w} \left| \hat{\boldsymbol{\beta}}_N(\varepsilon, c_w) - \boldsymbol{\beta} \right|, \tag{8.61}$$

where $\hat{\boldsymbol{\beta}}_N(\varepsilon, c_w)$ is the worst-case estimate, evaluated over all MC runs, for every given contamination probability ε. The contamination distribution $F(w)$ of the AO process w_t (see (8.8)), is given by $\Pr(w_t = -c_w) = \Pr(w_t = c_w) = 0.5$, and, in each MC run, the deterministic value c_w, is varied over a grid.

The MBC is generalized (Muma and Zoubir, 2017) by letting

$$\text{QBC}_\alpha(\varepsilon) = Q_\alpha \left\{ \left\| \hat{\boldsymbol{\beta}}_N(\varepsilon, c_w) - \boldsymbol{\beta} \right\| \right\}. \tag{8.62}$$

denote the *quantile bias curve*, which states that α percent of the sorted data is to the left of the α-quantile Q_α. Clearly, $\text{QBC}_{100}(\varepsilon) \equiv \text{MBC}(\varepsilon)$, and, for example, $\text{QBC}_{95}(\varepsilon)$ represents the MBC obtained in 95 percent of the MC runs, where c_w is varying and ε is fixed.

An example of the quantile bias curves of the BIP τ-estimator for a zero-mean AR(1) model with $\phi_1 = 0.5$ and independent AOs is provided in Figure 8.11. The asymptotic value was approximated using $N = 10,000$. Figure 8.11 shows that the MBC saturates at 0.5 for $\varepsilon \geq 0.38$. This breakdown, however, only occurs for a minority of the data, as can be seen from the $\text{QBC}_\alpha(\varepsilon)$ with $\alpha < 100$. It is observed that the bias curves redescend. This is easily explained by the fact, that, for large values of ε, the probability of obtaining patches of outliers increases. The effect of the patches is to increase the correlation and, therewith, to prevent a further shrinkage of the estimates toward zero.

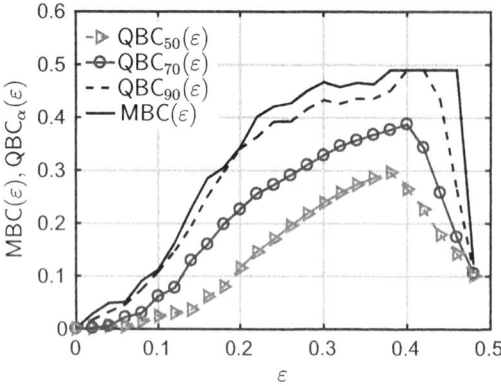

Figure 8.11 The maximum bias and quantile bias curves of the BIP τ-estimator for the AR(1) model with AOs.

8.5.3   Breakdown Point for Dependent Data

The definition of the BPs for dependent data parameter estimators must take into account the nature of the contaminating process. For example, i.i.d. outliers, on the one hand, and highly correlated or patchy outliers, on the other, may yield different BP. In the independent data setting, as discussed in Section 1.3.2, breakdown is defined as the fraction of outliers that drives a criterion function (bias) into a critical region (∞ or the edge of the parameter space). Unfortunately, this definition does not extend to the dependent data case (Genton and Lucas, 2003).

Consider the previously discussed example of a zero-mean AR(1) model that is contaminated by AOs, and where the contaminated observations $y_t^\varepsilon, t = 1, \ldots, N$, follow

$$y_t^\varepsilon = y_t + \xi_t^\varepsilon w_t \tag{8.63}$$

with

$$y_t = \phi_1 y_{t-1} + a_t \tag{8.64}$$

and all variables are defined as in (8.8). Figure 8.12 plots the empirical pdfs $f(\hat{\phi}_1|y^\varepsilon)$ based on 1000 MC experiments of the LSE for $\phi_1 = 0.5$, $N = 100$ and for the three cases:

1. $y^\varepsilon = y$: no outliers
2. $y_{\text{singleRO}}^\varepsilon$: contains a single RO of amplitude $c_w = 10,000$ at y_{50}^ε
3. $y_{\text{patchRO}}^\varepsilon$: contains a patch of two ROs of amplitude $c_w = 10,000$ at y_{50}^ε and y_{51}^ε

Clearly, in none of the cases does the bias of the estimator reach the maximum possible value, given that the parameter space of a stable AR(1) process is $-1 < \phi_1 < 1$. This example illustrates that driving a criterion function into a critical region will not serve to identify the BP of an ARMA model parameter estimator. Although it is clear that both $y_{\text{singleRO}}^\varepsilon$ and $y_{\text{patchRO}}^\varepsilon$ dominate the parameter estimator $\hat{\phi}_1$, the bias depends on the type of contamination. The intuition behind the classical (i.i.d.) definition of the BP,

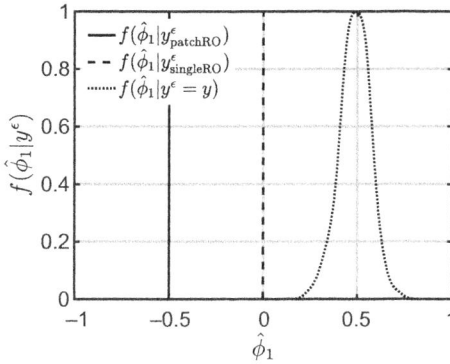

Figure 8.12 Empirical pdf of the LSE of an AR(1) model with $\phi_1 = 0.5$, given $N = 100$ observations that follow, respectively, the clean AR(1) model, the isolated RO model, and the patchy RO model.

namely, the point where the estimator no longer conveys any useful information on the data-generating process, and becomes trapped at the boundary of the parameter space, is not supported by the process defined by (8.64).

Genton and Lucas (2003) provided a definition for BPs that applies to dependent as well as independent observations. It also accommodates all of the earlier definitions of breakdown. The intuition for the BP relies on the crucial property of an estimator of taking on different values for different sample realizations. In the case of location estimation, the estimator may take on values $\hat{\mu} \in \mathbb{R}^N$, while in the case of AR(1) parameter estimation $-1 < \hat{\phi}_1 < 1$. The BP definition by Genton and Lucas is based on the fraction of contamination for which this property of the estimator is lost. The BP, therefore, is based on the fraction of outliers ε (and corresponding outlier configuration defined by ξ_t^ε) that cause a so-called measure of badness (e.g., bias) of an estimator only to take on a finite number of different values despite a continuum of possible uncontaminated sample realizations (Genton and Lucas, 2003). In the example that was illustrated in Figure 8.12, the distribution of the estimator for both $y_{\text{singleRO}}^\varepsilon$ and $y_{\text{patchRO}}^\varepsilon$ is concentrated at a single point, even though the estimator was given a multitude of different uncontaminated sample realizations in each MC run.

Let $\mathbf{y}^\varepsilon \triangleq (y_1^\varepsilon, \ldots, y_t^\varepsilon, \ldots, y_N^\varepsilon)^\top$, with $y_t^\varepsilon = y_t + \xi_t^\varepsilon w_t$, denote the observations that were generated consistent with (8.8) or (8.10). Then, with $\mathbf{y} \triangleq (y_1, \ldots, y_N)^\top$ being an observation of an ARMA process, by writing

$$\mathbf{y}^\varepsilon = \mathbf{y} + \mathbf{z}_M^{c_w}$$

the outliers are collected into an M-sparse contaminating sample $\mathbf{z}_M^{c_w} \in \mathbb{R}^{N \times 1}$. The index $c_w \in \bar{\mathbb{R}} \equiv \mathbb{R} \cup \{-\infty, \infty\}$ indicates the magnitude of the outliers and, for example, a single AO is specified by

$$\mathbf{z}_1^{c_w} = (0, \ldots, 0, c_w, 0, \ldots, 0)^\top$$

while an AO of patch length two, for example, is given by

$$\mathbf{z}_2^{c_w} = (0, \dots, 0, c_w, c_w, 0, \dots, 0)^\top.$$

In general, for multiple outliers with different amplitudes, for example, the elements of $\mathbf{z}_2^{c_w}$ may be multiplied by arbitrary constants. Let $\mathcal{Z}_M^{c_w}$ denote the set of allowable outlier constellations and let $\mathcal{Y}$ be the nondiscrete allowable set of the realizations of the uncontaminated samples. Then, a crucial property of an estimator is that it takes on different values for different elements of $\mathcal{Y}$. This results in a continuum of possible values of a so-called measure of badness (Genton and Lucas, 2003)

$$R(\hat{\boldsymbol{\beta}}, \mathbf{y}^\epsilon) \in \bar{\mathbb{R}}^+ \equiv \mathbb{R}^+ \cup \infty. \tag{8.65}$$

A simple example for $R(\hat{\boldsymbol{\beta}}, \mathbf{y}^\epsilon)$ is the distance of the estimate to the true parameter, that is,

$$R(\hat{\boldsymbol{\beta}}, \mathbf{y}^\epsilon) = |\hat{\boldsymbol{\beta}} - \boldsymbol{\beta}| \tag{8.66}$$

but other measures of model fit may also be chosen. Based on (8.65), the badness set is defined by

$$\mathcal{R}(\hat{\boldsymbol{\beta}}, \mathbf{y}^\epsilon, \mathbf{z}_M^{c_w}, \mathcal{Y}) = \bigcup_{\mathbf{y}' \in \mathcal{Y}} \left\{ R(\hat{\boldsymbol{\beta}}(\mathbf{y}' + \mathbf{z}_M^{c_w}), \mathbf{y}^\epsilon) \right\} \tag{8.67}$$

where $\mathbf{y}'$ denotes an alternative realization of $\mathbf{y}$, and $\hat{\boldsymbol{\beta}}(\mathbf{y}' + \mathbf{z}_M^{c_w})$ is an estimate based on observations $\mathbf{y}' + \mathbf{z}_M^{c_w}$. The BP is then defined by Genton and Lucas (2003) as the fraction of contamination for which $\mathcal{R}(\hat{\boldsymbol{\beta}}, \mathbf{y}^\epsilon, \mathbf{z}_M^{c_w}, \mathcal{Y})$ collapses from an uncountable to a finite set. This is formalized as follows:

DEFINITION 23 *Finite-Sample BP for Dependent Data*
The dependent data BP for an estimator $\hat{\boldsymbol{\beta}}$, based on observations $\mathbf{y}^\epsilon \in \mathbb{R}^{N \times 1}$, is defined by

$$\varepsilon_N^*(\hat{\boldsymbol{\beta}}, R(\hat{\boldsymbol{\beta}}, \mathbf{y}^\epsilon), \mathcal{Z}_M^{c_w}, \mathcal{Y}) = \min \left\{ \frac{M-1}{N} \,\middle|\, \exists\, \mathbf{z}_M^{c_w} \in \mathcal{Z}_M^{c_w} \text{ and } c_w \in \bar{\mathbb{R}} \text{ such that} \right.$$
$$\left. \mathcal{R}(\hat{\boldsymbol{\beta}}, \mathbf{y}^\epsilon, \mathbf{z}_M^{c_w}, \mathcal{Y}) \text{ is a finite set for every } \mathbf{y} \in \mathcal{Y} \right\}. \tag{8.68}$$

From Definition 23, it becomes clear that the BP of an estimator $\hat{\boldsymbol{\beta}}$, in the dependent data case, depends on the measure of badness $R(\hat{\boldsymbol{\beta}}, \mathbf{y}^\epsilon)$, the outlier constellation $\mathcal{Z}_M^{c_w}$, and the set of alternative uncontaminated samples $\mathcal{Y}$. An asymptotic version of $\varepsilon_N^*(\hat{\boldsymbol{\beta}}, R(\hat{\boldsymbol{\beta}}, \mathbf{y}^\epsilon), \mathcal{Z}_M^{c_w}, \mathcal{Y})$ is provided in Genton and Lucas (2003).

The finite-sample BP defined in (8.68) can also be evaluated by studying the quantile bias curve $\text{QBC}_\alpha(\varepsilon)$; see (8.62). A breakdown, according to Definition 23, occurs for the value $\varepsilon > \varepsilon^*$ where $\text{QBC}_\alpha(\varepsilon)$ is equal for all values of α. Note that such a definition captures the idea that a measure of badness collapses from an uncountable to a finite set without necessarily driving a criterion function, such as, for example, the bias, into a critical region. From Figure 8.11, it can be empirically deduced that the BP of the BIP-τ estimator for the AR(1) model, with an outlier constellation $\mathcal{Z}_M^{c_w}$ following a Bernoulli sequence, approaches $\varepsilon_N^*(\hat{\boldsymbol{\beta}}, R(\hat{\boldsymbol{\beta}}, \mathbf{y}^\epsilon), \mathcal{Z}_M^{c_w}, \mathcal{Y}) = 50\%$.

8.6 Algorithms

This section provides some algorithms to compute robust ARMA(p,q) parameter estimates. In principle, for example, M-estimates, S-estimates, MM-estimates, or τ-estimates could be found as the solution to a nonlinear LS problem. However, because the objective function is nonconvex, the performance of such iterative estimators crucially depends on the starting point of the iterative procedure. Due to the computational complexity, except for some very simple cases, such as ARMA models with $p + q \leq 2$, it is not possible to perform an exhaustive grid search to find a good starting point.

This section provides a computationally efficient algorithm to compute BIP-AR(p) or filtered AR(p) τ-parameter estimates (or S-estimates) based on a robust Levinson–Durbin procedure. By using a robust Levinson–Durbin procedure for the AR model, the use of iterative algorithms can be avoided. For the case of robust ARMA(p,q) parameter estimates, however, one must resort to iterative procedures. Therefore, a method to find a robust initialization, that is, a good starting point that is close to $\boldsymbol{\beta}$, is also discussed. Without loss of generality, it is assumed in this section that the observations are zero-mean, that is, $\mu = 0$. In practice, a robust location estimate, as discussed in Section 1.2.1, can be subtracted before applying the algorithms that are described in the following text.

8.6.1 Computing BIP-AR(p) or Filtered AR(p) τ- (or S-Estimates) Based on a Robust Levinson–Durbin Procedure

In the sequel, a robust Levinson–Durbin procedure, which does not require choosing a starting point, is detailed. Without loss of generality, a Robust Levinson–Durbin procedure for the BIP-AR(p) τ-estimator is considered. For S-estimators, the algorithm is nearly identical. The only difference is that the τ-scale (8.41) of the BIP-AR(p) innovations, that is,

$$\hat{\sigma}_\tau(a_N^b(\boldsymbol{\phi})) = \hat{\sigma}(a_N^b(\boldsymbol{\phi}))\sqrt{\frac{1}{N-p}\sum_{t=p+1}^{N}\rho_2\left(\frac{a_t^b(\boldsymbol{\phi})}{\hat{\sigma}(a_N^b(\boldsymbol{\phi}))}\right)}, \qquad (8.69)$$

where

$$a_t^b(\boldsymbol{\phi},\sigma) = y_t - \sum_{i=1}^{p}\phi_i y_{t-i} + \sum_{i=1}^{p}\left(\phi_i a_{t-i}^b(\boldsymbol{\phi},\sigma) - \phi_i\sigma\eta\left(\frac{a_{t-i}^b(\boldsymbol{\phi},\sigma)}{\sigma}\right)\right), \qquad (8.70)$$

is replaced by an M-scale, $\hat{\sigma}(a_N^b(\boldsymbol{\phi}))$, of the BIP-AR($p$) innovations, as specified by (1.19). To compute filtered AR(p) τ- or S-estimates, the filtered AR(p) innovations

$$a_t^f(\boldsymbol{\phi},\sigma_t) = y_t - \sum_{i=1}^{p}\phi_i y_{t-i}^f(\boldsymbol{\phi},\sigma_t) \qquad (8.71)$$

are used instead.

Algorithm 17: bipTauLevinson: computes the BIP-τ estimate of the AR(p) model parameters via a robust Levinson–Durbin procedure.

---

input : $\mathbf{y} \in \mathbb{R}^N$, p, grid spacing Δ_{ζ^0}
output : $\hat{\boldsymbol{\phi}}_\tau = (\hat{\phi}_1, \ldots, \hat{\phi}_p)^\top$, innovations τ-scale $\hat{\sigma}_\tau$
initialize: no initialization required

1 **if** $p > 1$ **then**
2   **for** $\zeta^0 = -0.99 : \Delta_{\zeta^0} : 0.99$ **do**
3   Compute AR(1) innovations $\mathbf{a}_N(\zeta^0)$ from (8.34) and BIP-AR(1) innovations $\mathbf{a}_N^b(\zeta^0, \hat{\sigma}(\zeta^0))$ from (8.70) and (8.26).
4   Compute τ-scales from (2.90) resulting in $\hat{\sigma}_\tau(\mathbf{a}_N(\zeta^0))$ and $\hat{\sigma}_\tau(\mathbf{a}_N^b(\zeta^0, \hat{\sigma}(\zeta^0)))$.

5   Fit polynomial to $(\zeta^0, \hat{\sigma}_\tau(\mathbf{a}_N(\zeta^0)))$, and $(\zeta^0, \hat{\sigma}_\tau(\mathbf{a}_N^b(\zeta^0, \hat{\sigma}(\zeta^0))))$ at $\zeta^0 = -0.99 : \Delta_{\zeta^0} : 0.99$.
6   Compute the values ζ_1, ζ_2, for which the objective functions that are approximated by the polynomials, is minimized. Use the estimates based on the model that provides the smaller minimum, that is,

$$\hat{\phi}_1 = \arg\min_{\zeta_1, \zeta_2} \left\{ \hat{\sigma}_{\tau,\min}(\mathbf{a}_N(\zeta_1)), \hat{\sigma}_{\tau,\min}(\mathbf{a}_N^b(\zeta_2, \hat{\sigma}(\zeta_2))) \right\}. \qquad (8.72)$$

7 **if** $p > 1$ **then**
8   **for** $m = 2, \ldots, p$ **do**
9   **for** $\zeta^0 = -0.99 : \Delta_{\zeta^0} : 0.99$ **do**
10   Compute AR(m) innovations $\mathbf{a}_N(\zeta^0)$ from (8.34) and BIP-AR(m) innovations $\mathbf{a}_N^b(\zeta^0, \hat{\sigma}(\zeta^0))$ from (8.70) and (8.26).
11   Compute τ-scales from (2.90) resulting in $\hat{\sigma}_\tau(\mathbf{a}_N(\zeta^0))$ and $\hat{\sigma}_\tau(\mathbf{a}_N^b(\zeta^0, \hat{\sigma}(\zeta^0)))$.

12   Fit polynomial to $(\zeta^0, \hat{\sigma}_\tau(\mathbf{a}_N(\zeta^0)))$, and $(\zeta^0, \hat{\sigma}_\tau(\mathbf{a}_N^b(\zeta^0, \hat{\sigma}(\zeta^0))))$ at $\zeta^0 = -0.99 : \Delta_{\zeta^0} : 0.99$.

13   Compute the values ζ_1, ζ_2, for which the objective functions that are approximated by the polynomials is minimized. Use the estimates based on the model that provides the smaller minimum, that is,

$$\hat{\phi}_{m,m} = \arg\min_{\zeta_1, \zeta_2} \left\{ \hat{\sigma}_{\tau,\min}(\mathbf{a}_N(\zeta)), \hat{\sigma}_{\tau,\min}(\mathbf{a}_N^b(\zeta, \hat{\sigma}(\zeta))) \right\} \qquad (8.73)$$

where the estimates from lower orders are incorporated by applying the Levinson–Durbin recursion

$$\hat{\phi}_{m,m} = \begin{cases} \zeta & \text{if } i = m \\ \hat{\phi}_{m-1,i} - \zeta \hat{\phi}_{m-1,m-i} & \text{if } 1 \leq i \leq m-1 \end{cases} \qquad (8.74)$$

(a)

(b)

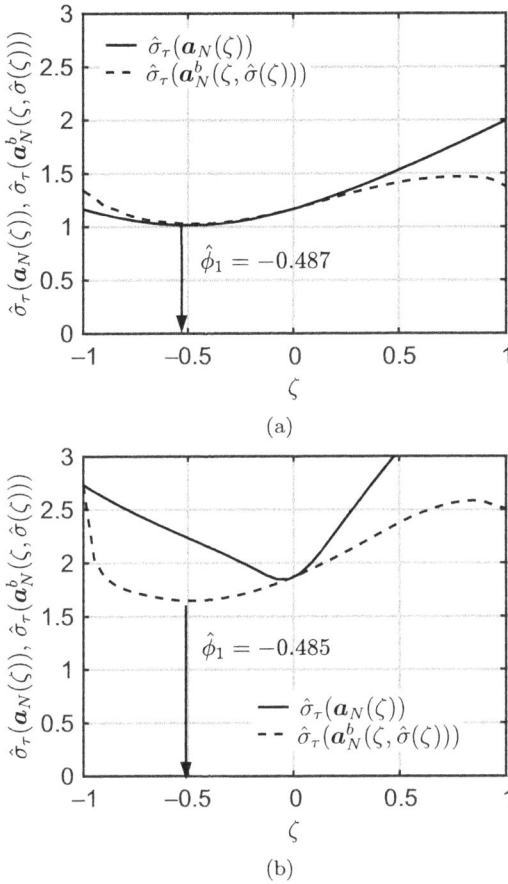

Figure 8.13 Example of finding $-1 < \zeta < 1$, which minimizes $\hat{\sigma}_\tau(\boldsymbol{a}_N(\zeta))$ and $\hat{\sigma}_\tau(\boldsymbol{a}_N^b(\zeta, \hat{\sigma}(\zeta)))$ for an AR(1) process with $\phi_1 = -0.5$ and $\sigma = 1$. (a) $y_t^\varepsilon = y_t$ clean data example. (b) 10 percent equally spaced AOs with amplitudes of 10.

Figure 8.13 details results for the robust Levinson–Durbin procedure that allow the computation of the BIP-τ estimate of ϕ_1, i.e., the AR(1) model parameter. The top graph depicts the results for $y_t^\varepsilon = y_t$ (no contamination/outliers) with $\phi_1 = -0.5$ for $\sigma = 1, N = 1000$. The bottom graph displays an illustrative AO example, where $\xi_t^\varepsilon w_t$ in (8.8) produces 10 percent equally spaced AOs of amplitude 10. The top plot confirms that when $y_t^\varepsilon = y_t$, the final estimate $\hat{\phi}_1$ is based on the AR(1) model rather than on the BIP-AR(1) model. This is in line with theorem 2 of Muma and Zoubir (2017), which states, under an ARMA model, the BIP-τ estimator is asymptotically equivalent to a τ-estimator. In the same work, the almost sure convergence of the τ-estimator of the innovations scale to the population value based on the expectation operator is also proven.

The bottom plot clearly shows the necessity of preventing the propagation of outliers. The minimum of the τ-scale objective function for the AR(1) model is shifted to zero, making it useless in terms of robust parameter estimation. The τ-scale objective function for the BIP-AR(1) model, by contrast, maintains its minimal value close to the true value of $\phi_1 = -0.5$ for 10 percent AOs.

The Algorithm 17: bipTauLevinson that computes the BIP-τ estimate of the AR(p) model parameters, via a robust Levinson–Durbin procedure, is included in the downloadable RobustSP toolbox. First evaluating $\hat{\sigma}_\tau(a_N(\zeta^0))$ and $\hat{\sigma}_\tau(a_N^b(\zeta^0, \hat{\sigma}(\zeta^0)))$ on a coarse grid (e.g., using a step size of $\Delta_{\zeta^0} = 0.05$) and then modeling the true curves by a polynomial is an optional step to speed up the algorithm when compared to choosing a fine grid. An alternative algorithm to bipTauLevinson for computing the BIP-τ estimates, which is based on a robust forward-backward algorithm, is detailed in Muma and Zoubir (2017).

8.6.2   Algorithms for Computing MA(q) and ARMA(p, q) Parameter Estimates

It has been shown that the M-estimators (Stockinger and Dutter, 1987), S- and MM-estimators (Muler et al., 2009), and τ-estimators (Muma and Zoubir, 2017) of the ARMA model parameters can be formulated as the solution to a nonlinear LS problem. An example is given in the sequel.

Example 17   Nonlinear LS formulation for the S-estimators of the ARMA(p, q) parameters.

From the definition of the M-scale, that is,

$$b = \frac{1}{N - p} \sum_{t=p+1}^{N} \rho\left(\frac{a_t^e(\boldsymbol{\beta})}{\hat{\sigma}(a_N(\boldsymbol{\beta}))}\right)$$

if follows immediately that

$$\hat{\sigma}^2(a_N(\boldsymbol{\beta})) = \sum_{t=p+1}^{N} \frac{\hat{\sigma}^2(a_N(\boldsymbol{\beta}))}{(N - p)b} \rho\left(\frac{a_t^e(\boldsymbol{\beta})}{\hat{\sigma}(a_N(\boldsymbol{\beta}))}\right).$$

Thus, by defining

$$r_t^2(\boldsymbol{\beta}) \triangleq \frac{\hat{\sigma}^2(a_N(\boldsymbol{\beta}))}{(N - p)b} \rho\left(\frac{a_t^e(\boldsymbol{\beta})}{\hat{\sigma}(a_N(\boldsymbol{\beta}))}\right)$$

it becomes evident that solving (8.36) is equivalent to minimizing a nonlinear LS problem that is defined by

$$\hat{\boldsymbol{\beta}} = \arg\min_{\boldsymbol{\beta}} \sum_{t=p+1}^{N} r_t^2(\boldsymbol{\beta}).$$

Performing similar steps, for example, as done in Stockinger and Dutter (1987) for the M-estimator, quickly reveals that M-estimators, MM-estimators, and τ-estimators are consistent with the minimization of a nonlinear LS problem. This means that they can be computed by generic nonlinear LS methods, such as the Levenberg–Marquardt algorithm. Naturally, this also holds for BIP-model based estimators and filtered estimators. Therefore, in principle, these minimization problems can be solved by using any nonlinear LS algorithm, for example, the Levenberg–Marquardt algorithm; see for example, Madsen et al. (2004).

However, determining an estimate for $\boldsymbol{\beta}$ with $q > 0$, in general, requires finding the $\boldsymbol{\beta}$ that minimizes a nonconvex problem. The crucial point is to find a starting point that is sufficiently close to the true $\boldsymbol{\beta}$. Due to the computational complexity, except for some very simple cases (e.g., $p + q \leq 2$), it is not possible to perform an exhaustive grid search.

Robust Starting Point Algorithm

In the sequel, a simple procedure to find a starting point $\hat{\boldsymbol{\beta}}_0$, based on the BIP-AR model, is described. From (8.20) it follows, for the AR model, that the one-step prediction of y_t can be computed recursively for $t \geq p + 1$ using:

$$\hat{y}_t = \sum_{i=1}^{p} \phi_i \left(y_{t-i} - a_{t-i}^b(\hat{\boldsymbol{\beta}}, \hat{\sigma}) + \hat{\sigma} \eta \left(\frac{a_{t-i}^b(\hat{\boldsymbol{\beta}}, \hat{\sigma})}{\hat{\sigma}} \right) \right). \tag{8.75}$$

From (8.75), outlier-cleaned observations are obtained for $t \geq p + 1$ by computing

$$y_t^* = y_t - a_t^b(\hat{\boldsymbol{\beta}}, \hat{\sigma}) + \hat{\sigma} \eta \left(\frac{a_t^b(\hat{\boldsymbol{\beta}}, \hat{\sigma})}{\hat{\sigma}} \right). \tag{8.76}$$

To find a starting point for the ARMA parameter estimation, the data is first cleaned from outliers using an AR(p) approximation, which can be computed with the bipTauLevinson algorithm. Note that the ARMA parameter estimation is performed on the observed data and the AR-approximation–based outlier cleaning is only used within the starting point algorithm to find $\hat{\boldsymbol{\beta}}_0$.

The choice of p to be used in the approximation is discussed in Muma and Zoubir (2017). The starting point $\hat{\boldsymbol{\beta}}_0$ for the BIP-τ ARMA parameter estimation algorithm can then be computed, based on y_t^*, and by using any classical ARMA parameter estimator, for example, Jones (1980).

8.6.3   Simulation and Real-Data Examples

As will be shown in Chapter 9, parametric spectral estimation on an ARMA model is equivalent to signal modeling. Therefore, all robust ARMA parameter estimators that were discussed in Section 8.3 also lead to robust spectral estimators. Section 9.3 provides a simulation example for an ARMA(4,3) process that is affected by AOs. A real-world example for robust data cleaning for photoplethysmography-based pulse-rate variability data, and based on robust AR parameter estimation, is given in Section 11.5.

8.7   Concluding Remarks

The objective of this chapter has been to discuss robust statistics for dependent data. We have introduced important signal and outlier models and have described the propagation of outliers and its effect on parameter estimators. The focus was on ARMA parameter estimators that involve robust ACM type filters or the BIP-ARMA model. Robust model order selection was briefly discussed. Robustness measures for the dependent data case, that is, the IF, BP, and MBC, were introduced, and the differences to the i.i.d. data case were highlighted. Algorithms to robustly estimate ARMA parameters are included as MATLAB$^{\copyright}$ code in the downloadable RobustSP toolbox. Due to limited space, this chapter did not include topics such as robust forecasting, multivariate ARMA models, and nonstationary data.

9   Robust Spectral Estimation

Spectral estimation refers to determining the distribution of power over frequency, that is, the spectral content based on an observed stationary time series. Spectral analysis has been applied in a wide range of areas and disciplines, including vibration monitoring, biomedicine, speech analysis, astronomy, and many more. Excellent textbooks exist that cover spectral estimation, for example Priestley (1981), Marple (1987), Brillinger (2001), and Stoica and Selen (2004). Approaches to spectral estimation can generally be classed as being either nonparametric or parametric. A further distinction is made between continuous spectra and line spectra. Both the nonparametric and the parametric approaches to spectral estimation are sensitive to outliers and heavy-tailed noise, which have been reported in many real-life applications (Zoubir et al., 2012).

This chapter treats both robust nonparametric and robust parametric spectral estimation. In particular, nonparametric spectral approaches, which are based on robustifying the periodogram, are discussed. Next, we discuss robust parametric approaches, which are based on an ARMA model, and where the robust spectral estimation task reduces to the robust ARMA parameter estimation task that was the subject of Chapter 8. Finally, we consider robust subspace frequency estimation. Subspace frequency estimation algorithms exploit the different properties of the noise subspace and the signal subspace. We will explain robust subspace frequency estimators, such as the robust MUSIC algorithm, similarly to the approach used in Chapter 5. We will show that robust subspace frequency estimation can be performed by applying robust scatter (covariance) estimators that were discussed in Chapter 4. Using Monte Carlo experiments, we compare the performance of different estimators. This chapter also contains an illustrating real-life example of spectral analysis for ECG signals with motion artifacts.

9.1   Robust Nonparametric Spectral Estimation

As illustrated in Figure 9.1, let $y_{t,i}$ with $t, i \in \mathbb{Z}$ denote the realization (or sample function) associated with the event ζ_i of a random process $y_t(\zeta)$. The set of all possible sample functions is called the ensemble and defines the random process. If we assume that the ith realization $y_{t,i}$, for $t \in \mathbb{Z}$, is truncated to a length $N = 2M + 1$, with $N, M \in \mathbb{Z}^+$, such that $y_{t,i} = 0$ for $|t| > M$, as long as $M < \infty$, it is reasonable to assume that

$$\sum_{t=-\infty}^{\infty} |y_{t,i}|^2 < \infty. \tag{9.1}$$

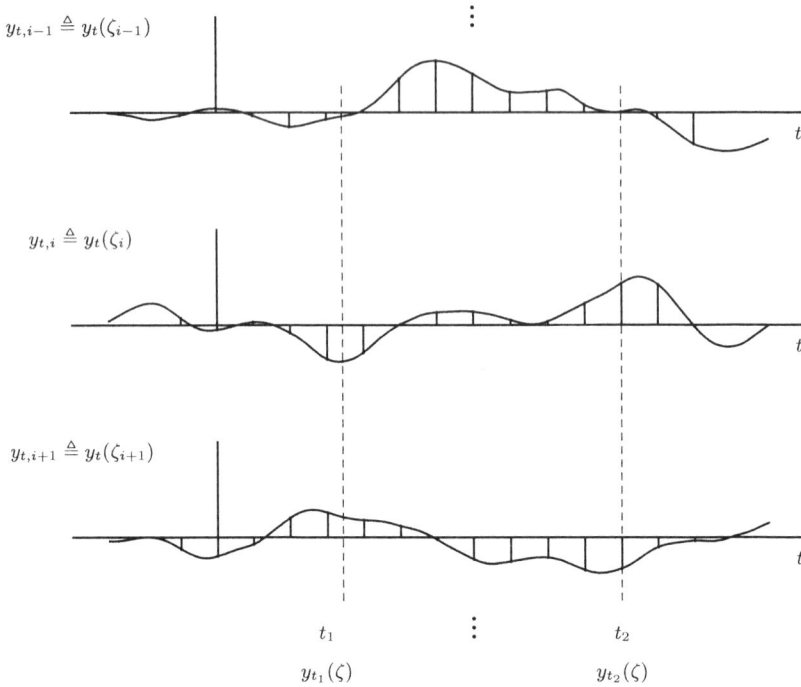

Figure 9.1 Illustration of a random process $y_t(\zeta)$ with realizations (sample functions) $y_{t,i} \triangleq y_t(\zeta_i)$ that are associated with the events $\zeta_i, i \in \mathbb{Z}$.

Then, under some additional regularity conditions (Stoica and Selen, 2004), each sample function has a discrete-time Fourier transform (DTFT), also called the finite Fourier transform (Brillinger, 2001), defined according to

$$Y_N(e^{J\omega}, \zeta_i) = \sum_{t=-M}^{M} y_{t,i} e^{-J\omega t}, \qquad (9.2)$$

where the normalized angular frequency ω is measured in radians per sample. The inverse DTFT is defined by

$$y_{t,i} = \frac{1}{2\pi} \int_{-\pi}^{\pi} Y_N(e^{J\omega}, \zeta_i) e^{J\omega t} d\omega. \qquad (9.3)$$

Then, the average power of the random process $y_t(\zeta)$ is defined as

$$P_{yy} = \lim_{M \to \infty} \frac{1}{2M+1} \sum_{t=-M}^{M} \mathsf{E}\left[y_t(\zeta)^2\right]$$

$$= \lim_{M \to \infty} \frac{1}{2\pi} \int_{-\pi}^{\pi} \frac{\mathsf{E}\left[|Y_N(e^{J\omega}, \zeta)|^2\right]}{2M+1} d\omega. \qquad (9.4)$$

Eq. (9.4) allows us to define the PSD of the random process:

DEFINITION 24   *The PSD of the random process $y_t(\zeta)$ is defined by*

$$S_{yy}(e^{J\omega}) = \lim_{M \to \infty} \frac{\mathsf{E}\left[|Y_N(e^{J\omega}, \zeta)|^2\right]}{2M + 1}. \tag{9.5}$$

The Wiener–Khintchine theorem provides a second definition of the PSD:

DEFINITION 25   *Wiener–Khintchine Theorem*
The PSD of the WSS random process $y_t(\zeta)$ is defined by the Fourier transform of the SOMF

$$S_{yy}(e^{J\omega}) = \sum_{m=-\infty}^{\infty} r_{yy}(m)e^{-J\omega m}, \tag{9.6}$$

where $r_{yy}(m)$ is the SOMF consistent with Definition 17.

The most frequently used nonparametric estimators of the PSD are based on the periodogram, which relies on (9.5) in Definition 24. The periodogram estimates $S_{yy}(e^{J\omega})$ based on an available sample $\mathbf{y}_N = (y_1, y_2, \ldots, y_N)^\top$ by means of

$$\hat{S}_{yy}(e^{J\omega}) = \frac{1}{N} \left| \sum_{t=1}^{N} y_t e^{-J\omega t} \right|^2. \tag{9.7}$$

The periodogram is asymptotically unbiased but it is not consistent because its variance does not decrease for an increasing N. However, its variance can be reduced by averaging periodograms of adjacent blocks, or by smoothing neighboring periodogram ordinates.

For example, Bartlett's estimator simply splits up the available sample of N observations into $K = \lfloor N/L \rfloor$ nonoverlapping blocks of length L each, and takes the sample average for each value of ω. To define Bartlett's estimator, let

$$y_t^{(k)} = y_{(k-1)L+t}, \quad t = 1, \ldots, L, \quad k = 1, \ldots, K$$

denote the samples of the kth block, and let

$$I_{yy}(e^{J\omega}, k) = \frac{1}{L} \left| \sum_{t=1}^{L} y_t^{(k)} e^{-J\omega t} \right|^2 \tag{9.8}$$

be the corresponding periodogram. Then, Bartlett's estimate is given by

$$\hat{S}_{yy}(e^{J\omega}) = \frac{1}{K} \sum_{k=1}^{K} I_{yy}(e^{J\omega}, k), \quad k = 1, \ldots, K. \tag{9.9}$$

9.1.1   Robust-Averaging–Based Nonparametric Estimators

A simple approach to robustify nonparametric spectral estimation is to replace the averaging in (9.9) by a robust average, such as the median, as suggested by Maronna et al. (2006). This yields a robust Bartlett estimator

$$\hat{S}_{yy}(e^{J\omega}) = \mathsf{med}\left(I_{yy}(e^{J\omega}, k)\right), \tag{9.10}$$

where the median is taken for each frequency ω over the available K periodograms. Of course, other robust location estimators can be adopted, not only the median. For example, one could choose an M-estimator that was discussed in Section 1.2.

The strategy of replacing the sample average by a robust location estimate can be straightforwardly applied to other periodogram-based estimators, such as the Welch and Daniell estimators (Stoica and Selen, 2004). Robust-averaging–based estimators are computationally efficient and can thus be used when processing large data sets. However, these estimators are limited in the sense that they only provide robustness, as long as the majority of the blocks do not contain outliers. Otherwise, they are not any more robust than the estimators that are based on the sample average. Nevertheless, they are useful in some situations, as illustrated in the following real-data example.

Example 18 Robust Bartlett Estimator for Spectral Analysis of ECG Signals

The ECG measures the electrical activity of the heart over a period using electrodes placed on the skin. Figure 9.2a shows an ECG recording, where at approximately the thirteenth second, a motion artifact is caused by an arm movement of the subject. Figure 9.2b displays the Bartlett and robust Bartlett estimates of the PSD based on the entire measurement while Figure 9.2c computes the same estimates, excluding the artifact at the thirteenth second.

Clearly, the Bartlett estimator is largely influenced by the motion artifact, while the robust Bartlett estimator provides very similar spectral density estimates in both cases. This example highlights the usefulness of robust Bartlett estimators in situations when most of the data segments do not contain outliers. However, if the outliers are not concentrated in a few patches but are spread evenly over the measurement, the robust Bartlett estimator does not provide an advantage over its classical counterpart.

If the majority of blocks contain outliers, it becomes necessary to consider more advanced estimators, such as the M-periodogram, which is described in the following section. Note that robust averaging and the M-periodogram can also be combined, which is beneficial if the temporal structure of the patchy outliers is such that the majority of the samples within some of the blocks are outliers. With this strategy, robust PSD estimators are obtained even if all blocks contain outliers, and long patches cause a minority of the K robust M-periodogram estimators to break down. However, it is still required that a majority of available samples, N, are uncontaminated.

9.1.2 The M-Periodogram and the ℓ_p-Periodogram

A more sophisticated nonparametric approach than the previously presented robust Bartlett estimator is to interpret the periodogram as the LSE solution of a nonlinear harmonic regression (trigonometric regression). Starting from (9.7)

$$I_{yy}(e^{J\omega}) = \frac{1}{N}\left|\sum_{t=1}^{N} y_t e^{-J\omega t}\right|^2,$$

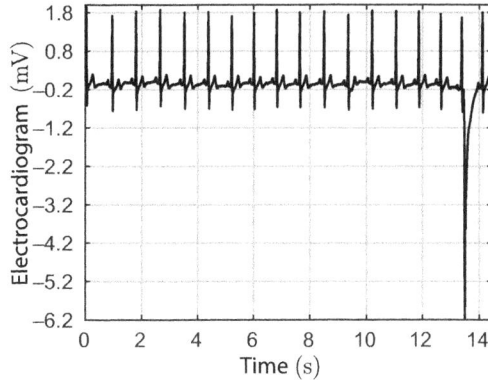

(a) ECG signal with a motion artifact at the thirteenth second.

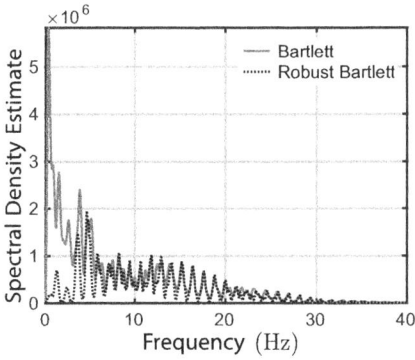

(b) Bartlett and Robust Bartlett estimates of the PSD based on the entire measurement.

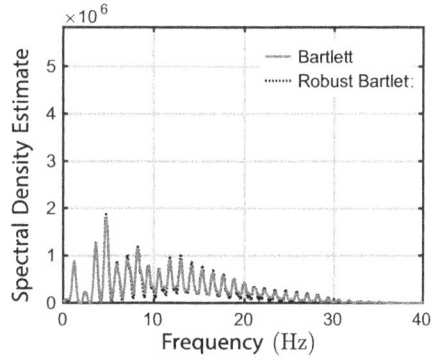

(c) Bartlett and Robust Bartlett estimates of the PSD when taking only the first 12 seconds (without the artifact).

Figure 9.2 Robust Bartlett based PSD estimation of an ECG signal with a single motion artifact that is caused by the arm movement of a subject.

and if $\omega = \frac{2\pi k}{N}$ for some integer number k, it follows that ω is a Fourier frequency and the periodogram can be expressed as

$$\frac{N}{4}\left|\hat{\boldsymbol{\beta}}_{\text{LS}}(\omega)\right|^2,\tag{9.11}$$

where $\hat{\boldsymbol{\beta}}_{\text{LS}}(\omega) \in \mathbb{R}^2$ is the LSE

$$\hat{\boldsymbol{\beta}}_{\text{LS}}(\omega) = \arg\min_{\boldsymbol{\beta}\in\mathbb{R}^2} \sum_{t=1}^{N} |y_t - \mathbf{x}_{[t]}^{\top}\boldsymbol{\beta}|^2\tag{9.12}$$

based on the harmonic regressor

$$\mathbf{x}_{[t]} = (\cos(\omega t), \sin(\omega t))^{\top}.\tag{9.13}$$

The fact that ω is a Fourier frequency implies that

$$\sum_{t=1}^{N} \mathbf{x}_{[t]}\mathbf{x}_{[t]}^{\top} = \frac{N}{2}\mathbf{I}_2,$$

(9.14)

where $\mathbf{I}_2 \in \mathbb{R}^{2\times 2}$ is the identity matrix. From (9.14), it follows that

$$\hat{\boldsymbol{\beta}}_{\mathrm{LS}}(\omega) = \left(\sum_{t=1}^{N} \mathbf{x}_{[t]}\mathbf{x}_{[t]}^{\top}\right)^{-1} \sum_{t=1}^{N} \mathbf{x}_{[t]}y_t$$

$$= \frac{2}{N}\sum_{t=1}^{N} \mathbf{x}_{[t]}y_t$$

and thus

$$\frac{N}{4}\left|\hat{\boldsymbol{\beta}}_{\mathrm{LS}}(\omega)\right|^2 = \frac{1}{N}\left|\sum_{t=1}^{N} y_t e^{-j\omega t}\right|^2 = I_{yy}(e^{j\omega}).$$

(9.15)

Starting from (9.12), and consistent with the M-estimation approach for linear regression discussed in Chapter 2, the so-called M-periodogram estimate (Katkovnik, 1998) can be computed using

$$\hat{\boldsymbol{\beta}}_{\mathrm{M}}(\omega) = \underset{\boldsymbol{\beta}\in\mathbb{R}^2}{\arg\min} \sum_{t=1}^{N} \rho\left(y_t - \mathbf{x}_{[t]}^{\top}\boldsymbol{\beta}\right),$$

(9.16)

where $\rho(x)$ is a robustifying objective function that defines the particular M-estimator, for example Huber's as given in (1.20).

The M-periodogram PSD estimate is then defined by

$$\hat{S}_{yy}(e^{j\omega}) = \frac{N}{4}\left|\hat{\boldsymbol{\beta}}_{\mathrm{M}}(\omega)\right|^2.$$

(9.17)

In Katkovnik (1998) a minimax optimal ρ is derived, which minimizes the worst-case variance of the M-periodogram over a class of distributions $\mathcal{F}$ that contains the nominal distribution $F(x)$. Katkovnik also established, under the following assumptions,

1. $F(x)$ is symmetric, that is, $F(x) = 1 - F(-x)$,
2. $\rho(x)$ is a symmetric, convex, piecewise twice differentiable function that satisfies $\rho(x) \geq 0$ and $\rho(0) = 0$,

that $\hat{\boldsymbol{\beta}}_{\mathrm{M}}(\omega)$ is asymptotically unbiased with a diagonal asymptotic covariance matrix given by

$$\frac{2}{N}V\left(\rho, F(x)\right)\mathbf{I}_2,$$

where

$$V(\rho, F(x)) = \frac{\int \psi^2(x)dF(x)}{(\psi'(x)dF(x))^2},$$

with $\psi(x) = \frac{d\rho(x)}{dx}$ and $\psi'(x) \triangleq \frac{d\psi(x)}{dx}$.

An interesting subclass of M-periodogram estimates defined by (9.16) are the ones that use $\rho(x) = |x|^p$ with $1 \leq p \leq 2$. These so-called ℓ_p-periodograms are defined as

$$\hat{S}_{yy}(e^{J\omega}) = \frac{N}{4} \left| \hat{\boldsymbol{\beta}}_{\ell_p}(\omega) \right|^2, \qquad (9.18)$$

where

$$\hat{\boldsymbol{\beta}}_{\ell_p}(\omega) = \arg \min_{\boldsymbol{\beta} \in \mathbb{R}^2} \sum_{t=1}^{N} |y_t - \mathbf{x}_{[t]}^\top \boldsymbol{\beta}|^p \qquad (9.19)$$

with $1 \leq p \leq 2$. Clearly, $p = 2$ yields the classical LSE periodogram, as defined by (9.12), which also coincides with the MLE, assuming that $F(x)$ is Gaussian. The well-known vulnerability of the LSE to heavy-tailed distributions $F(x)$ underpins the choice of $p < 2$. Most prominently, $p = 1$ results in the Laplace periodogram,

$$\hat{\boldsymbol{\beta}}_{\ell_1}(\omega) = \arg \min_{\boldsymbol{\beta} \in \mathbb{R}^2} \sum_{t=1}^{N} |y_t - \mathbf{x}_{[t]}^\top \boldsymbol{\beta}| \qquad (9.20)$$

which is the MLE, under the assumption that $F(x)$ is the Laplace distribution. Note that $\hat{\boldsymbol{\beta}}_{\ell_1}(\omega)$ is an LAD estimate and the Laplace periodogram has similar robustness properties as the linear regression LAD estimate that was discussed in Section 2.4. A trade-off between robustness and efficiency under the Gaussian distribution can be obtained by varying p. The asymptotic unbiasedness and normality of $\hat{\boldsymbol{\beta}}_{\ell_p}(\omega)$ has been established by Li (2010) for $p > 1$, while the same properties were obtained for $\hat{\boldsymbol{\beta}}_{\ell_1}(\omega)$ by Li (2008). A related approach is the ℓ_p-norm–based iterative adaptive approach for robust spectral analysis by Chen et al. (2014). Its main idea is to reformulate the nonlinear frequency estimation problem as a linear model whose parameters are updated iteratively. Robust periodograms can also be obtained from linear combinations of order statistics (L-estimates); see Djurovic et al. (2003) for details on the L-periodogram.

9.1.3   The Biweight Robust Fourier Transform

Related work, which may be regarded as a predecessor to the M-periodogram, is the method by Tatum and Hurvich (1993) who proposed a frequency-domain approach to the problem of cleaning time series of outliers. This approach, again, uses robust trigonometric regression to fit sine and cosine coefficients at each Fourier frequency. An available sample $\mathbf{y}_N = (y_1, \ldots, y_N)^\top$, can be represented as the sum of N cosines and sines at the Fourier frequencies $\omega_k = 2\pi k/N$, for $k = 0, \ldots, \lfloor N/2 \rfloor$, according to

$$y_t = a_0 + \sum_{0 < k < N/2} (a_k \cos(\omega_k t) + b_k \sin(\omega_k t)) + (-1)^t a_{N/2}, \quad t = 1, \ldots, N \quad (9.21)$$

where

$$a_k = \frac{2}{N} \sum_{t=1}^{N} y_t \cos(\omega_k t),$$

$$b_k = \frac{2}{N} \sum_{t=1}^{N} y_t \sin(\omega_k t),$$

$$a_0 = \frac{1}{N} \sum_{t=1}^{N} y_t,$$

and the last term

$$a_{N/2} = \frac{1}{N} \sum_{t=1}^{N} y_t (-1)^t$$

is included only if N is even.

In particular, Tatum and Hurvich (1993) proposed that the sine and cosine coefficients, $a_k, b_k, k = 1, \ldots, \lfloor N/2 \rfloor$, for each discrete frequency ω_k, be obtained from

$$\arg\min_{a_k, b_k} \sum_{t=1}^{N} \rho \left(\frac{r_t^{(k)}}{\hat{\sigma}_k} \right), \tag{9.22}$$

where $\hat{\sigma}_k$ is a robust scale estimate of $\mathbf{r}_k = (r_1^{(k)}, \ldots, r_N^{(k)})^\top$ with $r_t^{(k)} = y_t - a_k \cos(\omega_k t) - b_k \sin(\omega_k t)$; see Section 1.2.1, and $\rho(x)$ is Tukey's biweight function as given by (1.22). Because the optimization problem stated in (9.22) is nonconvex, Tatum and Hurvich proposed using the repeated median estimator by Siegel (1982) as an initializer.

Algorithm 18: RepMed details the pseudo-code for computing the initial repeated median estimates of $a_k, b_k, k = 1, \ldots, (N-1)/2$, for the case of N being prime. In this case, the estimator has a breakdown point of 50 percent. The full algorithm for nonprime N, which also computes a cleaned version of $\mathbf{y}_N$ by using the inverse Fourier transform of the robustly estimated Fourier coefficients, is given in the RobustSP toolbox. Unfortunately, Algorithm 18 is computationally complex.

9.2   Autoregressive Moving-Average Model–Based Robust Parametric Spectral Estimation

The nonparametric approaches discussed in the preceding text only assume signal stationarity. Parametric models allow to incorporate *a priori* knowledge of a statistical model into the estimation. This section deals with parametric methods for rational PSDs that form a dense set in the class of continuous PSDs (Stoica and Selen, 2004).

A rational PSD is a rational function of $e^{-J\omega}$, which means that it is the ratio of two polynomials. Motivated by the Weierstrass theorem which states that any continuous PSD can be approximated arbitrarily closely by a rational PSD provided that the degrees of the polynomials are chosen sufficiently large, there has been much interest in parametrizing the PSD in the following manner:

$$S_{yy}(e^{J\omega}) = \sigma^2 \left| \frac{\theta(e^{J\omega})}{\phi(e^{J\omega})} \right|^2, \tag{9.23}$$

Algorithm 18: RepMed: computes the repeated median estimate of the Fourier coefficients for a sample of length $N > 2$, with N prime.

---

    input   : $\mathbf{y}_N \in \mathbb{R}$ sample of length $N > 2$ and N prime, M
    output : $\hat{a}_k, \hat{b}_k, k = 1, \ldots, (N-1)/2$
    initialize: $\hat{a}_k, \hat{b}_k = 0, k = 1, \ldots, (N-1)/2$

1

$$a_0 = \mathsf{med}(\mathbf{y}_N)$$

$$\mathbf{y}_N \leftarrow \mathbf{y}_N - a_0$$

2 **for** $m = 1, \ldots, M$ **do**
3   **for** *kth strongest frequency,* $k \in \{1, \ldots, (N-1)/2\}$ **do**
4   **for** $u = 1, \ldots, N$ **do**
5   **for** $v = 1, \ldots, N$ *with* $v \neq u$ **do**
6

$$a_{kuv} \leftarrow \frac{y_u \sin(\omega_k v) - y_v \sin(\omega_k u)}{\sin(\omega_k(v-u))}$$

7

$$b_{kuv} \leftarrow \frac{y_v \sin(\omega_k u) - y_u \sin(\omega_k v)}{\sin(\omega_k(v-u))}$$

8

$$\tilde{a}_k \leftarrow \mathsf{med}_v(\mathsf{med}_u(a_{kuv}))$$

$$\tilde{b}_k \leftarrow \mathsf{med}_v(\mathsf{med}_u(a_{kuv}))$$

9

$$\mathbf{y}_N \leftarrow \mathbf{y}_N - \tilde{a}_k \cos(\omega_k t) - \tilde{b}_k \sin(\omega_k t)$$

10

$$\hat{a}_k \leftarrow \hat{a}_k + \tilde{a}_k$$

$$\hat{b}_k \leftarrow \hat{b}_k + \tilde{b}_k$$

---

where

$$\phi(e^{J\omega}) = 1 - \sum_{k=1}^{p} \phi_k e^{-J\omega k}, \tag{9.24}$$

and

$$\theta(e^{J\omega}) = 1 - \sum_{l=1}^{q} \theta_l e^{-J\omega l}. \tag{9.25}$$

Consistent with (8.2), $\boldsymbol{\phi} = (\phi_1, \ldots, \phi_p)^\top$, $\boldsymbol{\theta} = (\theta_1, \ldots, \theta_q)^\top$, respectively, are the autoregressive and moving-average parameters of an ARMA model, and σ^2 denotes the variance of the innovations sequence, as in defined in (8.6).

Based on (9.23) and (8.2), it is obvious that the ARMA model transforms the spectral estimation problem into a parameter estimation problem. Therefore, all the robust ARMA parameter estimators that were discussed in Section 8.3, lead also to robust PSD estimators if the parameter estimates are inserted into (9.23). This insertion yields

$$\hat{S}_{yy}(e^{J\omega}) = \hat{\sigma}^2 \left| \frac{\hat{\theta}(e^{J\omega})}{\hat{\phi}(e^{J\omega})} \right|^2, \tag{9.26}$$

with

$$\hat{\phi}(e^{J\omega}) = 1 - \sum_{k=1}^{p} \hat{\phi}_k e^{-J\omega k}, \tag{9.27}$$

and

$$\hat{\theta}(e^{J\omega}) = 1 - \sum_{l=1}^{q} \hat{\theta}_l e^{-J\omega l}. \tag{9.28}$$

To prevent the propagation of outliers (see Section 8.2), robust scale estimates $\hat{\sigma}$, e.g., the M-scale estimate (1.19), can be computed using bounded influence propagation (BIP) innovations, as defined in (8.22).

9.3   Simulation Example: Robust Spectral Estimation

To compare the different estimators that were detailed in the previous sections, a Monte Carlo simulation was conducted. The simulation is based on example 2 introduced by Moses et al. (1994), which considers the clean data case for an ARMA(4,3) process resulting in a rational PSD that follows (9.26) with

$$\phi(e^{J\omega}) = 1 + 1.3136e^{-J\omega t} + 1.4401e^{-J\omega 2t} - 1.0919e^{-J\omega 3t} + 0.83527e^{-J\omega 4t}, \tag{9.29}$$

$$\theta(e^{J\omega}) = 1 + 0.1792e^{-J\omega t} + 0.8202e^{-J\omega 2t} + 0.2676e^{-J\omega 3t} \tag{9.30}$$

and

$$\sigma^2 = 0.13137. \tag{9.31}$$

The true PSD, which is plotted as a dashed line in Figure 9.3, is given by

$$S_{yy}(e^{J\omega})$$

$$= 0.13137 \left| \frac{1 + 0.1792e^{-J\omega t} + 0.8202e^{-J\omega 2t} + 0.2676e^{-J\omega 3t}}{1 + 1.3136e^{-J\omega t} + 1.4401e^{-J\omega 2t} - 1.0919e^{-J\omega 3t} + 0.83527e^{-J\omega 4t}} \right|^2. \tag{9.32}$$

The displayed results are based on 100 Monte Carlo experiments. In each of these experiments, observations $\mathbf{y}_N$ of length $N = 512$ that follow a Gaussian ARMA(4,3)

process with parameters as given in (9.29)–(9.31) were generated. The outlier contami-
nated samples $\mathbf{y}^\varepsilon$ were generated by contaminating $\varepsilon = 2\%$ of the samples with isolated
additive outliers, see Section 8.1.2, sampled from a zero-mean contaminating Gaussian
distribution with variance $\kappa\sigma_y^2$, where $\kappa = 10$ and σ_y^2 is the variance of $\mathbf{y}_N$.

Figure 9.3 compares the following PSD estimators:

1.   Classical periodogram,
2.   Laplace periodogram (ℓ_1-periodogram) (Li, 2010),
3.   Biweight robust Fourier transform (Tatum and Hurvich, 1993),
4.   BIP-τ ARMA estimate (Muma and Zoubir, 2017),

and displays the mean values plus-minus one standard deviation of the PSD estimates
over 100 Monte Carlo runs. The left column shows the estimates for clean data samples,
whereas the right column shows the results for the additive outlier case.

As expected, the classical periodogram provides a nearly unbiased solution (with a
large variance) for clean data and breaks down completely in the presence of additive
outliers. The robust estimates, however, are similar in both cases. The ℓ_1-periodogram
has a slight bias (and a large variance) and is slightly outperformed by the biweight
robust Fourier transform. The nonparametric estimates are all outperformed by the
BIP-τ ARMA estimate, which largely benefits from incorporating the ARMA(4,3)
parametrization.

9.4   Robust Subspace-Based Frequency Estimation

Subspace-based frequency estimation is popular in the signal processing community,
and many approaches have been proposed. Commonly used approaches, such as MUSIC
and ESPRIT, rely on the eigendecomposition of the sample autocovariance matrix[1] as
defined in (9.34). However, as discussed extensively in Chapter 4, the sample covariance
matrix estimator experiences drastic performance degradation for the non-Gaussian
noise case. This, in turn, leads to inaccurate or even completely faulty frequency
estimation results.

This section considers robust subspace-based frequency estimation (Visuri et al.,
2000b). A performance comparison, for the case of ε-contaminated noise, through
Monte Carlo simulations, is provided. The first robust subspace frequency estimation
method that we consider has its basis in nonparametric statistics and is based on the
spatial sign concept that was introduced in Section 5.4.2 (Visuri et al., 2001). As
previously defined in (5.5), the spatial sign function is defined for a vector $\mathbf{z} \in \mathbb{C}^{p \times 1}$ as

$$\mathbf{s}(\mathbf{z}) = \begin{cases} \frac{\mathbf{z}}{||\mathbf{z}||}, & \mathbf{z} \neq \mathbf{0} \\ \mathbf{0}, & \mathbf{z} = \mathbf{0}. \end{cases}$$

---

[1] In spectral analysis, we use the term *autocovariance matrix* instead of *covariance matrix* to highlight its
Toeplitz structure, i.e., for the autocovariance matrix, each descending diagonal from left to right is
constant.

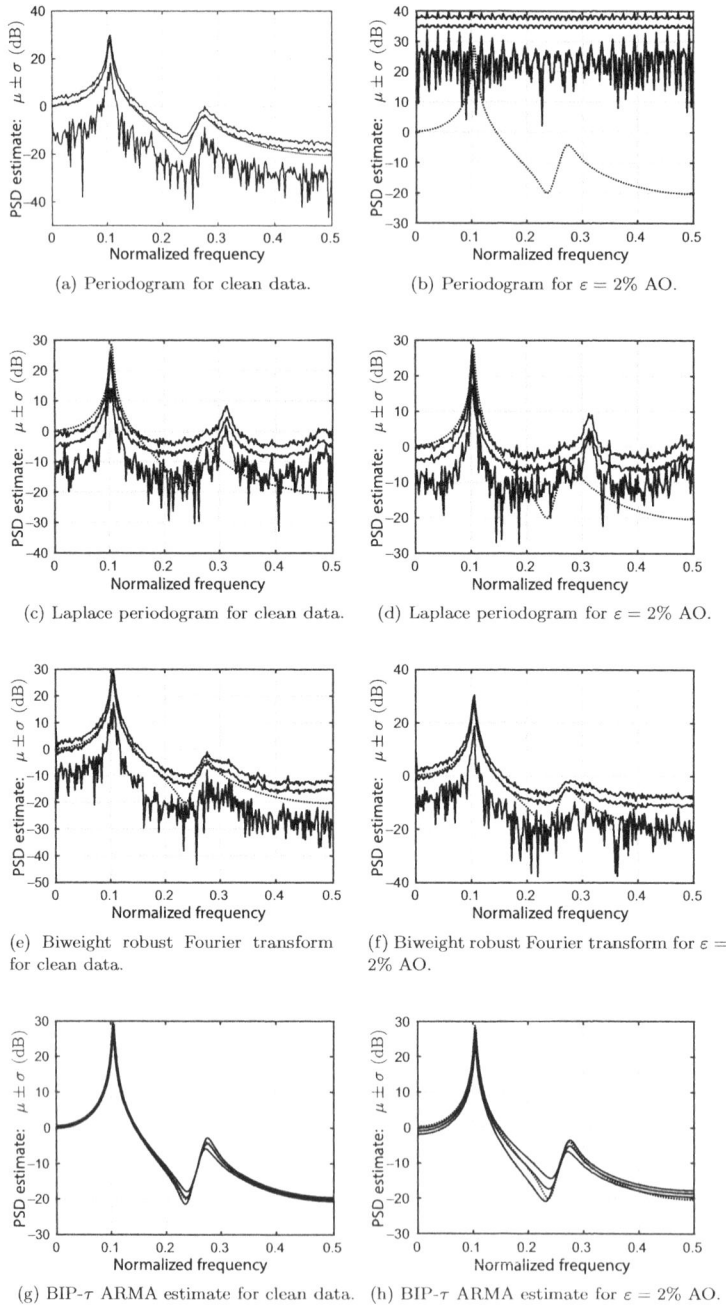

(a) Periodogram for clean data.

(b) Periodogram for $\varepsilon = 2\%$ AO.

(c) Laplace periodogram for clean data.

(d) Laplace periodogram for $\varepsilon = 2\%$ AO.

(e) Biweight robust Fourier transform for clean data.

(f) Biweight robust Fourier transform for $\varepsilon = 2\%$ AO.

(g) BIP-τ ARMA estimate for clean data.

(h) BIP-τ ARMA estimate for $\varepsilon = 2\%$ AO.

Figure 9.3 PSD for an ARMA(4,3) model underlying the observations of length $N = 512$. The dashed curve is the true PSD, the solid curves are estimates of the $\mu \pm \sigma$ bounds based on 100 Monte Carlo experiments. The left column shows the estimates for clean data samples, whereas the right column shows the additive outlier (AO) results for $\varepsilon = 2\%$ contamination and $\kappa = 10$.

Given an observed time series $\mathbf{y} = (y_1, y_2, \ldots, y_N)^\top$, by denoting

$$\mathbf{z}_t = (y_t, \ldots, y_{t+p-1})^\top, \quad t = 1, \ldots, N-p+1,$$

the *sample spatial sign autocovariance matrix* of size $p \times p$ is given by

$$\hat{\boldsymbol{\Sigma}}_{\text{SAM}} = \frac{1}{N-p+1} \sum_{t=1}^{N-p+1} \mathbf{s}(\mathbf{z}_t)\mathbf{s}(\mathbf{z}_t)^{\mathsf{H}}. \tag{9.33}$$

Here, p denotes the number of lags used in the autocovariance estimation, and it is assumed that $p < N$. Assuming that $\mathbf{z}_t$, $t = 1, \ldots, N-p+1$, is zero-mean, analogously to (4.1), the classical *sample autocovariance matrix* of size $p \times p$ is defined as

$$\hat{\boldsymbol{\Sigma}}_{\text{AM}} = \frac{1}{N-p+1} \sum_{t=1}^{N-p+1} \mathbf{z}_t \mathbf{z}_t^{\mathsf{H}}. \tag{9.34}$$

By noticing the equivalence of (9.34) and (4.1), an M-estimate of the autocovariance matrix can be obtained according to (4.22) as

$$\hat{\boldsymbol{\Sigma}} = \arg\min_{\boldsymbol{\Sigma}>0} \left\{ L(\boldsymbol{\Sigma}) = \sum_{t=1}^{N} \rho(\mathbf{z}_t^{\mathsf{H}} \boldsymbol{\Sigma}^{-1} \mathbf{z}_t) - \frac{N}{\gamma} \ln |\boldsymbol{\Sigma}^{-1}| \right\}.$$

See Section 4.3 for a detailed discussion.

Subspace-based frequency estimation utilizes the complex exponential signal model

$$y_t = \sum_{k=1}^{K} A_k e^{j t \omega_k} + v_t, \quad t = 1, \ldots, N, \tag{9.35}$$

to identify narrow and, potentially, closely-spaced spectral lines. Here, $K < p$ refers to the number of exponential signals, ω_k, $k = 1, \ldots, K$, are the frequencies, v_t, $t = 1, \ldots, N$, are complex-valued circular white noise samples with variance σ^2, and the complex amplitudes are given by

$$A_k = |A_k| e^{j\phi_k},$$

where ϕ_k denotes the phase.

By defining

$$\mathbf{v}_t = (v_t, \ldots, v_{t+p-1})^\top$$

$$\mathbf{z}_t = (y_t, \ldots, y_{t+p-1})^\top$$

$$\mathbf{a} = (A_1, \ldots, A_K)^\top$$

$$\mathbf{B} = \begin{pmatrix} 1 & 1 & \cdots & 1 \\ e^{j\omega_1} & e^{j\omega_2} & \cdots & e^{j\omega_K} \\ \vdots & \vdots & \ddots & \vdots \\ e^{j(p-1)\omega_1} & e^{j(p-1)\omega_2} & \cdots & e^{j(p-1)\omega_K} \end{pmatrix}$$

and

$$\mathbf{D}_t = \mathsf{diag}(e^{Jt\omega_1}, \ldots, e^{Jt\omega_K})$$

it follows from (9.35) that

$$\mathbf{z}_t = \mathbf{BD}_t\mathbf{a} + \mathbf{v}_t.$$

If the initial phases ϕ_k are independent, and uniformly distributed on $[-\pi, \pi)$, and with $\mathbf{z}_t$ being zero-mean, the $p \times p$ autocovariance matrix of the complex exponential signal $y_t, t = 1, \ldots, N$, becomes

$$\Sigma = \mathsf{E}\left[\mathbf{z}_t\mathbf{z}_t^{\mathsf{H}}\right] = \mathbf{B}\Sigma_{\mathbf{a}}\mathbf{B}^{\mathsf{H}} + \sigma^2\mathbf{I}, \qquad (9.36)$$

where $\Sigma_{\mathbf{a}} = \mathsf{diag}(|A_1|^2, \ldots, |A_K|^2)$ and σ^2 denotes the variance of the complex-valued circular white noise samples $\mathbf{v}_t$. The $p-K$ smallest eigenvalues of Σ are equal to σ^2, that is, the noise variance. The corresponding eigenvectors span the *noise subspace* whereas the eigenvectors associated with the K largest eigenvalues span the *signal subspace* that is orthogonal to the noise subspace. The signals lie in the signal subspace. The columns of the matrix $\mathbf{B}$ span the same signal subspace.

Subspace frequency estimation algorithms, such as MUSIC or ESPRIT, exploit different properties of the noise subspace and the signal subspace. In this section, the focus is on the MUSIC algorithm, which is based on the fact that the noise eigenvectors are orthogonal to the columns of $\mathbf{B}$. Using this orthogonality, the frequency estimates are chosen as the K largest peaks of the so-called MUSIC pseudospectrum

$$P_{\mathrm{MUSIC}}(e^{J\omega}) = \frac{1}{\mathbf{b}^{\mathsf{H}}(e^{J\omega})\hat{\mathbf{V}}\hat{\mathbf{V}}^{\mathsf{H}}\mathbf{b}(e^{J\omega})}, \qquad (9.37)$$

where $\mathbf{b}(e^{J\omega}) = (1, e^{J\omega}, \ldots, e^{J(p-1)\omega})^{\top}$ and $\hat{\mathbf{V}} = (\hat{\mathbf{v}}_{K+1}, \ldots, \hat{\mathbf{v}}_p)$ is the matrix composed of the eigenvectors corresponding to the $p - K$ smallest eigenvalues of the an autocovariance matrix estimate $\hat{\Sigma}$.

Robust MUSIC frequency estimation algorithms are constructed by replacing the nonrobust estimates $\hat{\Sigma}_{\mathrm{AM}}$ by their robust counterparts, for example, $\hat{\Sigma}_{\mathrm{SAM}}$ or $\hat{\Sigma}_{\mathrm{M}}$. Other robust subspace frequency estimation algorithms, such as robust ESPRIT, as well as the task of estimating the number of exponentials, that is, estimating K, are discussed in Visuri et al. (2001).

The following example illustrates robust MUSIC frequency estimation.

Example 19   Robust MUSIC frequency estimation

In this example, the difference between the MUSIC frequency estimation algorithm, which is based on nonrobust autocovariance estimates, that is, $\hat{\Sigma}_{\text{AM}}$, and its robust counterpart, which uses $\hat{\Sigma}_{\text{M}}$, are illustrated. In particular, results are presented for the $t_\nu M$-autocovariance matrix estimator, as defined in (4.22), with a t_ν loss function (4.28), where $\nu = 3$. The MATLAB$^{\copyright}$ function Mscat in the RobustSP toolbox is used to compute the M-estimator.

The following setup is considered. For each Monte Carlo run, $N = 2000$ observations are generated according to (9.35) with $K = 4$ sources, that is,

$$y_t = \sum_{k=1}^{4} A_k e^{Jt w_k} + n_t, \quad t = 1, \dots, 2000,$$

where $\phi_k \sim \mathcal{U}(-\pi, \pi)$, $|A_k| = \sqrt{50}$ and $\omega = (\omega_1, \omega_2, \omega_3, \omega_4)^\top = (\frac{89}{72}\pi, \frac{91}{72}\pi, \frac{13}{18}\pi, \frac{14}{18}\pi)^\top$. The number of lags for the autocovariance matrix computation is chosen to be $p = 30$. The frequency grid for $\mathbf{b}(e^{J\omega})$ is equidistant with a grid spacing of $2\pi/512$.

Figure 9.4 shows the superposition of the MUSIC pseudospectra arising from 10 MC experiments. The left column contains the results for the complex Gaussian noise case in which $n_t \sim \mathbb{C}\mathcal{N}_p(\mathbf{0}, \mathbf{I})$. For the right column, the noise is generated according to

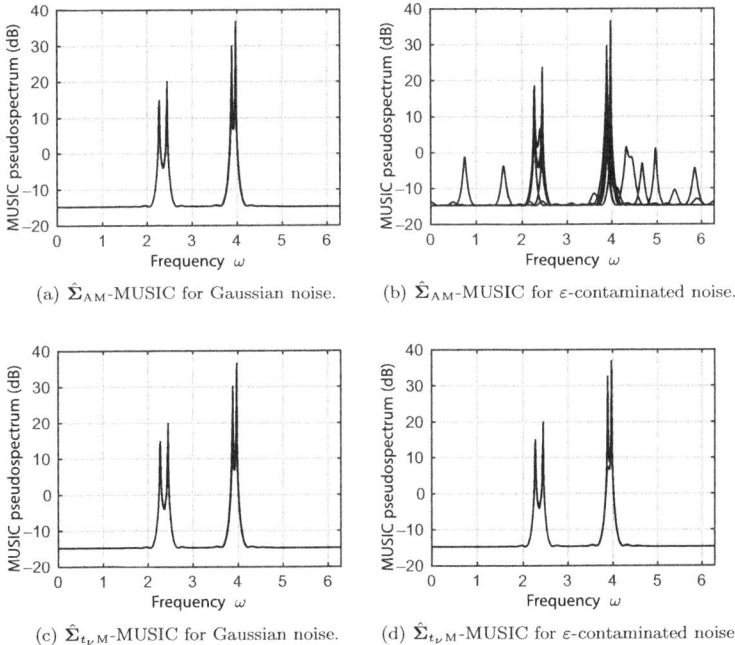

(a) $\hat{\Sigma}_{\text{AM}}$-MUSIC for Gaussian noise.

(b) $\hat{\Sigma}_{\text{AM}}$-MUSIC for ε-contaminated noise.

(c) $\hat{\Sigma}_{t_\nu \text{M}}$-MUSIC for Gaussian noise.

(d) $\hat{\Sigma}_{t_\nu \text{M}}$-MUSIC for ε-contaminated noise.

Figure 9.4 Superposition of the MUSIC pseudospectra arising from 10 MC experiments. The left column shows results for the Gaussian noise case, while the right column displays the results for the ε-contaminated noise case with $\varepsilon = 2\%$.

the ε-contaminated outlier model and where $\varepsilon = 2\%$ of the samples are replaced by outliers that follow $\mathbb{C}\mathcal{N}_p(\mathbf{0}, \kappa \mathbf{I})$, where $\kappa = 225$. While the robust MUSIC frequency estimator, based on the $t_\nu M$-autocovariance matrix estimator, maintains its excellent performance and can resolve the closely spaced frequency components, the classical MUSIC frequency estimator, based on the sample autocovariance matrix, is affected to a much greater extent by the outliers.

9.5   Concluding Remarks

In this chapter, both robust nonparametric and robust parametric spectrum estimation have been considered. Various nonparametric spectrum approaches, which are based on robustifying the periodogram, were discussed. When the outliers are concentrated in a few data segments, simple robust averaging of periodograms, for example, the robust Bartlett estimator, can be used. If the outliers are spread over the entire data set then, for example, robust M-periodograms can be used. As for the classical case, robust parametric approaches may provide an advantage because they can incorporate a model, such as the ARMA model. With such a model, the robust spectrum estimation task reduces to a robust ARMA parameter estimation task. For this case, robust ARMA parameter estimates, such as the ones discussed in Chapter 8, provide robust spectrum estimates. Finally, robust subspace frequency estimation was considered. Subspace frequency estimation algorithms exploit different properties of the noise subspace and the signal subspace. Robust subspace frequency estimators, such as the robust MUSIC, are based on the robust estimation of the autocovariance matrix. For this case, robust covariance matrix estimators, as discussed in Chapter 4, are required.

10 Robust Bootstrap Methods

10.1 Introduction

The chapter provides a brief treatment of the most recent developments of robust bootstrap methods. Although robust estimation, as discussed in previous chapters, has largely matured, there are many situations in which it is difficult, or even impossible, to do statistical inference for robust estimators, such as constructing confidence intervals (CIs) for the unknown parameters. Even if in some cases the asymptotic behavior of robust estimators is known, convergence to the limiting distribution may be slow. Contrary to asymptotic inference, the bootstrap provides better results in the small sample size regime and has become the method of choice (Efron, 1979; Efron and Tibshirani, 1994; Zoubir and Boashash, 1998; Zoubir and Iskander, 2004, 2007).

Bootstrap methods are based on the principle of resampling from the original sample, which has the drawback of propagating outlying samples that are likely to be present in collected measurements. It is therefore of utmost importance to apply robust bootstrap methods rather than to bootstrap robust estimators (Salibián-Barrera and Zamar, 2002; Amado and Pires, 2004; Müller and Welsh, 2005; Salibián-Barrera et al., 2008a; Amado et al., 2014; Basiri et al., 2014, 2016, 2017; Vlaski et al., 2014; Vlaski and Zoubir, 2014).

A further drawback of the bootstrap is that it cannot handle "big data" where statistical inference is prohibitively expensive and the presence of outlying observations is more than likely. This chapter reports not only on robust bootstrap methods, but also on the most recent advances on robust methods for large-scale data. In particular, fast, robust, and scalable bootstrap methods are introduced and discussed (Basiri et al., 2016, 2017).

This chapter includes real-life examples that highlight the efficacy of robust bootstrap methods. The first example is that of geolocation where the position of a mobile terminal in a harsh mixed LOS/NLOS environment is estimated (Vlaski et al., 2014; Vlaski and Zoubir, 2014). Furthermore, CIs for large-scale data based on a fast, robust, and scalable bootstrap (Basiri et al., 2016) are estimated. Finally, a hypothesis test is detailed that decides, based on a simplified version of the Million Song data set that is available on the UCI Machine Learning Repository with $N = 515,345$ music tracks from 1922 to 2011, whether a song is typical for its release year.

10.1.1   What Is the Bootstrap?

The bootstrap is a computational method for statistical inference, used in place of an approach based on asymptotics, particularly when the sample size is small and analytical inference is intractable. It is a powerful method used for the estimation of statistical characteristics, such as bias, variance, distribution functions, and, thus, CIs. It can also be used for hypothesis tests, for example, signal detection and model selection (Zoubir and Iskander, 2004). To understand robustness issues related to the bootstrap, we begin by briefly revisiting the nonparametric bootstrap principle.

Let $\mathcal{X} = \{x_1, x_2, \ldots, x_N\}$ denote a sample, that is, a collection of N numbers drawn at random from a completely unspecified distribution F. In this chapter, we assume that the x_is are realizations of i.i.d. random variables with distribution F. Let β denote an unknown characteristic of F, such as its mean or variance. The bootstrap enables us to find the distribution $G(\hat{\beta})$ of $\hat{\beta}$, an estimator of β, derived from the sample $\mathcal{X}$.

In the nonparametric bootstrap, a so-called bootstrap sample, also called a bootstrap resample, is generated by sampling *with* replacement from $\mathcal{X}$. Description 10.1 summarizes the nonparametric bootstrap, while Figure 10.1 illustrates its implementation using a schematic diagram. The underlying idea is to recalculate estimates for each of the B bootstrap samples to obtain bootstrapped estimates $\hat{\beta}^*$ whose empirical distribution $\widehat{G}(\hat{\beta})$ approximates the true distribution of $\hat{\beta}$.

Description 10.1 Principle of the nonparametric bootstrap.

Step 1.   Conduct the experiment to obtain the sample

$$\mathcal{X} = \{x_1, x_2, \ldots, x_N\}$$

and calculate the estimate $\hat{\beta}$ from the sample $\mathcal{X}$.

Step 2.   Construct the empirical distribution $\widehat{F}$, which puts equal mass $1/N$ at each observation

$$x_1, x_2, \ldots, x_N.$$

Step 3.   From the selected $\widehat{F}$, draw a bootstrap resample

$$\mathcal{X}^* = \{x_1^*, x_2^*, \ldots, x_N^*\}.$$

Step 4.   Compute the estimate $\hat{\beta}^*$ based on the bootstrap resample $\mathcal{X}^*$.

Step 5.   Approximate the distribution of $\hat{\beta}$ by the distribution of $\hat{\beta}^*$ derived from $\mathcal{X}^*$.

10.1.2   The Problem When Using the Bootstrap for Robust Estimators

On the one hand, the computational cost of bootstrapping robust estimators may become prohibitively large, especially when robust estimators involve solving nonconvex optimization problems. In particular, if one applies the classical bootstrap, the computational complexity is multiplied by a factor B, that is, the number of bootstrap samples, which usually is in the order of 10^2–10^3.

On the other hand, when taking a close look at the diagram of Figure 10.1, a second problem becomes immediately apparent. Through resampling (**Step 3** in Figure 10.1)

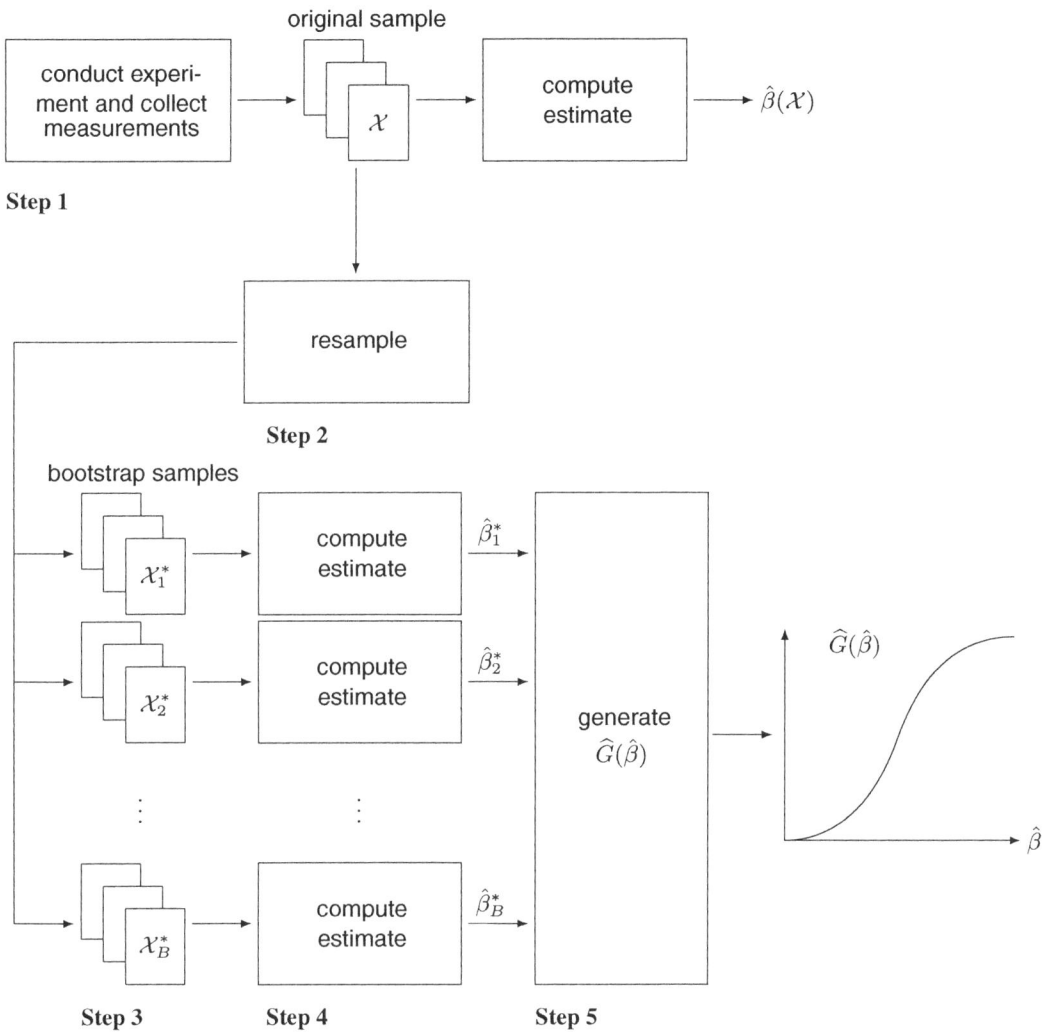

Figure 10.1 Implementation of the nonparametric bootstrap to estimate the distribution function $G(\hat{\beta})$ of $\hat{\beta}$, which is an estimator for the unknown parameter β.

an unduly large proportion of outliers might enter a significant number of bootstrap samples, which may severely affect the tails of $\widehat{G}(\hat{\beta})$. Unfortunately, this problem occurs independently of the estimator that is being bootstrapped. In fact, for large enough B, the asymptotic BP of the classical bootstrap is $\frac{1}{N}$ irrespective of the BP of the underlying estimator; see, for example, Singh (1998, theorem 1) or Stromberg (1997, appendix A).

10.2 Existing Robust Bootstrap Methods

In recent years, a few robust bootstrap methods have been developed and these are discussed in this section. To introduce and discuss the different methods, we consider the task of robustly estimating the CIs (detailed treatment in Section 10.3) based on robust estimators for the parameters in a simple linear regression model

$$y_i = x_i \beta + v_i, \quad i = 1, \ldots, N, \tag{10.1}$$

as discussed in Section 2.4.1. Of course, the bootstrap can be applied to a much larger class of problems, as discussed in Zoubir and Iskander (2004, 2007). Recently, it has also been extended to the big data regime (Kleiner et al., 2014; Basiri et al., 2016).

10.2.1   The Influence Function Bootstrap

As discussed earlier, for example, in Sections 1.3.1 and 2.7.2, robust estimators are characterized by bounded IFs. The influence function bootstrap (IFB) (Amado and Pires, 2004; Amado et al., 2014) assigns resampling probabilities based on the IF. This means that **Step 2** in Figure 10.1 is modified such that potential outliers are assigned small probabilities of appearing in any bootstrap resample. Because the harmful effect of outliers is eliminated through robust resampling, even a nonrobust estimator of the unknown parameter, β in (10.1), can be applied to the bootstrap resamples, yielding robust bootstrap estimates at an affordable computational complexity.

Let $F(\Omega)$ be some specified nominal parametric model, and let the actual distribution of the data belong to a contamination neighborhood in the sense of (1.42). Here, $\Omega = (\beta, \tau)$, where $\beta \in \mathbb{R}^p$ is the parameter vector of interest, while $\tau \in \mathbb{R}^s$ are the nuisance parameters. Then, following Amado and Pires (2004) and Amado et al. (2014), the standardized influence function (SIF) of some nonrobust estimator $\hat{\beta}^{nr}$, for example, the LSE, is defined as

$$\mathsf{SIF}\left(\mathbf{x}, \hat{\beta}^{nr}, F(\Omega)\right) = \left(\mathsf{IF}\left(\mathbf{x}, \hat{\beta}^{nr}, F(\Omega)\right)^{\top} \mathsf{var}^{-1}\left(\hat{\beta}_{\infty}^{nr}\right) \mathsf{IF}\left(\mathbf{x}, \hat{\beta}^{nr}, F(\Omega)\right) \right)^{\frac{1}{2}}, \tag{10.2}$$

with $\mathsf{var}^{-1}\left(\hat{\beta}_{\infty}\right) = \mathsf{E}\left[\mathsf{IF}\left(\mathbf{x}, \hat{\beta}, F(\Omega)\right) \mathsf{IF}\left(\mathbf{x}, \hat{\beta}, F(\Omega)\right)^{\top} \right]$ denoting the asymptotic variance of an estimator, and where the expectation is taken with respect to the nominal distribution $F(\Omega)$. Then, the robust standardized empirical influence function $\mathsf{RESIF}\left(\mathbf{x}, \hat{\beta}^{nr}, \hat{\Omega}^{r}\right)$ is obtained by substituting robust estimates $\hat{\Omega}^{r}$ of Ω into (10.2).

For the simple linear regression model, $\Omega = (\beta, \sigma)^{\top}$, where σ^2 denotes the variance of the residuals, it is the case that

$$r_i = y_i - x_i \beta, \quad i = 1, \ldots, N.$$

Recall from Chapter 2 that the LSE is an M-estimator with $\rho\left(\frac{r_i}{\sigma}\right) = \left(\frac{r_i}{\sigma}\right)^2, i = 1, \ldots, N$. Then, it follows that

$$\mathsf{SIF}(r_i, \sigma) = \left(\frac{r_i^2}{\sigma^2}\right)^{\frac{1}{2}} = \frac{|r_i|}{\sigma} \quad i = 1, \ldots, N.$$

The robust empirical SIF is then obtained by inserting robust estimates of β and σ, that is,

$$\text{RESIF}(\hat{r}_i, \hat{\sigma}) = \frac{|y_i - x_i \hat{\beta}|}{\hat{\sigma}}, \quad i = 1, \ldots, N. \tag{10.3}$$

Note that the computational cost of the IFB is reasonable because robust estimation is performed only once prior to bootstrapping. Every bootstrapped estimate $\hat{\beta}^*$ is obtained through the classical LSE as defined in Section 2.3.

Let $\mathcal{X} = \{(x_i, y_i), i = 1, \ldots, N\}$ be an available sample and let $\hat{\beta}$ and $\hat{\sigma}$ be the associated estimates of regression and scale, respectively. The resampling probabilities $\mathbf{p} = (p_1, p_2, \ldots, p_N)^\top$ are then obtained from (10.5), as described in Description 10.2. For more details, see Amado and Pires (2004). Note that the first condition on ϕ,

Description 10.2 The IFB obtains its robustness by resampling with unequal probabilities, that is, replacing **Step 2** and **Step 3** in Description 10.1 as described in the following text.

Step 2. Compute $\text{RESIF}(\hat{r}_i, \hat{\sigma})$ at each data point $i = 1, 2, \ldots, N$ by evaluating (10.3). Then, compute the weights w_i according to

$$w_i = \mathbb{1}_{[0,c]}(|\text{RESIF}(\hat{r}_i, \hat{\sigma})|) + \phi(c, |\text{RESIF}(\hat{r}_i, \hat{\sigma})|) \times \mathbb{1}_{(c,+\infty]}(|\text{RESIF}(\hat{r}_i, \hat{\sigma})|),$$
$$i = 1, 2, \ldots, N. \tag{10.4}$$

Here, $\mathbb{1}_{\mathcal{A}}$ is the indicator function of the set $\mathcal{A}$, $c > 0$ is a tuning constant and ϕ is a nonnegative function satisfying the two conditions:
Condition 1: $\lim_{t \to +\infty} t^2 \phi(c, t) = 0$ (for fixed c)
Condition 2: $\frac{\partial}{\partial t} \phi(c, t)|_{t=c} = 0$.
Step 3. Obtain bootstrap resamples by sampling with probabilities $\mathbf{p} = (p_1, p_2, \ldots, p_N)^\top$ where

$$p_i = \frac{w_i}{\sum_{n=1}^{N} w_n}. \tag{10.5}$$

---

as specified in Description 10.2, protects the bootstrap distribution from the harmful effects of outliers, while the second condition preserves the efficiency of the procedure with clean data (Amado and Pires, 2004). The bootstrap samples are then obtained by sampling with replacement from $\mathcal{X}$ and with the probabilities as specified by (10.5); see Description 10.2. The tuning constant c offers a trade-off between efficiency and robustness. A flexible family of functions from which ϕ can be chosen is the kernel of the pdf of the t-distribution and its limiting form, the normal distribution (Amado and Pires, 2004):

$$\phi(x; c, d, \gamma) = \begin{cases} \left(1 + \frac{(x-c)^2}{\gamma d^2}\right)^{-\frac{\gamma+1}{2}} & 0 < \gamma < \infty \\ \exp\left(-\frac{(x-c)^2}{2d^2}\right) & \gamma = \infty. \end{cases} \tag{10.6}$$

Here, c is the location parameter that is equal to the tuning constant in (10.4), d is the scale parameter, and γ is the shape parameter. Condition 1 is satisfied if $\gamma > 1$ and Condition 2 is satisfied if c is the tuning constant. To reduce the number of parameters to be adjusted to two (c and γ), the parameter d can further be set equal to c (Amado and Pires, 2004).

Through the weighting function defined in (10.4), outlying samples have a smaller probability of appearing in the bootstrap samples. As a result, the effective sample size is reduced. By sampling N times with replacement from a sample of effective size N_{eff}, variability is underestimated and consistency is not achieved. Vlaski et al. (2014) proposed that N be replaced by the effective sample size

$$
N_{\text{eff}} = \left\lfloor \sum_{i=1}^{N} w_i \right\rfloor,
$$

where $\lfloor \cdot \rfloor$ indicates the integer part. Note that for the case in which no outliers are present, every data point receives the weight $w = 1$ and $N_{\text{eff}} = N$.

10.2.2   The Stratified Bootstrap

The stratified bootstrap was proposed by Müller and Welsh (2005) in the context of robust model selection. By dividing data points into groups and bootstrapping each stratum individually, the fraction of contamination in every bootstrap sample is kept more representative, in terms of contamination level, of the original sample.

It was advocated by Müller and Welsh (2005) to group the samples, based on the residuals $r_i = y_i - x_i \hat{\beta}, i = 1, \ldots, N$, into eight strata of equal length. This groups potential outliers together and ensures that bootstrap resamples are more representative of the original sample in terms of fraction of contamination. Two issues arise when applying this approach:

1.   The number of strata offers a trade-off between robustness and consistency. A large number of strata leads to higher robustness at the cost of lower consistency because the size of the subset of the considered bootstrap resamples is reduced significantly. If, for example, we consider samples of size $N = 20$, dividing them into eight strata would lead to subsamples of size two or three, which would, in turn, result in a drastic underestimation of variability.

2.   Even a large number of strata does not guarantee that every single bootstrap resample is representative of the contamination in the original sample.

Vlaski et al. (2014) proposed an improvement to the stratified bootstrap that reme-dies these problems. Let ε_{max} denote the maximum number of contaminated samples before an estimator breaks down. Then the stratified sample is obtained as detailed in Description 10.3.

This stratification groups up to a total number of ε_{max} outliers, which corresponds to the maximum amount of contamination with which the underlying estimator can deal. By considering the BP of the underlying estimator, and allowing strata of unequal length, as few as two strata are sufficient to ensure that no bootstrap resample contains more than ε_{max} outliers. As a result, the robustness of the estimator is maintained for

Description 10.3 The stratified bootstrap divides the data into strata and bootstraps each stratum individually.

Step 1.   Compute the estimate $\hat{\beta}$ based on the original sample $\mathcal{X}$.

Step 2.   For every pair (x_i, y_i), compute the residual $r_i = y_i - x_i \hat{\beta}, i = 1, \ldots, N$, and obtain the ordered sample $\mathcal{X}_s = \{(x_{(1)}, y_{(1)}), \ldots, (x_{(N)}, y_{(N)})\}$, based on the associated ordered residuals $|r_{(1)}| \leq |r_{(2)}| \leq, \ldots, \leq |r_{(N)}|$.

Step 3.   Define the strata of $\mathcal{X}$ according to

$$\mathcal{X}_1 = \{(x_{(1)}, y_{(1)}), \ldots, (x_{(N-\varepsilon_{\max})}, y_{(N-\varepsilon_{\max})})\}$$
$$\mathcal{X}_2 = \{(x_{(N-\varepsilon_{\max}+1)}, y_{(N-\varepsilon_{\max}+1)}), \ldots, (x_{(N)}, y_{(N)})\}$$

---

every bootstrap resample. Furthermore, the sample is divided into only two strata, which improves accuracy, especially for small sample sizes.

10.2.3   The Fast and Robust Bootstrap

The fast and robust bootstrap (FRB) was proposed by Salibián-Barrera and Zamar (2002). Consider any estimate $\hat{\beta}$ that can be represented as a solution of a fixed-point equation

$$\hat{\beta} = g(\hat{\beta}; \mathcal{X}), \tag{10.7}$$

where $g : \mathbb{R}^p \to \mathbb{R}^p$ and $g(\cdot)$ is a continuous and differentiable function. This includes a large class of estimators, such as the M-, MM-, or τ-estimators, discussed in Chapter 2. The bootstrap estimator $\hat{\beta}^*$ then solves

$$\hat{\beta}^* = g(\hat{\beta}^*; \mathcal{X}^*). \tag{10.8}$$

The computation of $\hat{\beta}^*$ for every bootstrap sample is both computationally expensive and non-robust because an unduly large fraction of contamination may enter the bootstrap sample. A significant reduction in computational complexity can be achieved by using the FRB method (Salibián-Barrera et al., 2008a), which does not require recomputing the estimates in each stage. It is applicable for estimators $\hat{\beta} \in \mathbb{R}^p$ that can be expressed as a solution to a system of smooth fixed-point estimation equations. Instead of computing $\hat{\beta}^*$ as previously, the following approximation is used:

$$\hat{\beta}^{1*} = g(\hat{\beta}; \mathcal{X}^*). \tag{10.9}$$

Because the distribution of $\hat{\beta}^{1*}$ typically underestimates the sampling variability of $\hat{\beta}$, a linear correction term based on a Taylor-series approximation of the function g is applied as follows:

$$\hat{\beta}^{R*} = \hat{\beta} + \left[\mathbf{I} - \nabla g(\hat{\beta}; \mathcal{X})\right]^{-1} (\hat{\beta}^{1*} - \hat{\beta}), \tag{10.10}$$

where $\nabla g(\hat{\beta}; \mathcal{X}) \in \mathbb{R}^{p \times p}$ is the matrix of partial derivatives w.r.t. $\hat{\beta}$. The FRB method has the following advantages: first, it is fast to compute because the initial estimate $\hat{\beta}$ is computed only once, for example, for the distinct data subset, as in Section 10.4, or for the full data, if feasible. Second, the one-step improvement step $\hat{\boldsymbol{\beta}}^{1*}$ requires only one iteration of the fixed-point estimation equation. Further, the estimator is also robust in the face of outliers if a robust fixed-point estimator, such as the MM-estimator, is used. Under conditions detailed in Salibián-Barrera and Zamar (2002), the obtained solution $\hat{\boldsymbol{\beta}}^{R*}$ estimates the same limiting distribution as the actual bootstrap calculations $\hat{\boldsymbol{\beta}}^{*}$. The estimate $\hat{\boldsymbol{\beta}}^{R*}$ is not only much faster to compute than $\hat{\boldsymbol{\beta}}^{*}$, it is also more robust because outlying points carry their small weights into the bootstrap recalculations.

10.2.4 The Robust Starting Point Bootstrap

The robust starting point bootstrap (RSPB) was proposed by Vlaski et al. (2014) and Vlaski and Zoubir (2014). Consider a high-BP regression estimator, such as the S-estimator, the τ-estimator, or the MM-estimator, for which the objective function, in general, is nonconvex. As discussed in Chapter 2, approximate solutions are found by choosing a sufficiently large set of starting points for an iterative algorithm that converges to a set of solutions (local minima). These solutions will usually contain the global minimum of the objective function, as long as the level of contamination is below the BP of the estimator. However, the bootstrap resamples may contain a large fraction of outliers such that the level of contamination is greater than the BP of the underlying estimator. For this reason, the smallest local minimum may follow the contaminated data, which leads to a breakdown of the estimator. When bootstraping such contaminated data, it is thus not desirable to find the global minimum of a loss function of a bootstrap sample $\mathcal{X}^{*}$, but rather to find a local minimum in the neighborhood of $\hat{\boldsymbol{\beta}}$, the solution for the original sample $\mathcal{X}$. Thus, the RSPB modifies **Step 2** and **Step 4** of the implementation of the nonparametric bootstrap procedure, as specified in Description 10.1.

Description 10.4 The RSPB modifies the computation of the bootstrap estimates by finding the local minimum of the associated objective function starting from the original robust estimate.

> **Step 2.** Obtain $\hat{\beta}$ and $\hat{\sigma}$ for the original sample $\mathcal{X}$ by evaluating the full iterative algorithm that solves the minimization problem of the robust estimator. This is done only once prior to bootstrapping.
>
> **Step 4a.** For every bootstrap resample $\mathcal{X}^{*}$ choose the estimate $\hat{\beta}$ with associated scale estimate $\hat{\sigma}$ that is based on $\mathcal{X}$ as the only candidate point for the minimization problem.
>
> **Step 4b.** Apply an iterative algorithm, which reduces the respective objective function, until the estimate converges to a solution of (10.8), which is close to that of (10.7). Then, $\mathcal{X}^{*}$ $\hat{\beta}^{*}$ and $\hat{\sigma}^{*}$ can be obtained.

By choosing a robust starting point for the iterative algorithm, $\hat{\beta}^{*}$ is likely to converge to the local minimum of the objective function for $\mathcal{X}^{*}$, which is close to $\hat{\beta}$, even when the absolute minimum is shifted due to a level of contamination that exceeds the BP of the estimator. An additional advantage of this strategy is that, as $\hat{\beta}$ is a good starting point, faster convergence is usually obtained.

10.3 Robust Bootstrap Confidence Interval Estimation in Linear Regression

In any estimation task, we should not be interested in point estimates only, but also in their accuracy. A CI for an unknown parameter, say β, provides a measure of accuracy of the estimator for β, $\hat{\beta}$, based on the sample $\mathcal{X}$. A CI $\text{CI}_\alpha = (a, b)$ with confidence level $1 - \alpha$ indicates that, with stochastic bounds a, b, the parameter β lies within the interval (a, b) with probability $1 - \alpha$, the so-called confidence level. The confidence length $l = b - a$ is used as a measure for the accuracy of $\hat{\beta}$. The bootstrap allows the estimation of CIs through the estimation of the distribution of the estimator $\hat{\beta}$ (Zoubir and Iskander, 2004).

10.3.1 Breakdown of Confidence Intervals

There are two distinct cases for the breakdown of a CI. Let $\text{CI}_\alpha = (a, b)$ be the confidence with confidence level $1 - \alpha$ and length $l = b - a$ centered around $m = \frac{a+b}{2}$ as shown in Figure 10.2.

No information about β can be retrieved from CI_α, if either

Case 1: $(m - \beta) \to \pm\infty$ (bias toward infinity), or if
Case 2: $l \to \infty$ (interval length toward infinity).

Maximum-Bias and Interval Length Curves

MBCs are a well-established measure of robustness, and they facilitate distinguishing between the performance of estimators with similar BPs; see Chapter 1. For CIs, two distinct cases of breakdown are possible, namely, a breakdown in bias and a breakdown in interval length. In this section, the maximum CI bias and the maximum CI length are estimated based on the worst case arising from all MC iterations. The worst-case results can be used to assess the BP of the robust bootstrap estimators for the CI.

Empirical Coverage Probability

The MBC alone does not yield sufficient insight about the quality of a CI. While a small bias is generally desirable, a short CI does not necessarily imply a good CI. To measure the accuracy of a CI, a second measure, the empirical coverage probability (ECP) (Amado and Pires, 2004), is used. The ECP of a CI measures the probability that a CI contains the true value of a statistic of interest. Ideally, for large sample sizes and clean data, the ECP of a CI tends to its confidence level $1 - \alpha$. It is important to note that both higher and lower ECP are undesirable. In this section, the ECP is estimated, based on MC experiments, as the fraction of CIs that contain the true value of the statistic of interest.

$$\text{CI}_\alpha = (a, b)$$

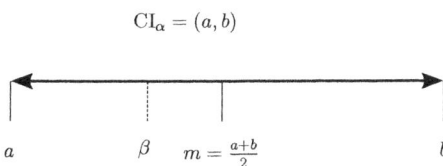

Figure 10.2 CI with confidence level $1 - \alpha$ and length $l = b - a$ centered around m.

Connection between Confidence Interval Bias, Length and Empirical Coverage Probability

The CI bias, the confidence length, and the ECP are related as follows:

1.   An *increase* in the CI bias leads to a *decrease* in the coverage probability.
2.   An *increase* in a CI length leads to an *increase* in the coverage probability

Generally, as the level of contamination increases, estimators suffer from increased bias and increased variability, which leads to increased CI bias and CI length. Whether the coverage probability increases or decreases depends on the rate of increase in bias and interval length.

10.3.2   Comparison of Robust Bootstrap Methods

To compare the previously discussed robust bootstrap methods in terms of their CI bias, CI length, and ECP, a simulation study based on 1,000 MC realizations of data from a contaminated normal distribution was conducted.

Simulation Setup

A clean sample of size N is generated according to the simple linear regression model

$$y_i = x_i \beta + v_i, \quad i = 1, \ldots, N,$$

where the regressors x_i and noise terms v_i, $i = 1, \ldots, N$, are i.i.d. and distributed according to $\mathcal{U}(0, x_{\max})$ and $\mathcal{N}(0, \sigma_v^2)$, respectively, and β is the deterministic parameter of interest. Then, a fraction ε of randomly selected data points (y_j, x_j) are corrupted by additive outliers that are distributed according to a bivariate normal distribution $\mathcal{N}_2(\boldsymbol{\mu}, \boldsymbol{\Sigma})$ with mean vector $\boldsymbol{\mu} = (0, 0)^\top$ and covariance matrix $\boldsymbol{\Sigma} = \begin{pmatrix} \sigma_e^2 & 0 \\ 0 & \sigma_e^2 \end{pmatrix}$.

For the simulations, the parameters were chosen as: $N = 20$, $x_{\max} = 60$, $\beta = 5$, $\sigma_v = 1$, $\sigma_e = 100$, $\alpha = 10\%$. A large choice of σ_e is suitable because worst-case scenarios are desirable to test the robust bootstrap methods under extreme circumstances. For all bootstrap methods, robust *MM*-estimators, as defined in (2.93), were used. The estimates, which are approximate solutions to (2.93), were computed with the algorithm proposed by Salibián-Barrera and Yohai (2012).

Simulation Results

Figure 10.3 shows the ECP, which is estimated, based on 1000 MC iterations, as the fraction of CIs that contain the true value of the statistic of interest. Ideally, for large sample sizes and clean data, the ECP of a CI tends to its confidence level, in this case $1 - \alpha = 90$ percent. It is important to note that, for good CIs, both high and low ECP are undesirable. For this simulation example, all robust bootstrap methods were able to provide similar and accurate ECPs.

As discussed in Section 10.3.1, CIs can break down in terms of their length and bias. Figure 10.4a, therefore, displays the maximum CI bias, whereas Figure 10.4b shows the maximum CI length. All methods perform similarly, except for the FRB, which is more sensitive to increasing levels of contamination.

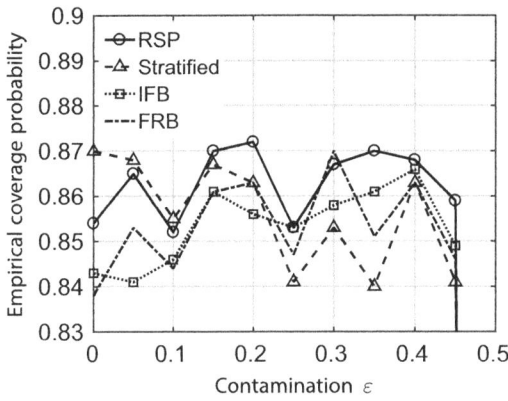

Figure 10.3 ECP of the CIs for the RSPB, stratified (stratified bootstrap), IFB, and FRB.

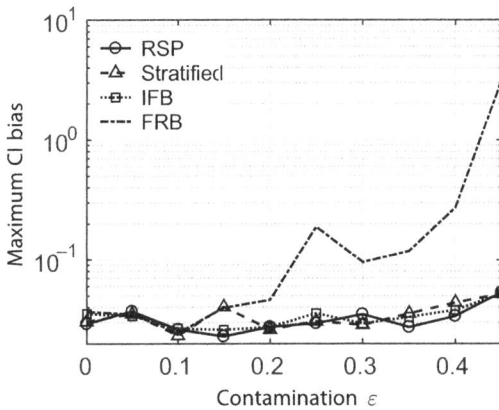

(a) CI bias for the robust bootstrap methods.

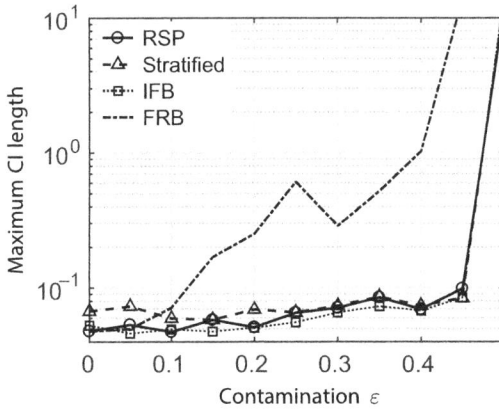

(b) CI length for the robust bootstrap methods.

Figure 10.4 Bias and length of the CIs for the RSPB, stratified (stratified bootstrap), IFB, and FRB.

Example 20   Example: Localization of UE

The bootstrap methods described previously can be directly applied to TOA-based localization when using the linearized regression-based approach that was discussed in Section 2.9.2. For this example, an *MM*-estimator of the regression parameters was used. Figure 10.5a displays the RSPB distribution estimate (contour plot) of the *MM*-estimate, the *MM*-point estimate, and the true position of the UE for the LOS case ($\varepsilon = 0$). Anchors are located at $(2500, 5000)$, $(1500, 4000)$, $(3000, 4500)$, $(4000, 3500)$, $(2000, 250)$, and $(1000, 1000)$ meters.

Figure 10.5b shows the RSPB distribution estimate in the mixed LOS/NLOS case with NLOS probability $\varepsilon = 0.4$. The distribution estimate adapts to the *MM*-estimate, which, due to the strong contamination, performs slightly worse than in the LOS case. Nonetheless, the quality of the distribution estimate is comparable to the LOS case. Unlike the classical bootstrap, the RSPB distribution estimate inherits the robustness

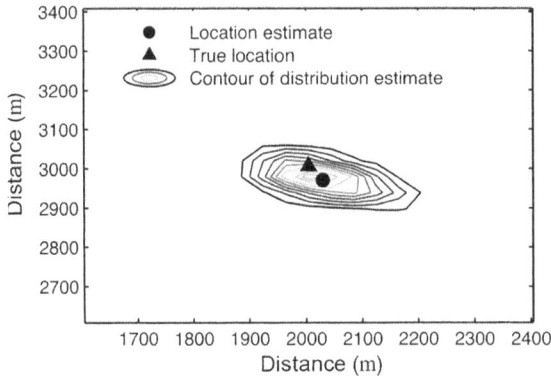

(a) LOS environment.

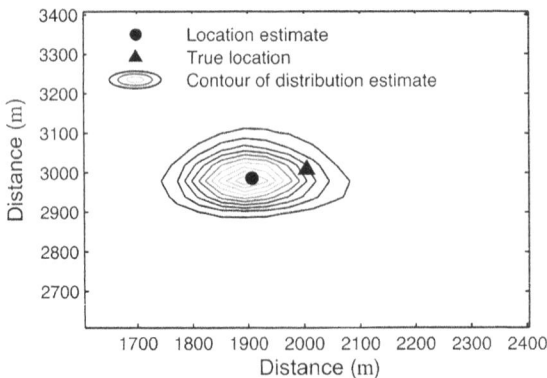

(b) Mixed LOS/NLOS environment.

Figure 10.5 The RSP bootstrap distribution estimate of the *MM*-estimator for the localization example.

properties of the underlying *MM*-estimate and is similar, in terms of area, to the LOS case.

The distribution estimate was extracted solely through the RSPB and without additional measurements or assumptions. This is crucial in geolocation and in many other applications because the experiment cannot be repeated under the same conditions. In particular, in geolocation, the UE is moving and the environment changes continuously.

The distribution estimate shows that the uncertainty in the horizontal direction is larger than the uncertainty in the vertical direction. The reason for this is the position of the fixed terminals relative to the true position of the UE and the structure of outliers in this particular example. This additional information could be utilized, for example, in a tracking scenario, where distribution estimates from past measurements can be combined to improve new position estimates.

10.4   Robust and Scalable Bootstrap for Large-Scale Data

10.4.1   Introduction

We live in an era of data deluge. Both the volume and the dimensionality of the data have grown to an extent that it cannot be effectively analyzed by traditional statistical inference methods. In fact, the data can be so large that it cannot be processed or stored in one computation or storage unit. Hence, the processing and storage of massive data sets may require the use of parallel and distributed computing architectures. Performing inference on massive data sets, using distributed and parallel platforms, requires fundamental changes in statistical signal processing and inference methods. Even a simple parameter estimation task, which uses all the data in a massive data set, can be prohibitively expensive. In addition, characterizing the uncertainty (e.g., CIs) in parameter estimation may not be computationally feasible using conventional bootstrap methods, for example. The lack of scalability of conventional signal processing, inference, and learning techniques forms a bottleneck in extracting relevant information from the data. For example, for the case of the bootstrap for large-scale data, one would have to generate a large number of bootstrap resamples of size N from the original massive data set that is of size N as well. Consequently, even more storage is required than for the original full data set. Furthermore, processing a good number of bootstrap resamples would further increase the computational complexity because the inference equations need to be solved for each bootstrap resample rather than only once for the original data set.

Variants of the resampling approach commonly used in the bootstrap have been proposed, for example subsampling and the m-out-of-N bootstrap (Bickel et al., 2012). The goal in this approach is to reduce the computational cost of the bootstrap by computing the parameter estimates using subsamples of size m, which is smaller than the size of the original data set, N. Implementing and analyzing the performance of such methods is problematic because the output is sensitive to the size of the subsamples m. Furthermore, using $m < N$ subsamples causes the variability of the bootstrap resamples to differ from the sampling variability of the estimator. Thus, additional analytical work is needed to compensate for the difference in scales of the outputs.

The bag of little bootstraps (BLB) (Kleiner et al., 2014) resampling approach was proposed to make bootstrapping scalable to large-scale data. In the BLB method, the massive data set is first resampled, without replacement, to create smaller disjoint subsets where each subset has distinct data items. An assumption of statistical independence is necessary in this step. The size of each disjoint subset is chosen to be $b = \lfloor N^\gamma \rfloor$ where $\gamma \in [0.5, 0.9]$.

Such smaller distinct data sets may be stored in distributed storage units and computations can be done in parallel distributed processing units, which makes the approach very scalable and attractive for large-scale data analysis problems. The bootstrap is employed for each distinct data set and bootstrap resamples of size N are drawn. It is important to note that this will not cause a storage problem because in each unit, bootstrap resamples are constructed by assigning a random weight vector $\mathbf{n}^* = (n_1^*, \ldots, n_b^*)^\top$, from a multinomial distribution $\mathcal{M}(N, (1/b)\mathbf{1}_b)$, to each distinct data point of the subsample, where $\sum_{i=1}^{b} n_i^* = N$, that is, the size of the original full data set. These weights describe the multiplicity of each data item in the bootstrap resample. Hence, the only additional storage needed is for the weights for b data items in each bootstrap resample, but no additional storage for the actual data is required. Unfortunately, one still needs to compute estimates for a large number of bootstrap resamples that may be prohibitively expensive, for example, if it requires solving a complicated optimization problem or inverting large matrices. Moreover, as discussed previously, the bootstrap is even more sensitive to outliers than conventional estimators because an outlier may appear with multiplicity in the bootstrap resample making it even more contaminated than the original data.

10.4.2   Making the Bootstrap Method Scalable

In this section, a scalable variation of the bootstrap method, called the bag of little bootstraps (BLB) is described (Kleiner et al., 2014). It facilitates using bootstrap analysis, as well as finding CIs and point estimates for large-scale data. It is scalable to very large volume and high-dimensional data sets, that is, big data. Moreover, it is compatible with distributed data storage systems and distributed and parallel computing architectures. The method is comprised of three key steps. First, the original big data is resampled *without* replacement to form multiple smaller distinct data sets. Second, the bootstrap is applied to each distinct data set so that the smaller and distinct resamples have N data items. However, storage and computation is needed only for b data items and their weights describing their multiplicity in a bootstrap resample. Third, for each resample, estimates of the parameters of interest are computed. Moreover, the empirical distribution of the parameter estimates is formed for each distinct subset of data. The CIs may then be estimated, for example, by using the bootstrap percentile method. In this method, to find the bounds on the CI for a parameter, the point estimates are rank ordered and the 5 and 95 percentile points of the empirical distribution are determined.

The point estimates, and their CIs, for the full large-scale data are found by combining the results from each distinct subset, for example by averaging the results over all the distinct subsets.

The BLB procedure is illustrated in Figure 10.6. It is described more formally in Description 10.5.

Description 10.5 The BLB procedure facilitates using bootstrap analysis, as well as finding CIs and point estimates for large-scale data.

Step 1.   Draw s subsamples $\check{\mathcal{X}} = \{\check{\mathbf{x}}_1, \ldots, \check{\mathbf{x}}_b\}$ of smaller size $b = \lfloor N^{\gamma} \rfloor |, \gamma \in [0.5, 0.9]$
by randomly sampling *without* replacement from $\mathcal{X} = \{\mathbf{x}_1, \ldots, \mathbf{x}_N\}$.

for   each subsample $\check{\mathcal{X}}$

   Step 2.   Generate B bootstrap samples $\mathcal{X}^* = \{\check{\mathcal{X}}, \mathbf{n}^*\}$ by assigning a random weight
   vector $\mathbf{n}^* = (n_1^*, \ldots, n_b^*)^{\top}$ from a multinomial distribution $\mathcal{M}(N, (1/b)\mathbf{1}_b)$ to the data
   points of $\check{\mathcal{X}}$.

   Step 3.   Compute the estimate $\hat{\boldsymbol{\beta}}^*$ based on each $\mathcal{X}^*$.

   Step 4.   Use the population of B bootstrap resamples $\hat{\boldsymbol{\beta}}^*$ to estimate the bootstrap CI
   $\hat{\boldsymbol{\xi}}^*$ (e.g., by using the bootstrap percentile method).

end

Step 5.   Average the computed values of $\hat{\boldsymbol{\xi}}^*$ over the subsamples, i.e., $\hat{\boldsymbol{\xi}}^* = \frac{1}{s} \sum_{k=1}^{s}$
$\hat{\boldsymbol{\xi}}^{*(k)}$.

A few clarifying remarks are in order:

- The distinct data of the subsamples generated in **Step 1** allows the original big data to be stored in distributed storage systems.
- In **Steps 2–4**, the processing of subsamples $\check{\mathcal{X}} = \{\check{\mathbf{x}}_1, \ldots, \check{\mathbf{x}}_b\}$ can be done in parallel using different computing units.
- $\mathcal{X}^* = \{\check{\mathcal{X}}, \mathbf{n}^*\}$ resembles a conventional bootstrap sample of size N with at most $b = \lfloor N^{\gamma} \rfloor |, \gamma \in [0.5, 0.9]$ distinct data points. Element n_i^* of $\mathbf{n}^* = (n_1^*, \ldots, n_b^*)^{\top}$ is a weight that defines the multiplicity of original subsample data item $\check{\mathbf{x}}_i$, $i = 1, \ldots, b$, for the bootstrap resample $\mathcal{X}^*$.
- The BLB is computationally less complex than the conventional bootstrap. Let us consider simple averaging to provide a concrete example of the computational advantages of the BLB scheme. In BLB, the averaging is computed by b summations $(+)$ and b multiplications $(\times)$ as $\hat{\mu}(\mathcal{X}^*) = \frac{1}{N} \sum_{i=1}^{b} n_i^* \check{\mathbf{x}}_i$, whereas in the conventional bootstrap N *summations* $(+)$ are needed as $\hat{\mu}(\mathcal{X}^*) = \frac{1}{N} \sum_{i=1}^{N} \mathbf{x}_i^*$.

In Figure 10.6, $\check{\mathcal{X}}^{(k)}$, $k = 1, \ldots, s$, denote the disjoint subsamples and $\mathcal{X}^{*(kj)}$, $j = 1, \ldots, B$, corresponds to the jth BLB sample generated based on the subsample k. In the original BLB method, a least square (LS) estimator was used for each bootstrap sample. It is well known that the LS estimator is not robust in the face of outlying observations. Moreover, the same outliers may appear many times in a bootstrap resample making the method even more sensitive than conventional least squares estimation. The BLB resampling approach is not limited to using least squares estimation.

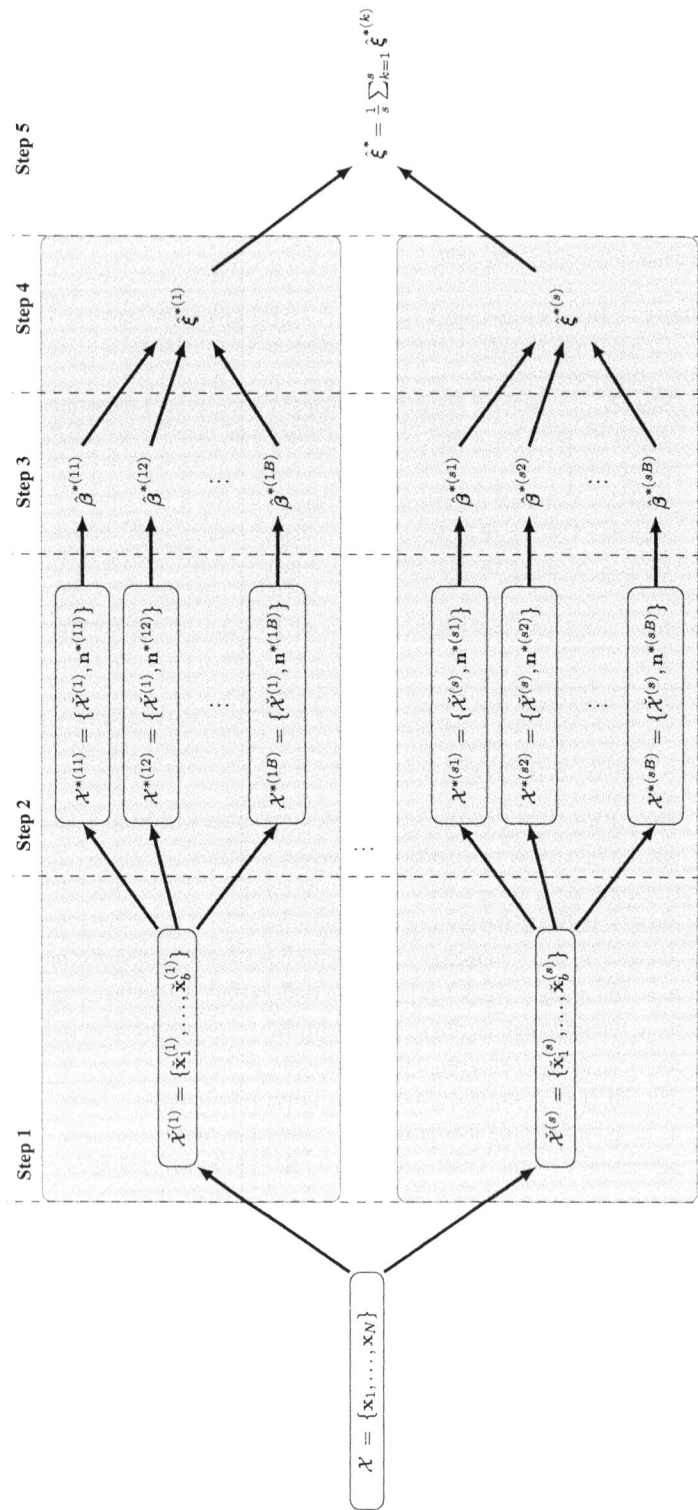

Figure 10.6 A block diagram of the BLB method. In Step 1, distinct data sets are created by resampling without replacement. Step 2 draws bootstrap resamples using multinomial weighting. In Step 3, point estimates are computed for each resample. Finally, in Steps 4 and 5, the results from different resamples and distinct subsets are combined to determine the final point estimates and their CIs.

10.4.3 A Fast, Robust, and Scalable Bootstrap Method

The FRB estimator is not scalable for large-scale data analysis problems, and it is difficult to parallelize it across distributed computing nodes. However, if it is used as an estimator for the resamples in the BLB, both scalability and statistical robustness can be achieved simultaneously. The main computational burden of the BLB scheme is in **Step 3** of the algorithm in Description 10.5 because the estimating equations need to be solved, potentially, many times for each bootstrap sample $\mathcal{X}^*$. The use of the FRB partially solves this problem.

The bag of little, fast, and robust bootstraps (BLFRB) method proposed by Basiri et al. (2016) combines the desirable properties of the BLB and FRB methods. The problem of recomputing estimates for each resample potentially multiple times is alleviated by using (fast to compute) fixed-point estimating equations that do not need to be (re)solved for each bootstrap resample. Many estimators, including many of the robust estimators presented in this book, can be expressed as a solution to fixed-point estimation equations. Prime examples include M-estimators, MM-estimators, S-estimators, and the Fast-ICA method for independent component analysis. If statistically robust estimators, such as MM-estimators, are employed for the bootstrap resamples, the complete bootstrap analysis method inherits the robustness of the employed estimator. Just like the BLB method (Kleiner et al., 2014), the BLFRB method is scalable to large volume data sets and is compatible with distributed data storage and processing architectures. Moreover, it is fast to compute because the estimating equations need to be computed only once for each subset. The bootstrap sample data are just used to find a one-step refinement of the initial estimate. Moreover, a linear correction term is applied to ensure that the variability is correct. The reduction in computation may be seen in the procedure described in detail:

- Let $\hat{\boldsymbol{\beta}}_b$ be a solution to $\hat{\boldsymbol{\beta}} = g(\hat{\boldsymbol{\beta}}; \mathcal{X})$, for subsample $\check{\mathcal{X}} \in \mathbb{R}^{d \times b}$:

$$\hat{\boldsymbol{\beta}}_b = g(\hat{\boldsymbol{\beta}}_b; \check{\mathcal{X}}). \tag{10.11}$$

- Let $\mathcal{X}^* \in \mathbb{R}^{d \times N}$ be a bootstrap sample of size N that is randomly resampled, with replacement, from the distinct data subset $\check{\mathcal{X}}$ of size b;
- The FRB resampling of $\hat{\boldsymbol{\beta}}_b$ can be obtained by

$$\hat{\boldsymbol{\beta}}_b^{R*} = \hat{\boldsymbol{\beta}}_b + \left[\mathbf{I} - \nabla g(\hat{\boldsymbol{\beta}}_b; \check{\mathcal{X}})\right]^{-1} (\hat{\boldsymbol{\beta}}_b^{1*} - \hat{\boldsymbol{\beta}}_b), \tag{10.12}$$

where $\hat{\boldsymbol{\beta}}_b^{1*} = g(\hat{\boldsymbol{\beta}}_b; \mathcal{X}^*)$ is the one-step estimator and $\nabla g(\cdot) \in \mathbb{R}^{p \times p}$ is the matrix of partial derivatives w.r.t. $\hat{\boldsymbol{\beta}}_b$.

The initial estimate $\hat{\boldsymbol{\beta}}_b$ and the correction term $\left[\mathbf{I} - \nabla g(\hat{\boldsymbol{\beta}}_b; \check{\mathcal{X}})\right]^{-1}$ are computed only once for each distinct data subset, which further reduces complexity.

Statistical robustness in the face of outliers is achieved by employing fixed-point robust estimators, for example, S-estimators or MM-estimators as described in Section 2.8. A block diagram of the BLFRB method is illustrated in Figure 10.7. The procedure

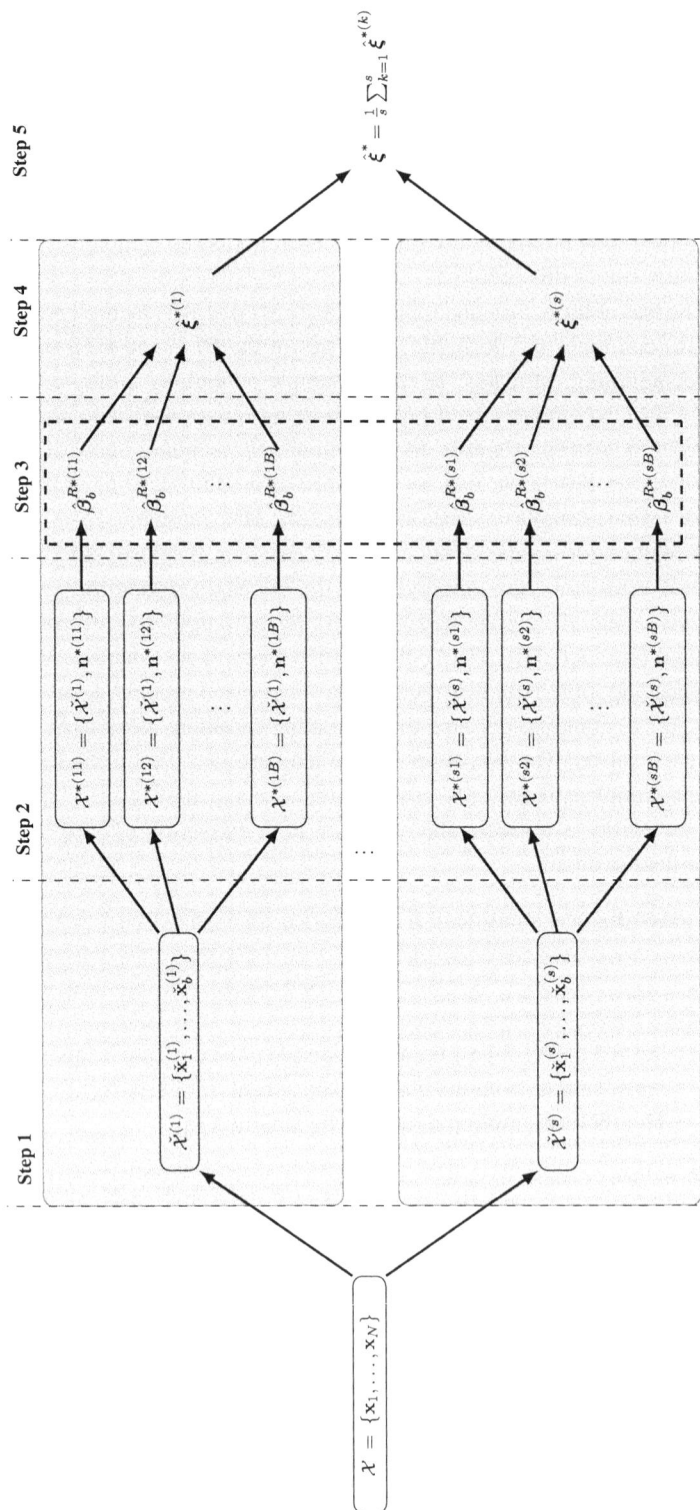

Figure 10.7 Block diagram of the BLFRB method. Significantly lower computational complexity, compared to the BLB, is achieved by using fixed-point estimation equations and one-step improvement of the initial estimate combined with a correction term in Step 3.

for performing estimation and inference using the BLFRB method is described, in detail, in Description 10.6.

Description 10.6 The BLFRB procedure enables robust bootstrap analysis, as well as finding CIs and point estimates, for large-scale data.

Step 1.   Draw s disjoint subsamples $\check{\mathcal{X}} = \{\check{x}_1, \ldots, \check{x}_b\}$ of smaller size $b = \lfloor N^\gamma \rfloor |$, $\gamma \in [0.5, 0.9]$.
for   each subsample $\check{\mathcal{X}}$
    Step 2.   Generate B bootstrap samples $\mathcal{X}^* = \{\check{\mathcal{X}}, \mathbf{n}^*\}$ according to the BLB procedure.
    Step 3.

    i)   Find the estimate $\hat{\boldsymbol{\beta}}_b$ based on $\check{\mathcal{X}}$.
    ii)   For each bootstrap sample $\mathcal{X}^*$ compute the FRB resample $\hat{\boldsymbol{\beta}}_b^{R*}$ from (10.12)
        using $\hat{\boldsymbol{\beta}}_b$.

    Step 4.   Compute the bootstrap CIs $\hat{\boldsymbol{\xi}}^*$ based on the population of B FRB resamples
    $\hat{\boldsymbol{\beta}}_b^{R*}$.
end
Step 5.   Average the computed values of $\hat{\boldsymbol{\xi}}^*$ over the subsamples, i.e., $\hat{\boldsymbol{\xi}}^* = \frac{1}{s}\sum_{k=1}^{s}$
$\hat{\boldsymbol{\xi}}^{*(k)}$.

10.4.4   *MM*-estimation Equations in the Bag of Little Fast and Robust Bootstraps Method

To provide a concrete example of robust estimation with the BLFRB procedure, consider the fixed-point *MM*-estimation equations for the linear regression problem of (2.7)

$$y_i = \mathbf{x}_{[i]}^\top \boldsymbol{\beta} + v_i, \quad i = 1, \ldots, N,$$

with $v_i = \sigma_0 e_i$. As described in Section 2.8.3, highly robust *MM*-estimators (Yohai, 1987) are defined in (2.93) as the solution to

$$\underset{\boldsymbol{\beta} \in \mathbb{R}^p}{\text{minimize}}\left\{ \sum_{i=1}^{N} \rho\left(\frac{r_i(\boldsymbol{\beta})}{\hat{\sigma}_0} \right) \right\}. \tag{10.13}$$

Here, $r_i(\boldsymbol{\beta}) \triangleq y_i - \mathbf{x}_{[i]}^\top \boldsymbol{\beta}, i = 1, \ldots, N$. The minimization problem can be solved by evaluating the system of *M*-equations that is given by

$$\frac{1}{N} \sum_{i=1}^{N} \psi\left(\frac{r_i(\boldsymbol{\beta})}{\hat{\sigma}_0} \right) \mathbf{x}_{[i]}^\top = \mathbf{0} \tag{10.14}$$

with $\psi(x) = \frac{d\rho(x)}{dx}$ and $\hat{\sigma}_0$ being the *M*-scale

$$\frac{1}{N} \sum_{i=1}^{N} \rho_0\left(\frac{r_i(\hat{\boldsymbol{\beta}}_0)}{\hat{\sigma}_0} \right) = m, \tag{10.15}$$

of the residuals $r_i(\hat{\boldsymbol{\beta}}_0)$, $i = 1, \ldots, N$, of an S-estimator $\hat{\boldsymbol{\beta}}_0$, and where $m < \rho_0(\infty)$ is a constant. See also Section 1.2.1.

Simple manipulations yield the following fixed-point formulation of the *MM*-estimator that is defined through (10.14) and (10.15):

$$\hat{\boldsymbol{\beta}} = \left(\sum_{i=1}^{N} \omega_i \mathbf{x}_{[i]} \mathbf{x}_{[i]}^{\top} \right)^{-1} \sum_{i=1}^{N} \omega_i \mathbf{x}_{[i]} y_i, \tag{10.16}$$

$$\hat{\sigma}_0 = \sum_{i=1}^{N} \upsilon_i(r_i(\hat{\boldsymbol{\beta}}_0)), \tag{10.17}$$

where

$$r_i(\hat{\boldsymbol{\beta}}) = y_i - \mathbf{x}_{[i]}^{\top} \hat{\boldsymbol{\beta}}, \qquad r_i(\hat{\boldsymbol{\beta}}_0) = y_i - \mathbf{x}_{[i]}^{\top} \hat{\boldsymbol{\beta}}_0,$$

$$\omega_i = \frac{\psi(r_i(\hat{\boldsymbol{\beta}})/\hat{\sigma}_0)}{r_i(\hat{\boldsymbol{\beta}})} \quad \text{and} \quad \upsilon_i = \frac{\hat{\sigma}_0}{Nm} \cdot \frac{\rho_0(r_i(\hat{\boldsymbol{\beta}}_0)/\hat{\sigma}_0)}{r_i(\hat{\boldsymbol{\beta}}_0)}.$$

Let $\mathcal{X}^* = \{\check{\mathcal{X}}, \mathbf{n}^*\}$ denote a BLFRB bootstrap resample based on the subsample

$$\check{\mathcal{X}} = \{(\check{y}_1, \check{\mathbf{x}}_{[1]}^{\top})^{\top}, \ldots, (\check{y}_b, \check{\mathbf{x}}_{[b]}^{\top})^{\top}\},$$

$\check{\mathbf{x}}_{[i]} \in \mathbb{R}^p$ and a weight vector $\mathbf{n}^* = (n_1^*, \ldots, n_b^*)^{\top} \in \mathbb{R}^b$. Then, the one-step *MM*-estimator $\hat{\boldsymbol{\beta}}_b^{1*}$ is given by

$$\hat{\boldsymbol{\beta}}_b^{1*} = \left(\sum_{i=1}^{b} n_i^* \check{\omega}_i \check{\mathbf{x}}_{[i]} \check{\mathbf{x}}_{[i]}^{\top} \right)^{-1} \sum_{i=1}^{b} n_i^* \check{\omega}_i \check{\mathbf{x}}_{[i]} \check{y}_i, \tag{10.18}$$

$$\hat{\sigma}_{0,b}^{1*} = \sum_{i=1}^{b} n_i^* \check{\upsilon}_i(\check{r}_i(\hat{\boldsymbol{\beta}}_{0,b})), \tag{10.19}$$

where

$$\check{r}_i(\hat{\boldsymbol{\beta}}_b) = \check{y}_i - \check{\mathbf{x}}_{[i]}^{\top} \hat{\boldsymbol{\beta}}_b, \qquad \check{r}_i(\hat{\boldsymbol{\beta}}_{0,b}) = \check{y}_i - \check{\mathbf{x}}_{[i]}^{\top} \hat{\boldsymbol{\beta}}_{0,b}$$

$$\check{\omega}_i = \frac{\psi(\check{r}_i(\hat{\boldsymbol{\beta}}_b)/\hat{\sigma}_{0,b})}{\check{r}_i(\hat{\boldsymbol{\beta}}_b)} \quad \text{and} \quad \check{\upsilon}_i = \frac{\hat{\sigma}_{0,b}}{Nm} \cdot \frac{\rho_0(\check{r}_i(\hat{\boldsymbol{\beta}}_{0,b})/\hat{\sigma}_{0,b})}{\check{r}_i(\hat{\boldsymbol{\beta}}_{0,b})}.$$

Here, $\hat{\boldsymbol{\beta}}_{0,b}$ denotes the S-estimate based on the subsample $\check{\mathcal{X}}$ and $\hat{\sigma}_{0,b}$ is the corresponding scale for the fixed-point equation. The BLFRB resamples of $\hat{\boldsymbol{\beta}}_b$ are obtained from the linearly corrected version of the one-step approximation in (10.18) and (10.19):

$$\hat{\boldsymbol{\beta}}_b^{R*} = \hat{\boldsymbol{\beta}}_b + \mathbf{M}_b(\hat{\boldsymbol{\beta}}_b^{1*} - \hat{\boldsymbol{\beta}}_b) + \mathbf{d}_b(\hat{\sigma}_{0,b}^{1*} - \hat{\sigma}_{0,b}), \tag{10.20}$$

where

$$
\mathbf{M}_b = \hat{\sigma}_{0,b} \left(\sum_{i=1}^{b} \psi'(\check{r}_i(\hat{\boldsymbol{\beta}}_b)/\hat{\sigma}_{0,b}) \check{\mathbf{x}}_{[i]} \check{\mathbf{x}}_{[i]}^{\top} \right)^{-1} \sum_{i=1}^{b} \check{\omega}_i \check{\mathbf{x}}_{[i]} \check{\mathbf{x}}_{[i]}^{\top},
$$

$$
\mathbf{d}_b = k_b^{-1} \left(\sum_{i=1}^{b} \psi'(\check{r}_i(\hat{\boldsymbol{\beta}}_b)/\hat{\sigma}_{0,b}) \check{\mathbf{x}}_{[i]} \check{\mathbf{x}}_{[i]}^{\top} \right)^{-1} \sum_{i=1}^{b} \psi'(\check{r}_i(\hat{\boldsymbol{\beta}}_b)/\hat{\sigma}_{0,b}) \check{r}_i(\hat{\boldsymbol{\beta}}_b) \check{\mathbf{x}}_{[i]},
$$

$$
\text{and} \qquad k_b = \frac{1}{Nm} \sum_{i=1}^{b} \left(\psi_0(\check{r}_i(\hat{\boldsymbol{\beta}}_{0,b})/\hat{\sigma}_{0,b}) \check{r}_i(\hat{\boldsymbol{\beta}}_{0,b})/\hat{\sigma}_{0,b} \right).
$$

As shown in Basiri et al. (2016), the BLFRB method, described previously, results in a convergent estimator (convergence in distribution) and the estimator has a high BP. The BLB method, however, is not robust because LS estimators that have a BP of zero are used in Kleiner et al. (2014). For large-scale data, and even for high model orders, the BP of the BLFRB method was found to be close to the maximally possible value of 50 percent. At the same time, the asymptotic relative efficiency was close to 95 percent for the nominal Gaussian case.

10.4.5   Simulation Example

We will first consider the performance of the BLFRB estimator in terms of standard error. The bootstrap estimate for the standard deviation of $\hat{\boldsymbol{\beta}}$ for distinct data subset (bag) k is defined as follows:

$$
\hat{\xi}_l^{*(k)} = \hat{\sigma}([\hat{\boldsymbol{\beta}}_b^{(k)}]_l) = \left(\sum_{j=1}^{B} \frac{\left([\hat{\boldsymbol{\beta}}_b^{*(kj)}]_l - \hat{\mu}_{b,l}^{*(k)} \right)^2}{B-1} \right)^{1/2}
$$

where $[\hat{\boldsymbol{\beta}}_b]_l$ denotes the lth element, $l = 1, \dots, p$, of $\hat{\boldsymbol{\beta}}_b$ and $\hat{\mu}_{b,l}^{*(k)} = \frac{1}{B} \sum_{j=1}^{B} [\hat{\boldsymbol{\beta}}_b^{*(kj)}]_l$. Furthermore,

$$
\hat{\xi}_l^* = \hat{\sigma}([\hat{\boldsymbol{\beta}}]_l) = \frac{1}{s} \sum_{k=1}^{s} \hat{\sigma}([\hat{\boldsymbol{\beta}}_b^{(k)}]_l), \quad l = 1, \dots, p.
$$

The performance of the BLB and BLFRB methods is assessed by computing a relative error defined as:

$$
e_{\mathrm{rel}} = \frac{\left| \hat{\sigma}(\hat{\boldsymbol{\beta}}) - \overline{\sigma}_o(\hat{\boldsymbol{\beta}}) \right|}{\overline{\sigma}_o(\hat{\boldsymbol{\beta}})}, \tag{10.21}
$$

where $\hat{\sigma}(\hat{\boldsymbol{\beta}}) = \frac{1}{p} \sum_{l=1}^{p} \hat{\sigma}([\hat{\boldsymbol{\beta}}]_l)$ and $\overline{\sigma}_o(\hat{\boldsymbol{\beta}}) = \sigma_0/\sqrt{N\mathcal{O}}$, with $\mathcal{O} = 0.95$ for the *MM*-estimator and $\mathcal{O} = 1$ for the LSE, which is an approximation of the average standard deviation based on the asymptotic covariance matrix; see Basiri et al. (2016) for a detailed explanation.

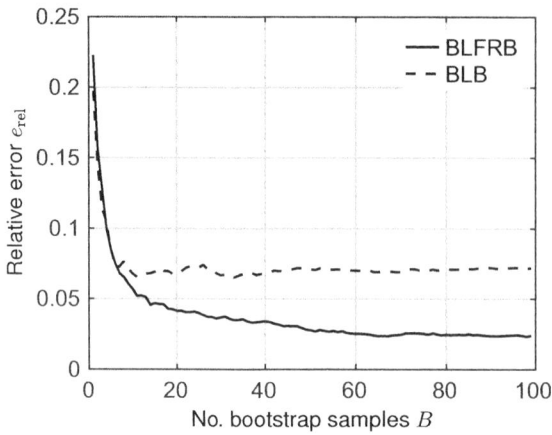

Figure 10.8 Graph of the relative error, as defined in (10.21), as a function of the number of bootstrap samples, for the BLB method (dashed line) and the BLFRB method (solid line). Both methods perform well when there are no outliers in the data.

The parameters of the bootstrap were selected as follows: The number of distinct data subsets (bags) was chosen as $s = 25$, the size of each subsample was $b = \lfloor N^\gamma \rfloor = 1946$ with $N = 50000$, $\gamma = 0.7$, and the maximum number of bootstrap samples in each subsample module as $B_{\max} = 300$.

Starting from $B = 2$, a new bootstrap resample is continuously added (while $B < B_{\max}$) to each of the subsample modules. It can be seen in Figure 10.8, from the relative error curves of the BLB and BLFRB methods, that both methods perform well, as the number of bootstrap resamples grows.

Next, based on simulations, the statistical robustness of the BLFRB method is compared to that of the BLB method. Outliers are introduced into the data by randomly choosing one of the data points and multiplying it by a large constant κ. First, the lack of robustness of the BLB, employing the LS estimator, is illustrated in Figure 10.9. Only one outlier is enough to increase the relative error significantly. A more demanding case is considered for the BLFRB method, where 40 percent of the data points are replaced by outliers based on a large value of $\kappa = 1,000$. The robustness of the BLFRB is demonstrated in Figure 10.10. The relative error remains small despite the high proportion of outliers in the contaminated data set.

The BLFRB method described in this chapter allows for using bootstrap methods for parameter estimation and inference in case of large-scale data. It lends itself to parallel computing and distributed storage systems. Moreover, the employed fixed-point estimation equations reduce the computational burden significantly because there is no need to recompute them repeatedly. Instead, one-step improvement of an initial estimate and a Taylor-series based correction term are used. The method is also nonparametric because empirical distribution models are used to find the bounds of the CI. The method is highly robust and also statistically highly efficient.

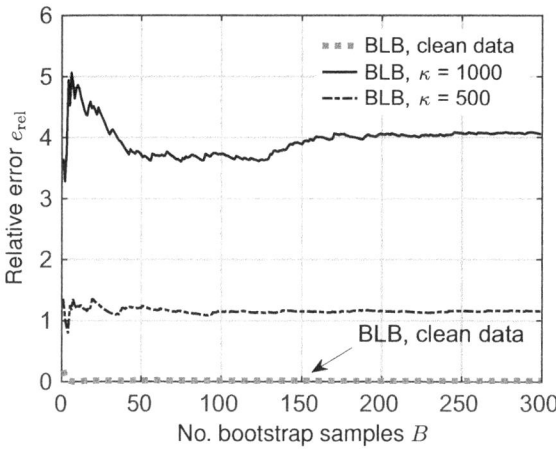

Figure 10.9 Graph of the relative error, as defined in (10.21), as a function of the number of bootstrap samples, for the BLB method. Only one outlier is enough to cause large errors in the BLB estimation. The outlier is introduced by randomly choosing one of the data points and multiplying it by a large constant κ.

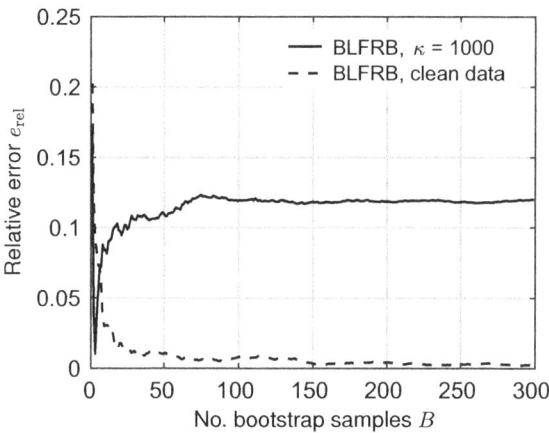

Figure 10.10 Graph of the relative error, as defined in (10.21), as a function of the number of bootstrap samples, for the BLFRB method, when outliers are introduced in the data. The BLFRB method is highly robust and results in a small relative error despite the high proportion, 40 percent, of outliers in the data set. Note the scale of the ordinate compared to Figure 10.9.

10.4.6   Million Song Data Set Example

Finally, an inference example using the BLFRB method with real-world data is considered. A simplified version of the Million Song data set, available on the UCI Machine Learning Repository, is used as our full data. The data set $\mathcal{X} = \{(y_1, \mathbf{x}_{[1]}^\top)^\top, \ldots, (y_N, \mathbf{x}_{[N]}^\top)^\top\}$ contains $N = 515,345$ music tracks, where y_i, $i = 1, \ldots, N$, represents the release or publication year of the ith song. Songs in the database are from the era spanning the years 1922 to 2011. Let $\mathbf{x}_{[i]} \in \mathbb{R}^p$ denote a vector of $p = 90$ different audio features characterizing the music content of each song. The characteristic features are the average and nonredundant covariance values of the timbre vectors of the song. The goal is to use linear regression as a tool to predict the release year of a song based on its audio features. The linear regression model is given by

$$y_i = \mathbf{x}_{[i]}^\top \boldsymbol{\beta} + v_i,$$

where $v_i = \sigma_0 e_i$ is an error term. We use the BLFRB method here to conduct a fast, robust, and scalable hypothesis test based on the estimated CIs. Let the null and alternative hypotheses be defined as:

$$\mathcal{H}_0 : [\boldsymbol{\beta}]_l = 0 \quad \text{vs}$$

$$\mathcal{H}_1 : [\boldsymbol{\beta}]_l \neq 0,$$

where $[\boldsymbol{\beta}]_l, l = 1, \ldots, p$, denotes the lth element of $\boldsymbol{\beta}$. The BLFRB hypothesis test, based on a level α, rejects the null hypothesis if the computed $100(1 - \alpha)\%$ CI of that particular parameter or feature coefficient does not contain zero. We perform a test on each feature coefficient β_l and discard it as being atypical for music from a particular year if its CI contains zero.

Using the BLFRB inference method, the BLFRB hypothesis test, at level $\alpha = 0.05$, is applied based on the following parameters:

- The number of distinct data subsamples being set to $s = 51$,
- The size of each subsample being set to $b = \lfloor N^\gamma \rfloor = 9,964$ with $\gamma = 0.7$, $N = 515345$, and

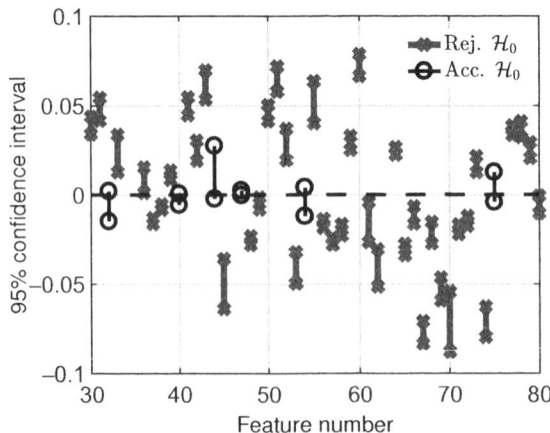

Figure 10.11 The 95 percent CIs, computed with the BLFRB method, for selected audio features of the Million Song data set. The null hypothesis is accepted for those features having 0 inside the CI.

- The number of bootstrap samples drawn for each distinct subsample module being set to $B = 500$.

An example result of the inference is depicted in Figure 10.11. It can be seen from this figure that the null hypothesis is accepted for six features with indexes: 32, 40, 44, 47, 54, and 75, so these features are not common for the music from the particular year. Obtained results can be exploited in reducing the dimension of the data by excluding the ineffective explanatory variables from the regression analysis. Moreover, they can be used for recommending music from a certain era.

10.5   Concluding Remarks

Robust bootstrap methods are required as computational tools for statistical inference, for example, to approximate the distribution of an estimator in the presence of outliers because the BP of the classical bootstrap tends to zero, even when bootstrapping a robust estimator. Different approaches were reviewed, with the focus being on eliminating the effects of outliers, in different ways, within the bootstrap procedure. To compare the different methods, the example of CI estimation, for the simple linear regression case, was considered. A localization example was used to highlight the usefulness of robust bootstrap methods in practical situations. Finally, the application of robust bootstrap methods to large-scale data sets, that is, big data, was discussed and examples were given.

11 Real-Life Applications

This chapter is devoted to real-life applications of robust methods. Here, we give several examples of how the methods that have been discussed in the book can be applied in areas as diverse as short-term load forecasting, diabetes monitoring, heart-rate variability analysis by means of PPG, inverse atmospherical problems, and indoor localization. These comprise of only a few examples; robust methods have shown to be beneficial in a much larger range of applications. See, for example, Zoubir et al. (2012) and references therein.

11.1 Localization of User Equipment in an Indoor Environment

This example applies the robust regression methods of Chapter 2 to the problem of indoor localization. The last decade has seen a significant increase in the number of sensors and in the processing power of stationary or mobile devices such as handheld or wearable devices. When equipped with communication capabilities, these devices can form a wireless sensor network (WSN) where the nodes cooperate to solve one (or multiple) signal processing tasks. Estimating the geographical position of a mobile device with a WSN has become of increasing importance in a variety of applications, such as indoor navigation, local search, mobile advertisement, tourism, and assistive healthcare. According to Markets and Markets (2016), the overall market is expected to grow from USD 15.04 billion in 2016 to USD 77.87 billion in 2021 at a Compound Annual Growth Rate of 38.9 percent.

In this example, we compare different robust geolocation estimators, based on TOA measurements from an indoor positioning experiment conducted by Patwari et al. (2003). The measurement environment is an 14×13 m office area that is partitioned by cubicle walls and wideband battery-powered direct-sequence spread-spectrum transmitters and receivers were used. The TOA measurements are based on the synchronized transmitters and receivers and are severely affected by NLOS-multipath propagation. Further details on the data collection and setup are available at http://web.eecs.umich.edu/~hero/localize/.

To conduct geolocation, TOA measurements between each pair of sensors, for a set of $N = 44$ sensors, are available. In our example, we consider Sensor 19 to be the agent, and the goal is to determine its position. Two different subsets of the remaining $N = 43$ sensors are chosen to be agents of known positions so as to simulate a LOS and

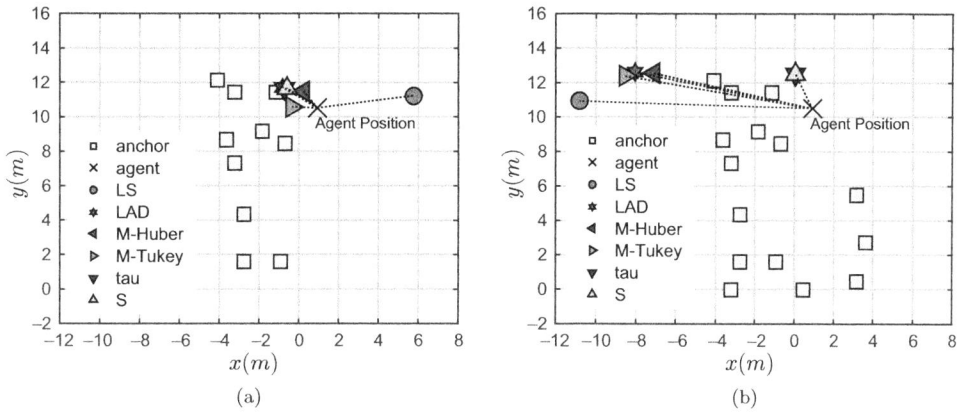

Figure 11.1 TOA-based position estimates of six estimators in two different setups. In (a), a LOS environment is considered, whereas (b) is a challenging mixed LOS/NLOS environment. In this example, only the τ- and S-estimators are able to achieve results that are similar to those obtained for the LOS case.

a mixed LOS/NLOS environment. The two setups and resulting estimates are depicted in Figure 11.1.

To obtain position estimates, as in (2.94), the ith nonlinear range measurement at the mth anchor is modeled by

$$r_{m,i} = h_m(\boldsymbol{\beta}) + v_{m,i}, \quad m = 1, \ldots, M, \quad i = 1, \ldots, N,$$

where $\boldsymbol{\beta} = (x_{\mathrm{UE}}, y_{\mathrm{UE}})^\top$ is the unknown position of the user equipment (agent). As in Section 2.9.2, the measurement model is linearized by introducing a range variable $R = \sqrt{x_{\mathrm{UE}}^2 + y_{\mathrm{UE}}^2}$.

We examine the performance of the following six estimators that have been introduced in Chapter 2:

1. The LSE defined by (2.14).
2. The LAD estimator defined by (2.15) and computed using Algorithm 3.
3. Huber's M-estimator defined by (2.52), (2.54) with $c_{.95} = 1.345$ and computed using Algorithm 4.
4. Tukey's M-estimator defined by (2.52), (2.55) with $c_{.95} = 4.685$ and computed using Algorithm 4.
5. S-estimator defined by (2.88) and computed using the IRWLS algorithm by Salibián-Barrera and Yohai (2012).
6. τ-estimator defined by (2.89) and computed using the IRWLS algorithm by Salibián-Barrera et al. (2008b).

Figure 11.1a shows the results for the LOS environment. All robust estimators yield a reasonable estimate for the position. Only the LSE is strongly biased, which suggests, even for this simple setup, that the distribution of errors is non-Gaussian. Figure 11.1b shows the results for the mixed LOS/NLOS case in which there is a high percentage of outliers. For this challenging setting, only the τ- and S-estimators are able to achieve results that are similar to those obtained in the LOS case.

11.2   Blood Glucose Concentration in Photometric Handheld Devices

Diabetics rely heavily on handheld blood glucose measurement devices to manage their disease efficiently. These devices are typically invasive and require extracting a blood drop from the patient's finger (Tamada et al., 2002). Reducing the blood sample volume lowers the self-monitoring inhibition for the patient and increases the usability of these devices (Tamada et al., 2002). This is crucial for diabetes care: Regular self-management has been proven to significantly delay or prevent complications of diabetes (Guerci et al., 2003). In this example, the robust location and scale M-estimators, introduced in Chapter 1, are utilized to estimate the glucose concentration in a blood sample.

Figure 11.2 illustrates a photometric setup that is used for blood glucose measurement that requires much smaller blood samples than is typical in state-of-the-art devices (Demitri and Zoubir, 2017). The blood drop extracted from the patient is placed on a chemical test strip. A camera observes the chemical reaction between the glucose in the blood sample and the chemical test strip, producing a set of images describing the reaction over time. The images are then segmented to identify the region of interest (ROI) and the intensity $\hat{R}$ in the ROI is estimated. This value is mapped to the underlying glucose value to obtain a glucose estimate $\hat{G}$. The measurement procedure results in a set of frames describing the chemical reaction at the different time instances. Figure 11.2 shows an example of an obtained glucose image after the chemical reaction has converged to its final state. The ROI is indicated by the two vertical stripes.

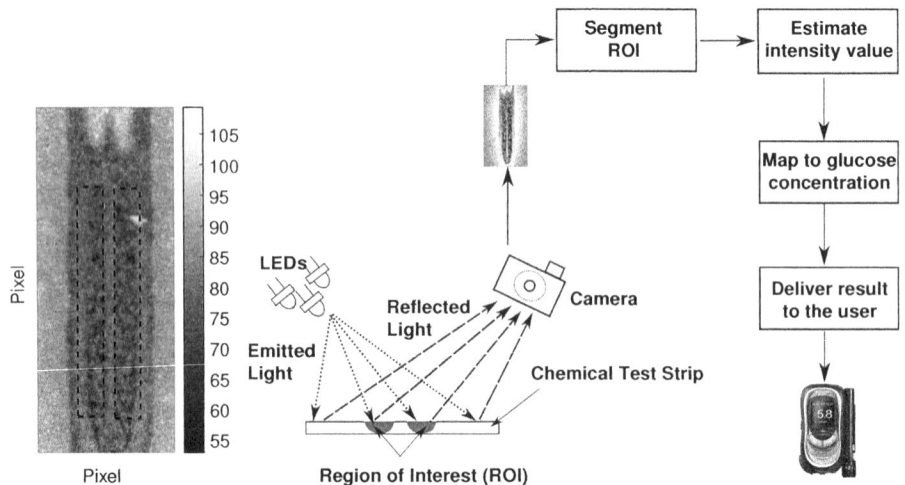

Figure 11.2 Example of a glucose image and a photometric setup that is used for blood glucose measurement.

The challenges with this problem are twofold: The ROI has to be extracted and the intensity value in the ROI has to be estimated. The mean-shift (MS) algorithm is appropriate for these problems, as it simultaneously performs the segmentation and the subsequent estimation of the intensity value of the ROI (Demitri and Zoubir, 2013). It does this by first estimating the pdf of the intensity values in an image with the help of a kernel density estimator (KDE). We assume that a vectorized image $\mathbf{x}$ contains N pixels given by the intensities $x_i, i = 1, \ldots, N$. The KDE of intensities is

$$\hat{f}_K(\mathbf{x}) = \frac{1}{Nh} \sum_{i=1}^{N} K\left(\frac{\mathbf{x} - x_i}{h}\right) = \langle \Phi(\mathbf{x}), \sum_{i=1}^{N} \frac{1}{Nh} \Phi(x_i) \rangle, \qquad (11.1)$$

where $K(\mathbf{x})$ is a radially symmetric kernel function with a strictly decreasing profile for $x \geq 0$ and h is a bandwidth parameter. The right form of the equation is an alternative way to express the KDE, where the kernel is considered to be an inner product in the Hilbert space $\mathcal{H}$, where $\Phi : \mathbb{R} \to \mathcal{H}$ is a mapping function and $\langle \cdot \rangle$ denotes the inner product. To estimate the mode locations of the KDE, the zeros of the gradient are calculated

$$\nabla \hat{f}_K(\mathbf{x}) = 0. \qquad (11.2)$$

Reformulating this term results in the so-called MS vector

$$\mathbf{m}_{h,K'}^{(MS)}(\mathbf{x}) = \left[\frac{\sum_{i=1}^{N} x_i K'\left(\frac{\mathbf{x} - x_i}{h}\right)}{\sum_{i=1}^{N} K'\left(\frac{\mathbf{x} - x_i}{h}\right)} - \mathbf{x} \right], \qquad (11.3)$$

where $K' = \frac{dK}{d\mathbf{x}}$. Each data point x_i is shifted by its MS vector in the direction of steepest ascent until it reaches its nearest mode location. After several iterations, all data points have converged to their nearest mode locations.

In some cases, however, the intensity estimate is not accurate due to occurring artifacts such as granularity of the chemical substance, air bubbles, or particles contaminating the measurement area. By utilizing a robust mean shift (R-MS) algorithm, the estimation accuracy in the presence of artifacts can be improved. This is achieved by substituting the mean of vectors in the MS formulation by a robust M-estimate

$$\hat{\mu}_\Phi = \arg\min_{\mu_\Phi} \sum_{i=1}^{N} \rho\left(\frac{\Phi(x_i) - \mu_\Phi}{\hat{\sigma}}\right), \qquad (11.4)$$

where $\rho(\cdot)$ is a monotone differentiable loss function, such as Huber's loss function, and the scale $\hat{\sigma}$ is initialized with a robust estimate based on the mean absolute deviation from the median, as defined in (1.13). This leads to the robust KDE

$$\hat{f}(\mathbf{x}) = \langle \Phi(\mathbf{x}), \hat{\mu}_\Phi \rangle = \frac{1}{h} \sum_{i=1}^{N} w_i K\left(\frac{\mathbf{x} - x_i}{h}\right), \qquad (11.5)$$

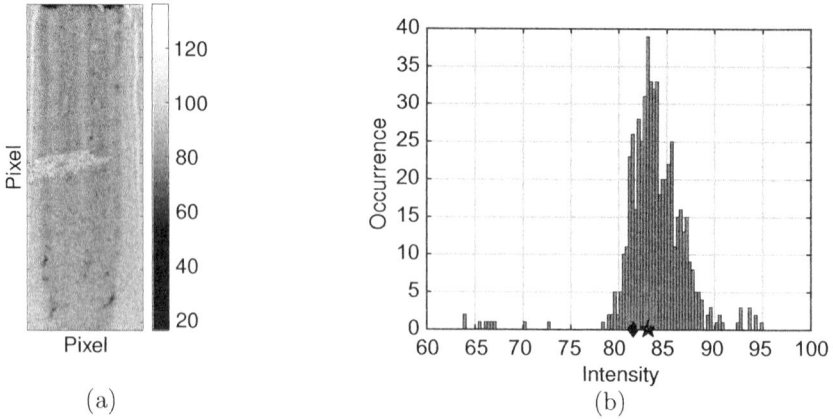

Figure 11.3 Example of a contaminated image (a) of the chemical reaction for a blood glucose sample of 150 mg/dl and (b) its histogram, as well as mode estimates based on the MS (diamond-shaped marker) and the R-MS (star-shaped marker) algorithm.

where the robust weights w_i can be determined using an IRWLS algorithm (Demitri and Zoubir, 2014), resulting in

$$w_i = \begin{cases} \dfrac{\hat{\sigma} \cdot \psi\left(\frac{\Phi(x_i) - \mu_\Phi}{\hat{\sigma}}\right)}{\Phi(x_i) - \mu_\Phi} & \text{if } \dfrac{\Phi(x_i) - \mu_\Phi}{\hat{\sigma}} \neq 0 \\ \psi'(0) & \text{if } \dfrac{\Phi(x_i) - \mu_\Phi}{\hat{\sigma}} = 0 \end{cases}, \qquad (11.6)$$

where $\psi = \rho'$. The RMS vector is given by

$$\mathbf{m}_{h,K'}^{(\text{R-MS})}(\mathbf{x}) = \left[\frac{\sum_{i=1}^{N} w_i x_i K'\left(\frac{\mathbf{x} - x_i}{h}\right)}{\sum_{i=1}^{N} w_i K'\left(\frac{\mathbf{x} - x_i}{h}\right)} - \mathbf{x} \right]. \qquad (11.7)$$

This example illustrates how the basic concepts, such as robust location and scale estimation, that were considered in Chapter 1 can be creatively applied to solve complex problems. For a more detailed description of the R-MS, the reader is referred to Demitri and Zoubir (2014). Figure 11.3 presents illustrative results for the performance of the standard MS, and the RMS algorithm, for the case of a contaminated image. Clearly, the image is disturbed by a low-intensity contamination in the middle of the image. This leads to the standard MS algorithm to produce a mode estimate that is biased toward the low-intensity region. The robust MS algorithm manages to down-weight the area of contamination and this results in a more accurate mode estimate.

11.3   European Tracer Experiment Source Estimation

This example details the application of robust penalized regression methods, as discussed in Chapter 3, to the problem of source estimation in an atmospheric inverse problem. For this example, the regression model is ill-posed and the parameter vector is (block) sparse and represents the emissions of a pollutant. Quantifying the emissions

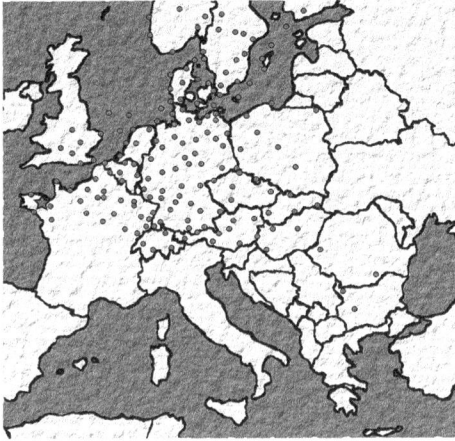

Figure 11.4 For the ETEX, 168 ground-level sampling stations in 17 European countries were used to estimate the temporal progression of the emission of a pollutant in Monterfil, France, whose location is represented by a star.

of a pollutant into the atmosphere is important, for example, in the case of nuclear power plant accidents, volcano eruptions, or tracking the releases of greenhouse gases. Penalized robust estimation is used to determine the temporal releases of the particles, at the source location, during the European Tracer Experiment (ETEX). During the ETEX experiment, tracers (perfluorocarbons) were released into the atmosphere in Monterfil, France, in 1994. Hourly measurements were taken at 168 ground-level sampling stations in 17 European countries as illustrated in Figure 11.4. Here, the source location, that is, Monterfil, France, is highlighted by a star.

The task is to estimate the temporal releases of the particles at the source location. Atmospherical dispersion models, such as the Lagrangian Particle Dispersion Model (LPDM), allow the problem to be modeled according to (2.7), that is,

$$y_i = \mathbf{x}_{[i]}^{\mathsf{H}}\boldsymbol{\beta} + v_i, \quad i = 1, \dots, N,$$

where every regression parameter $\beta_j, j = 1, \dots, p$, corresponds to the amount of perfluorocarbon that is released by the source at time instant j. The time resolution is one hour for this experiment and there are 120 time instances, resulting in $p = 120$ unknown regression variables. The LPDM allows to formulate the source estimation problem as a linear inverse problem according to (2.8) as follows: The regression matrix $\mathbf{X}$ is estimated as detailed in Martinez-Camara et al. (2014) by using the Flexible Particle Dispersion Model and is formed by

$$\mathbf{X} = (\tilde{\mathbf{X}}_1, \dots, \tilde{\mathbf{X}}_k, \dots, \tilde{\mathbf{X}}_K)^{\top},$$

where the kth matrix describes the distribution of the particles from the source to the kth sensor. The responses $\mathbf{y} = (y_1, \dots, y_N)^{\top} = (\tilde{\mathbf{y}}_1^{\top}, \dots, \tilde{\mathbf{y}}_k^{\top}, \dots, \tilde{\mathbf{y}}_K^{\top})^{\top}$ are a set of stacked observations of the $K = 168$ sensors.

The ETEX data is sparse in $\boldsymbol{\beta}$ because only 12 of the regression parameters, that is, 10 percent, are non-zero. The residuals $\mathbf{r} = \mathbf{y} - \mathbf{X}\boldsymbol{\beta}$, given the ground truth values of $\boldsymbol{\beta}$, are non-Gaussian and the regression matrix $\mathbf{X}$ contains outliers. Furthermore, $\mathbf{X}$ is sparse because most of the time, the perfluorocarbon particles do not reach a sensor, resulting in a regression matrix, which is dominated by zero-valued cells.

The following preprocessing steps are applied to the data:

1.   Remove data points $(\mathbf{x}_j, y_j)^\top$, where all entries of the predictor $\mathbf{x}_j$ are equal to zero.

2.   Center and normalize the data robustly:

$$\mathbf{y} \leftarrow \frac{\mathbf{y} - \mathrm{med}(\mathbf{y})}{\mathrm{mad}(\mathbf{y})}$$

$$\mathbf{x}_j \leftarrow \frac{\mathbf{x}_j - \mathrm{med}(\mathbf{x}_j)}{\mathrm{mad}(\mathbf{x}_j)}.$$

3.   When applying the median or the mad to the predictors $\mathbf{x}_j$, only use samples that are greater than zero.[1]

4.   Apply a robust PCA (Croux et al., 2007) and reconstruct $\mathbf{X}$ using only N_p principal components, such that the mad of the N_p principal components corresponds to 90 percent of the total mad.[2]

5.   Further, because we know that the number of particles omitted by the source can only be positive $\beta_j \geq 0$, we impose a nonnegativity constraint on the parameters $\beta_j, j \in \{1, \ldots, p\}$.

For the nonnegative data case, the Lasso estimator can be modified according to the following definition:

DEFINITION 26   *(The Positive Lasso Estimator)*

$$\hat{\boldsymbol{\beta}}_{\mathrm{pos\,Lasso}} = \arg\min_{\boldsymbol{\beta}} ||\mathbf{y} - \mathbf{X}\boldsymbol{\beta}||_2^2,$$

subject to $||\boldsymbol{\beta}||_1 \leq t$ *and* $\beta_j \geq 0 \quad j \in \{1, \ldots, p\}.$ (11.8)

The positive Lasso can be calculated using a modified CCD algorithm detailed in Algorithm 19.

To obtain robust M-Lasso estimates of regression and scale, using Huber's loss function, Algorithm 9 can be simply modified to encompass the positivity constraint, as detailed in Algorithm 19.

---

[1]   The predictors contain an overwhelming number of components that are zero because $\mathbf{X}$ is highly sparse. Robust estimates like the median or the mad will result in a value of zero if more than 50 percent of the entries of $\mathbf{x}_j$ are zero. Obviously, taking only the positive components into account is consistent with considering the data we are interested in, and this leads to estimates that are of interest.

[2]   The reason for applying a robust PCA is to provide, for all algorithms, a matrix that is not as badly conditioned as the original regression matrix. The value 90 percent has been empirically determined from a range of possible values between 85 and 99 percent.

Algorithm 19: posLasso: computes the positive Lasso solution using the CCD algorithm for (centered and) standardized data

---

input   : $\mathbf{y} \in \mathbb{R}^N, \mathbf{X} \in \mathbb{R}^{N \times p}, \hat{\boldsymbol{\beta}}_{\text{init}}, \lambda > 0$
output   : $\hat{\boldsymbol{\beta}} = \hat{\boldsymbol{\beta}}(\lambda) \in \mathbb{R}^p$, a solution to the positive Lasso optimization problem
initialize : $\hat{\boldsymbol{\beta}} \leftarrow \hat{\boldsymbol{\beta}}_{\text{init}}$

1 **while** *stopping criteria not met* **do**

2   **for** $j = 1$ **to** p **do**

3        $\mathbf{r} \leftarrow \mathbf{y} - \mathbf{X}\hat{\boldsymbol{\beta}}$

4

$$\hat{\beta}_j = \max \left\{ \text{soft}_\lambda \left(\sum_{i=1}^N x_{ij}(y_i - \sum_{k \neq j} x_{ik}\hat{\beta}_k), \lambda \right), 0 \right\},$$

       where soft_λ is defined in (3.12).

Performance Metrics

For this example, the MSE, the false-positive rate (FPR), and the false-negative rate (FNR), respectively, are defined according to:

$$\text{MSE}(\hat{\boldsymbol{\beta}}) = \frac{1}{p} \sum_{j=1}^p (\beta_j - \hat{\beta}_j)^2, \tag{11.9}$$

$$\text{FPR}(\hat{\boldsymbol{\beta}}) = \frac{|\{j \in \{1, \ldots, p\} : \beta_j = 0 \wedge \hat{\beta}_j \neq 0\}|}{|j \in \{1, \ldots, p\} : \beta_j = 0|} \tag{11.10}$$

$$\text{FNR}(\hat{\boldsymbol{\beta}}) = \frac{|\{j \in \{1, \ldots, p\} : \beta_j \neq 0 \wedge \hat{\beta}_j = 0\}|}{|j \in \{1, \ldots, p\} : \beta_j \neq 0|} \tag{11.11}$$

The FPR measures how many time instances are flagged as containing a source emission, while there was none. The FNR measures how many time instances containing a source emission are falsely left out.

Table 11.1 displays the MSE, FPR, and FNR for the ETEX data, while Figure 11.5 shows the estimated source emissions for the positive Lasso and positive robust *M*-Lasso. For this example, even the positive LS-Lasso results in a reasonable estimation,

Table 11.1 MSE, FPR, and FNR of the ETEX data using the LS-Lasso and the robust *M*-Lasso.

	MSE	FPR	FNR	1-(FPR+FNR)
LS-Lasso (Tibshirani, 1996)	70.66	0.046	0.36	0.594
M-Lasso (Ollila, 2016a)	35.51	0	0.25	0.75

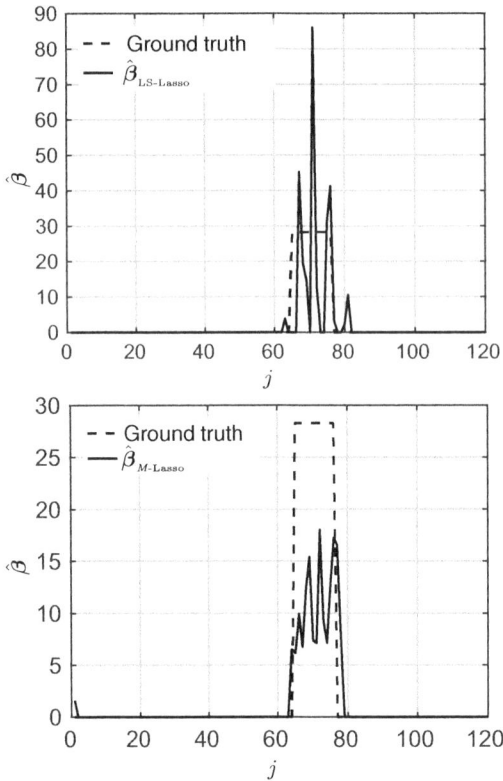

Figure 11.5 Estimated source emissions for the Lasso and the M-Lasso.

mainly thanks to robust PCA. However, the LS-Lasso estimate still has poor model selection properties.

11.4   Robust Short-Term Load Forecasting

This example treats the practical problem of short-term electricity consumption forecasting. To forecast electricity consumption, the load time series is first corrected for the weather influence by using a regression model where the explanatory variables are the temperature and the nebulosity. The nebulosity is a measure of the cloud cover in real time. In this step, robust regression estimation, such as described in Chapter 2, should be considered.

The obtained residual series encompasses some seasonalities (daily, weekly, yearly) and is generally modeled by a seasonal autoregressive integrated moving average (SARIMA) model. After removing the daily and weekly seasonalities, double differencing leads to a stationary process, that can be modeled using an ARMA model as defined in Chapter 8. Figure 11.6 displays the one-week differenced consumption in France at 07:00 in 2005. While the majority of data points can be represented

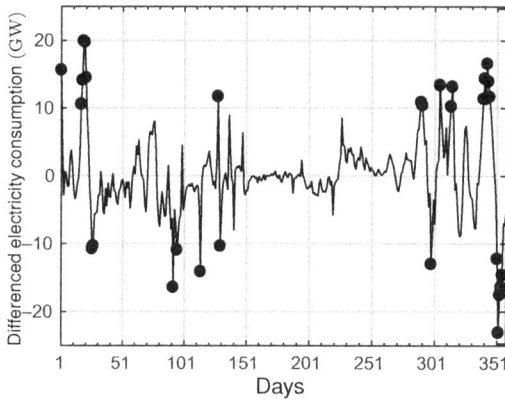

Figure 11.6 To forecast electricity consumption, the load time series is first corrected for weather influences, then daily and weekly seasonalities are removed. The figure shows the one-week differenced residual series of the electrical consumption in France at 07:00 in 2005. Outliers that arise, when extraordinary events influence the usual electricity consumption, are highlighted by circular markers.

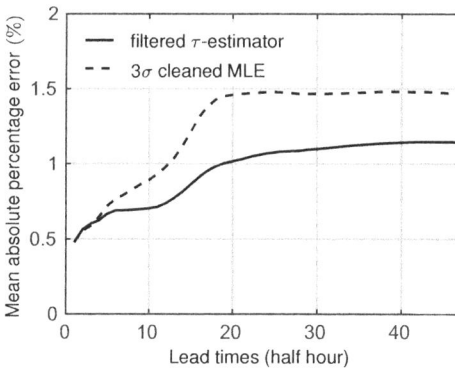

Figure 11.7 A plot of the MAPE versus the 48 half-hours of a day $(00\!:\!00, \ldots, 23\!:\!30)$ for France (200 days postsample period). The univariate cleaned maximum likelihood is the method that is used by the majority of electricity companies. It is outperformed by the robust and efficient multivariate filtered τ-estimator, which exploits the correlation structure of the multichannel data.

by an ARMA model, the series also contains outliers, which are highlighted by the circular markers in Figure 11.6. These outliers arise from special events, which have a massive influence on the electricity consumption, such as holidays, major strikes, the soccer World Cup, or storms. Robust forecasting methods are necessary to maintain the forecasting quality in the presence of outliers (Chakhchoukh et al., 2010).

The evaluation of the forecasting quality is generally assessed by plotting a robust version of the Mean Absolute Percentage Error (MAPE) (Chakhchoukh, 2010a) with respect to the forecasting horizon. Figure 11.7 (Chakhchoukh et al., 2010; Zoubir et al., 2012) shows the MAPE (%) of the univariate 3-sigma cleaned MLE, which is used by the majority of electricity companies, and that of the multivariate filtered τ-estimator, which is a robust multivariate ARMA estimator; see Chakhchoukh (2010a) for the full details. The 3σ cleaned MLE simply rejects all observations that exceed three times the robustly estimated standard deviation and then applies a Gaussian MLE that can handle missing data.

A further interesting criterion is the error between the observation at day N and its forecast from day $N - 1$. The MSE is evaluated over *normal* days of the 200 out-of-sample one-day-ahead forecast error for different robust methods for the French electrical consumption at 08:00. The univariate 3σ cleaned MLE, which has an MSE of 1.632×10^6 MW^2, is clearly outperformed by the filtered τ-estimator, which has an MSE of 0.708×10^6 MW^2.

11.5   Robust Data Cleaning for Photoplethysmography-Based Pulse-Rate Variability Analysis

This example considers the estimation of the pulse-rate variability from PPG signals. PPG sensors can be found in a number of different devices. Not only are they built into consumer goods, such as wrist-type fitness trackers, but also into devices used by medical professionals. Recently, Schäck et al. (2017) proposed a PPG-based method for atrial fibrillation detection using smartphones. However, PPG sensors are mostly used to either estimate the pulse rate (Schäck et al., 2015) or the oxygen saturation in the blood. They are becoming increasingly popular because a PPG signal can be obtained comfortably, continuously, and at low cost.

A plethysmograph is an instrument that measures changes in the volume of an organ, hence the name: "plethys" and "graphos" are Greek words for "fullness" and "to write." A photoplethysmograph is an optical sensor. The sensor consists of a light-emitting diode (LED), which emits light onto the skin and a photodiode that detects the reflected light. This diode is usually placed next to the LED to optimize the level of detected light. In this example, we used a PPG sensor that is placed at the finger tip and operates at a wavelength of 950 mm.

The PPG signal displays volume changes in small arteries and arterioles beneath the skin. Those changes occur because the human heart does not pump blood continually but in cycles. Every cardiac cycle, which consists of systole and diastole, results in one pulse wave in the PPG signal.

In this example, the heart rate variability (HRV), which describes the variation in the time interval between heartbeats from the PPG signal, is derived. For simplicity, it is assumed that the pulse rate variability (PRV), estimated from the time interval between pulses measured at the PPG sensors at the wrist and fingertip, is the same as the HRV.

As ground truth reference, the HRV is estimated by using the detected R-peaks from an ECG signal. The PRV from the PPG signals are obtained by the locations of the systolic feet. Figure 11.8 shows the detected peaks for both signals for the first 15 seconds of a recording that was taken in the biomedical laboratory of the Signal Processing Group at Technische Universität Darmstadt in Darmstadt, Germany.

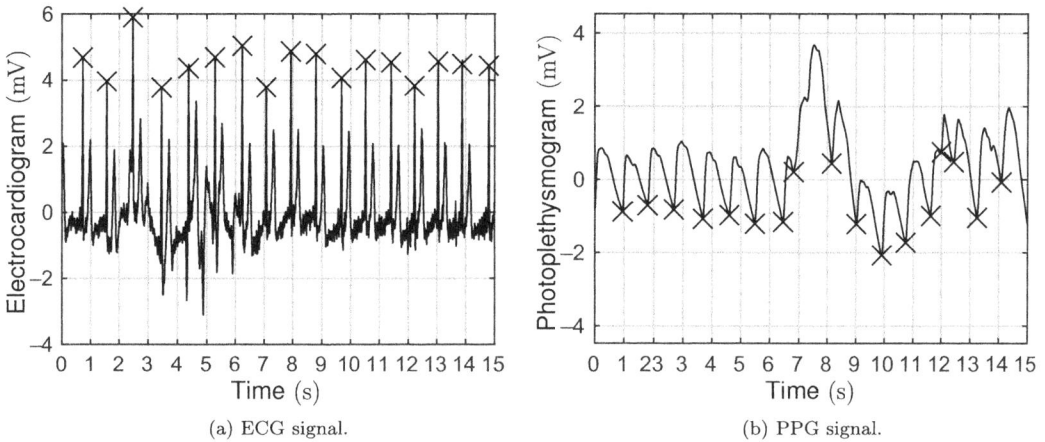

(a) ECG signal.

(b) PPG signal.

Figure 11.8 First 15 seconds of the measured signals with peak position estimates highlighted by crosses.

The PRV and HRV analysis is performed based on the peak-to-peak time difference between successive estimated peaks. Missed detections, or false alarms, which mainly occur due to subject motion, lead to a signal that contains outliers, as shown in the dashed line of Figure 11.10a. This line represents the peak-to-peak time difference estimates using a simple peak detection method that is described in Schäck et al. (2017).

Unfortunately, even a few of these outliers have a large effect on standard PRV metrics, such as the standard deviation of intervals (SDNN), measured in milliseconds, and the root mean square of successive differences (RMSSD), which is the square root of the mean of the squares of the successive differences between adjacent time intervals.

A robust cleaning of the corrupted peak-to-peak time difference series was performed, by applying the robust one-step ahead prediction of the BIP τ estimator (Muma and Zoubir, 2017), as defined in (8.76). Because in real-data applications the model order is unknown, model order selection needs to be performed. In this example, we considered the family of models to be the AR(p) models, where the candidate range of orders is given by $1 \le p \le 20$.

Figure 11.9 shows the outcome of applying AIC

$$\log(\hat{\sigma}_\tau(a_N^b(\boldsymbol{\phi}))^2) + 2p/N \tag{11.12}$$

and the BIC

$$\log(\hat{\sigma}_\tau(a_N^b(\boldsymbol{\phi}))^2) + p\log(N)/N \tag{11.13}$$

based on the τ scale estimate of the innovations that is provided by the BIP τ-estimator as defined in Equation (8.69). Both criteria show a minimum at $p = 5$, which is thus chosen for the data cleaning.

Figure 11.10a displays the original PPG-based estimate of the peak-to-peak time difference series (dashed line) and the BIP τ cleaned series using an AR(5) model with associated parameter vector estimate $\hat{\boldsymbol{\phi}} = (-0.6494, 0.3295, -0.5625, 0.2737, 0.0860)^\top$.

Table 11.2 SDNN, in milliseconds, and the RMSSD for PPG- and ECG-based HRV analysis.

	Original PPG	Robust BIP τ PPG	ECG
SDNN	136.3	61.6	58.0
RMSSD	178.1	58.3	50.5

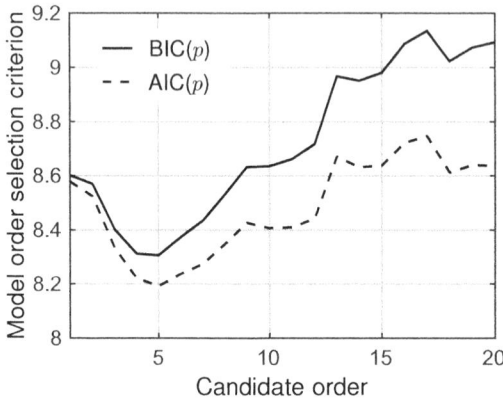

Figure 11.9 Model order selection for the BIP τ-estimator.

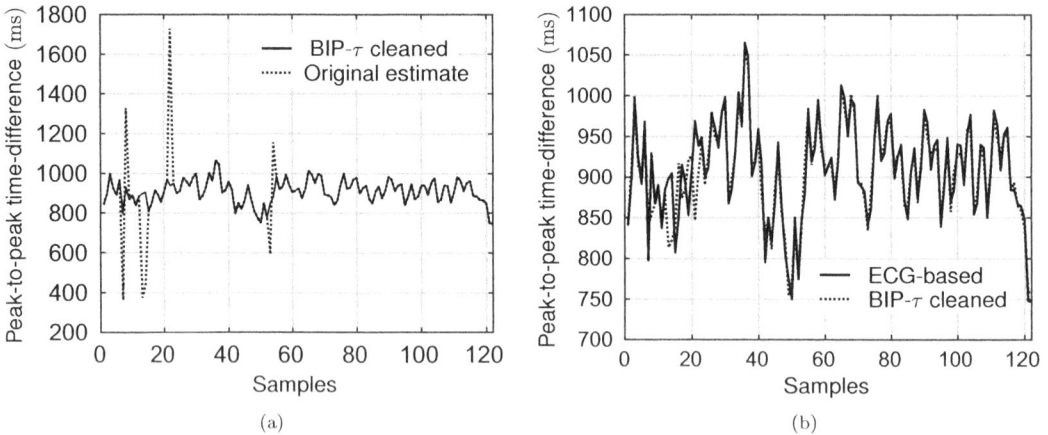

(a)   (b)

Figure 11.10 The left plot (a) shows peak-to-peak time difference series for the PPG signal-based estimates. The right plot (b) shows the ECG signal-based estimates (ground truth reference) and the BIP τ-cleaned series.

Both series coincide most of the time. In fact, they differ only in the positions that contain outliers.

Figure 11.10b compares the BIP τ cleaned series to the estimate of the peak-to-peak time difference series that is obtained from the ECG reference device. Clearly, the outliers have been successfully removed and replaced by predicted values that match the ones derived from the ECG signal. Table 11.2 lists the SDNN and RMSSD values, which are standard metrics for HRV analysis. The difference to the ECG-based estimates that was present when using the original series is largely reduced by the BIP τ cleaning operation, which demonstrates the practical applicability of robust methods.

Bibliography

Abramovich, Y. I. and Spencer, N. K. (2007). Diagonally loaded normalised sample matrix inversion (LNSMI) for outlier-resistant adaptive filtering. In *Proc. IEEE Int. Conf. Acoust. Speech Signal Process.*, 1105–1108.

Abramovich, Y. I. and Turcaj, P. (1999). Impulsive noise mitigation in spatial and temporal domains for surface-wave over-the-horizon radar. *DTIC document*. Cooperative Research Centre for Sensor Signal and Information Processing.

Adali, T., Schreier, P. J., and Scharf, L. L. (2011). Complex-valued signal processing: The proper way to deal with impropriety. *IEEE Trans. Signal Process.*, 59(11): 5101–5125.

Agostinelli, C. (2004). Robust Akaike information criterion for ARMA models. *Rendiconti per gli Studi Economici Quantitativi*, 1: 1–14.

Aittomaki, T. and Koivunen, V. (2004). Recursive householder-based space-time processor for jammer mitigation in navigation receivers. In *Proc. 38th Annu. CISS2004 Conf. Inform. Sci. Syst.*, 1–4.

Alexander, D. C., Barker, G. J., and Arridge, S. R. (2002). Detection and modeling of non-Gaussian apparent diffusion coefficient profiles in human brain data. *Magn. Reson. Med.*, 48(2): 331–340.

Al-Sayed, S., Zoubir, A. M., and Sayed, A. H. (2016). Robust adaptation in impulsive noise. *IEEE Trans. Signal Process.*, 64(11): 2851–2865.

Amado, C. and Pires, A. M. (2004). Robust bootstrap with non random weights based on the influence function. *Commun. Stat. Simul. Comput.*, 33(2): 377–396.

Amado, C., Bianco, A. M., Boente, G., and Pires, A. M. (2014). Robust bootstrap: An alternative to bootstrapping robust estimators. *REVSTAT Statist. J.*, 12(2): 169–197.

Andersen, H. H., Hojbjerre, M., Sorensen, D., and Eriksen, P. S. (1995). *Linear and graphical models for the multivariate complex normal distribution*. Lecture Notes in Statistics, vol. 101. Springer-Verlag, New York.

Andrews, B. (2008). Rank-based estimation for autoregressive moving average time series models. *J. Time Ser. Anal.*, 29(1): 51–73.

Astola, J., Haavisto, P., and Neuvo, Y. (1990). Vector median filters. *Proc. IEEE*, 78(4): 678–689.

Aysal, T. C. and Barner, K. E. (2007). Meridian filtering for robust signal processing. *IEEE Trans. Signal Process.*, 55(8): 3949–3962.

Barner, K. E. and Arce, G. R. (2003). *Nonlinear signal and image processing: Theory, methods, and applications*. CRC Press, Boca Raton, FL.

Barrodale, I. and Roberts, F. D. K. (1973). An improved algorithm for discrete ℓ_1 linear approximation. *SIAM J. Numerical Anal.*, 10(5): 839–848.

Basiri, S., Ollila, E., and Koivunen, V. (2014). Fast and robust bootstrap method for testing hypotheses in the ICA model. In *Proc. IEEE Int. Conf. Acoust. Speech Signal Process.*, 6–10, Florence, Italy.

Basiri, S., Ollila, E., and Koivunen, V. (2016). Robust, scalable, and fast bootstrap method for analyzing large scale data. *IEEE Trans. Signal Process.*, 64(4): 1007–1017.

Basiri, S., Ollila, E., and Koivunen, V. (2017). Enhanced bootstrap method for statistical inference in the ICA model. *Signal Process.*, 138: 53–62.

Becker, C., Fried, R., and Kuhnt, S., editors (2013). *Robustness and complex data structures: Festschrift in honour of Ursula Gather*. Springer, New York.

Belloni, F. and Koivunen, V. (2003). Unitary root-MUSIC technique for uniform circular array. In *Proc. 3rd IEEE Int. Symp. Signal Process. Inform. Technol.*, 451–454, Darmstadt, Germany.

Belloni, F., Richter, A., and Koivunen, V. (2007). DOA estimation via manifold separation for arbitrary array structures. *IEEE Trans. Signal Process.*, 55(10): 4800–4810.

Bénar, C. G., Schön, D., Grimault, S., Nazarian, B., Burle, B., Roth, M., Badier, J. M., Marquis, P., Liegeois-Chauvel, C., and Anton, J. L. (2007). Single-trial analysis of oddball event-related potentials in simultaneous EEG-fMRI. *Hum. Brain Map.*, 28(7): 602–613.

Bernoulli, D. (1777). Dijudicatio maxime probabilis plurium observationum discrepantium atque verisimillima inductio inde formanda. *Acta Acad. Sci. Petropolit.*, 1: 3–33.

Bickel, P. J., Götze, F., and van Zwet, W. R. (2012). Resampling fewer than *n* observations: Gains, losses, and remedies for losses. In *Selected Works of Willem van Zwet*, 267–297. Springer, New York.

Blankenship, T. K., Kriztman, D. M., and Rappaport, T. S. (1997). Measurements and simulation of radio frequency impulsive noise in hospitals and clinics. In *Proc. IEEE 47th Veh. Technol. Conf.*, vol. 3, 1942–1946, Phoenix, AZ.

Box, G. E. P. (1953). Non-normality and tests on variances. *Biometrika*, 40(3–4): 318–335.

Box, G. E. P. (1979). Robustness in the strategy of scientific model building. *Robustness Stat.*, 1: 201–236.

Boyd, S. and Vandenberghe, L. (2004). *Convex optimization*. Cambridge University Press, Cambridge.

Brandwood, D. H. (1983). A complex gradient operator and its applications in adaptive array theory. *IEE Proc. F and H*, 130(1): 11–16.

Brillinger, D. R. (2001). *Time series: Data analysis and theory*. SIAM, Philadelphia, PA.

Brodsky, B. E. and Darkhovsky, B. S. (2008). Minimax methods for multihypothesis sequential testing and change-point detection problems. *Seq. Anal.*, 27(2): 141–173.

Bustos, O. H. and Yohai, V. J. (1986). Robust estimates for ARMA models. *J. Am. Statist. Assoc.*, 81(393): 155–168.

Calvet, L. E., Czellar, V., and Ronchetti, E. (2015). Robust filtering. *J. Am. Stat. Assoc.*, 110: 1591–1606.

Candes, E. J. and Wakin, M. B. (2008). An introduction to compressive sampling. *IEEE Signal Proc. Mag.*, 25(2): 21–30.

Carlson, B. D. (1988). Covariance matrix estimation errors and diagonal loading in adaptive arrays. *IEEE Trans. Aerosp. Electron. Syst.*, 24(4): 397–401.

Carroll, J. D. and Chang, J.-J. (1970). Analysis of individual differences in multidimensional scaling via an *n*-way generalization of Eckart-Young decomposition. *Psychometrika*, 35(3): 283–319.

Chakhchoukh, Y. (2010a). *Contribution to the estimation of SARIMA (application to short-term forecasting of electricity consumption)*. PhD thesis, Université de Paris-Sud, Faculté des Sciences d'Orsay, Essonne.

Chakhchoukh, Y. (2010b). A new robust estimation method for ARMA models. *IEEE Trans. Signal Process.*, 58(7): 3512–3522.

Chakhchoukh, Y., Panciatici, P., and Bondon, P. (2009). Robust estimation of SARIMA models: Application to short-term load forecasting. In *Proc. IEEE Workshop Statist. Signal Process.* 77–80, Cardiff, UK.

Chakhchoukh, Y., Panciatici, P., and Mili, L. (2010). Electric load forecasting based on statistical robust methods. *IEEE Trans. Power Syst.*, 26(3): 982–991.

Chen, Y., So, H. C., and Sun, W. (2014). ℓ_p-norm based iterative adaptive approach for robust spectral analysis. *Signal Process.*, 94(1): 144–148.

Chen, Y., Wiesel, A., and Hero, A. O. (2011). Robust shrinkage estimation of high-dimensional covariance matrices. *IEEE Trans. Signal Process.*, 59(9): 4097–4107.

Chen, Y., Wiesel, A., Eldar, Y. C., and Hero, A. O. (2010). Shrinkage algorithms for MMSE covariance estimation. *IEEE Trans. Signal Process.*, 58(10): 5016–5029.

Cheung, K. W., So, H.-C., Ma, W.-K., and Chan, Y.-T. (2004). Least squares algorithms for time-of-arrival-based mobile location. *IEEE Trans. Signal Process.*, 52(4): 1121–1130.

Chi, E. and Kolda, T. G. (2011). Making tensor factorizations robust to non-Gaussian noise. Technical Report Tech. Rep. No. SAND2011-1877, Sandia National Laboratories, Livermore, CA.

Cichocki, A., Mandic, D., De Lathauwer, L., Zhou, G., Zhao, Q., Caiafa, C., and Phan, A.-H. (2015). Tensor decompositions for signal processing applications: From two-way to multiway component analysis. *IEEE Signal Process. Mag.*, 32(2): 145–163.

Conte, E., De Maio, A., and Ricci, G. (2002). Recursive estimation of the covariance matrix of a compound-Gaussian process and its application to adaptive CFAR detection. *IEEE Trans. Signal Process.*, 50(8): 1908–1915.

Costa, M. and Koivunen, V. (2014). Application of manifold separation to polarimetric capon beamformer and source tracking. *IEEE Trans. Signal Process.*, 62(4): 813–827.

Costa, M., Koivunen, V., and Viberg, M. (2013). *Array processing in the face of nonidealities, signal processing reference (handbook of signal processing)*, vol. 3, ch. 19. Academic Press Library in Signal Processing.

Costa, M., Richter, A., and Koivunen, V. (2010). Unified array manifold decomposition based on spherical harmonics and 2-D Fourier basis. *IEEE Trans. Signal Process.*, 58(9): 4634–4645.

Couillet, R. and McKay, M. (2014). Large dimensional analysis and optimization of robust shrinkage covariance matrix estimators. *J. Multivariate Anal.*, 131: 99–120.

Croux, C. (1998). Limit behavior of the empirical influence function of the median. *Stat. Prob. Lett.*, 37(4): 331–340.

Croux, C. and Exterkate, P. (2011). Robust and sparse factor modelling. Technical Report Tech. Rep. No. KBI 1120, KU Leuven, Faculty of Business and Economics, Flanders, Belgium.

Croux, C., Filzmoser, P., and Oliveira, M. R. (2007). Algorithms for projection–pursuit robust principal component analysis. *Chemom. Intell. Lab. Syst.*, 87(2): 218–225.

Dabak, A. G. and Johnson, D. H. (1993). Geometrically based robust detection. In *Proc. Inform. Science Syst. Conf.*, 73–77.

Davies, P. L. (1987). Asymptotic behaviour of *S*-estimates of multivariate location parameters and dispersion matrices. *Ann. Stat.*, 15(3): 1269–1292.

Davies, P. L. (1993). Aspects of robust linear regression. *Ann. Stat.*, 21(4): 1843–1899.

DeGroot, M. H. (1960). Minimax sequential tests of some composite hypotheses. *Ann. Math. Stat.*, 31(4): 1193–1200.

Dehling, H., Fried, R., and Wendler, M. (2015). A robust method for shift detection in time series. *arXiv preprint arXiv:1506.03345*.

de Luna, X. and Genton, M. G. (2001). Robust simulation-based estimation of ARMA models. *J. Comput. Graph. Stat.*, 10(2): 370–387.

Demitri, N. and Zoubir, A. M. (2013). Mean-shift based algorithm for the measurement of blood glucose in hand-held devices. In *Proc. 21st Eur. Signal Process. Conf.*, 1–5, Marrakech, Morocco.

Demitri, N. and Zoubir, A. M. (2014). A robust kernel density estimator based mean-shift algorithm. In *Proc. IEEE Int. Conf. Acoust. Speech Signal Process.*, 7964–7968, Florence, Italy.

Demitri, N. and Zoubir, A. M. (2017). Measuring blood glucose concentrations in photometric glucometers requiring very small sample volumes. *IEEE Trans. Biomed. Eng.*, 64(1): 28–39.

De Moivre, A. (1733). Approximatio ad summam terminorum binomii $(a+b)$ n in seriem expansi. *Suppl. Miscellanea Analytica*.

Deutsch, S. J., Richards, J. E., and Swain, J. J. (1990). Effects of a single outlier on ARMA identification. *Commun. Stat. Theory*, 19(6): 2207–2227.

Djurovic, I., Stankovic, L., and Böhme, J. F. (2003). Robust l-estimation based forms of signal transforms and time-frequency representations. *IEEE Trans. Signal Process.*, 51(7): 1753–1761.

Dong, H., Wang, Z., and Gao, H. (2010). Robust $H\infty$ filtering for a class of nonlinear networked systems with multiple stochastic communication delays and packet dropouts. *IEEE Trans. Signal Process.*, 58(4): 1957–1966.

Donoho, D. L. (2006). Compressive sensing. *IEEE Trans. Inf. Theory*, 52(2): 5406–5425.

Donoho, D. L. and Huber, P. J. (1983). The notion of breakdown point. *A festschrift for Erich L. Lehmann*, 157–184, Wadsworth, Belmont, CA.

Dryden, I. L. and Mardia, K. V. (1998). *Statistical shape analysis*. Wiley, Chichester, UK.

Duncan, D. B. and Horn, S. D. (1972). Linear dynamic recursive estimation from the viewpoint of regression analysis. *J. Am. Statist. Assoc.*, 67(340): 815–821.

Durovic, Z. M. and Kovacevic, B. D. (1999). Robust estimation with unknown noise statistics. *IEEE Trans. Autom. Control*, 44(6): 1292–1296.

Dürre, A., Fried, R., and Liboschik, T. (2015). Robust estimation of (partial) autocorrelation. *Wiley Interdisc. Rev. Comput. Statist.*, 7(3): 205–222.

Efron, B. (1979). Bootstap methods: Another look at the jackknife. *Ann. Stat.*, 7(1): 1–26.

Efron, B. and Tibshirani, R. J. (1994). *An introduction to the bootstrap*. CRC Press, Boca Raton, FL.

Eriksson, J., Ollila, E., and Koivunen, V. (2009). Statistics for complex random variables revisited. In *Proc. IEEE Int. Conf. Acoust. Speech Signal Process.*, 3565–3568, Taipei, Taiwan.

Eriksson, J., Ollila, E., and Koivunen, V. (2010). Essential statistics and tools for complex random variables. *IEEE Trans. Signal Process.*, 58(10): 5400–5408.

Etter, P. C. (2003). *Underwater acoustic modeling and simulation*. Taylor & Francis, Boca Raton, FL.

Fang, K.-T., Kotz, S., and Ng, K. W. (1990). *Symmetric multivariate and related distributions*. Chapman and Hall, London.

Fante, R. L. and Vaccaro, J. J. (2000). Wideband cancellation of interference in a GPS receive array. *IEEE Trans. Aerosp. Electron. Syst.*, 36(2): 549–564.

Fauß, M. (2016). *Design and analysis of optimal and minimax robust sequential hypothesis tests*. PhD thesis, Technische Universität Darmstadt, Darmstadt, Germany.

Fauß, M. and Zoubir, A. M. (2015). A linear programming approach to sequential hypothesis testing. *Seq. Anal.*, 34(2): 235–263.

Fauß, M. and Zoubir, A. M. (2016). Old bands, new tracks – Revisiting the band model for robust hypothesis testing. *IEEE Trans. Signal Process.*, 64(22): 5875–5886.

Fellouris, G. and Tartakovsky, A. G. (2012). Nearly minimax one-sided mixture-based sequential tests. *Seq. Anal.*, 31(3): 297–325.

Fisher, R. A. (1925). Theory of statistical estimation. *Math. Proc. Cambridge Philosoph. Soc.*, 22(5): 700–725.

Frahm, G. (2004). *Generalized elliptical distributions: Theory and applications*. PhD thesis, Universität zu Köln.

Frank, L. E. and Friedman, J. H. (1993). A statistical view of some chemometrics regression tools. *Technometrics*, 35(2): 109–135.

Friedman, J., Hastie, T., Höfling, H., and Tibshirani, R. (2007). Pathwise coordinate optimization. *Ann. Appl. Stat.*, 1(2): 302–332.

Fu, W. J. (1998). Penalized regressions: The bridge versus the Lasso. *J. Comput. Graph. Stat.*, 7(3): 397–416.

Galilei, G. (1632). *Dialogue concerning the two chief world systems*, Folio Society, London.

Gandhi, M. A. and Mili, L. (2010). Robust Kalman filter based on a generalized maximum-likelihood-type estimator. *IEEE Trans. Signal Process.*, 58(5): 2509–2520.

Gao, X. and Huang, J. (2010). Asymptotic analysis of high-dimensional LAD regression with Lasso. *Stat. Sin.*, 1: 1485–1506.

Gauss, C. F. (1809). *Theoria Motus Corporum Celestium*. Perthes et Besser.

Genton, M. G. and Lucas, A. (2003). Comprehensive definitions of breakdown points for independent and dependent observations. *J. R. Statist. Soc. B*, 65(1): 81–94.

Gershman, A. B. (2004). Robustness issues in adaptive beamforming and high-resolution direction finding, ch. 2. In Hua et al., *High-resolution and robust signal processing* (2004).

Gerstoft, P., Xenaki, A., and Mecklenbräuker, C. F. (2015). Multiple and single snapshot compressive beamforming. *J. Acoust. Soc. Am.*, 138(4): 2003–2014.

Gini, C. (1921). Sull'interpolazione di una retta quando i valori della variabile indipendente sono affetti da errori accidentali. *Metron*, 1: 63–82.

Gini, F. (1997). Sub-optimum coherent radar detection in a mixture of K-distributed and Gaussian clutter. *IEE Proc. Radar, Sonar Navig.*, 144(1): 39–48.

Gini, F. and Greco, M. (2002). Covariance matrix estimation for CFAR detection in correlated heavy tailed clutter. *Signal Process.*, 82(12): 1847–1859.

Godara, L. C. (2004). *Smart antennas*. CRC Press, Boca Raton, FL.

Goldstein, J. S., Reed, I. S., and Scharf, L. L. (1998). A multistage representation of the Wiener filter based on orthogonal projections. *IEEE Trans. Inf. Theory*, 44(7): 2943–2959.

Gonzalez, J. G. and Arce, G. R. (2001). Optimality of the myriad filter in practical impulsive-noise environments. *IEEE Trans. Signal Process.*, 49(2): 438–441.

Goodman, N. R. (1963). Statistical analysis based on certain multivariate complex Gaussian distribution (an introduction). *Ann. Math. Stat.*, 34(1): 152–177.

Guerci, B., Drouin, P., Grangé, V., Bougnères, P., Fontaine, P., Kerlan, V., Passa, P., Thivolet, C., Vialettes, B., and Charbonnel, B. (2003). Self-monitoring of blood glucose significantly improves metabolic control in patients with type 2 diabetes mellitus: The Auto-Surveillance Intervention Active (ASIA) study. *Diabetes Metab.*, 29(6): 587–594.

Gül, G. (2017). *Robust and distributed hypothesis testing*. Lecture Notes in Electrical Engineering. Springer International Publishing, New York.

Gül, G. and Zoubir, A. M. (2016). Robust hypothesis testing with α-divergence. *IEEE Trans. Signal Process.*, 64(18): 4737–4750.

Gül, G. and Zoubir, A. M. (2017). Minimax robust hypothesis testing. *IEEE Trans. Inf. Theory*, 63(9): 5572–5587.

Guvenc, I. and Chong, C.-C. (2009). A survey on TOA based wireless localization and NLOS mitigation techniques. *IEEE Commun. Surveys Tuts.*, 11(3): 107–124.

Hammes, U. and Zoubir, A. M. (2011). Robust MT tracking based on M-estimation and interacting multiple model algorithm. *IEEE Trans. Signal Process.*, 59(7): 3398–3409.

Hammes, U., Wolsztynski, E., and Zoubir, A. M. (2008). Transformation-based robust semiparametric estimation. *IEEE Signal Process. Lett.*, 15: 845–848.

Hammes, U., Wolsztynski, E., and Zoubir, A. M. (2009). Robust tracking and geolocation for wireless networks in NLOS environments. *IEEE J. Sel. Topics Signal Process.*, 3(5): 889–901.

Hampel, F. R. (1968). *Contributions to the theory of robust estimation*. PhD thesis, University of California, Berkeley.

Hampel, F. R. (1974). The influence curve and its role in robust estimation. *J. Am. Statist. Assoc.*, 69(346): 383–393.

Hampel, F. R., Ronchetti, E. M., Rousseeuw, P. J., and Stahel, W. A. (2011). *Robust statistics: The approach based on influence functions*, vol. 114. John Wiley & Sons, Toronto, ON.

Han, B., Muma, M., Feng, M., and Zoubir, A. M. (2013). An online approach for intracranial pressure forecasting based on signal decomposition and robust statistics. In *Proc. IEEE Int. Conf. Acoust. Speech Signal Process.*, 6239–6243, Vancouver, BC.

Hannan, E. J. and Quinn, B. G. (1979). The determination of the order of an autoregression. *J. R. Statist. Soc. B*, 41(2): 190–195.

Harshman, R. A. (1970). Foundations of the PARAFAC procedure: Models and conditions for an explanatory multi-modal factor analysis. *UCLA working papers phonetics*, 16: 1–84.

Hastie, T., Tibshirani, R., and Friedman, J. (2001). *The elements of statistical learning*. Springer, New York.

Hastie, T., Tibshirani, R., and Wainwright, M. (2015). *Statistical learning with sparsity: The Lasso and generalizations*. CRC Press, Boca Raton, FL.

Hjorungnes, A. and Gesbert, D. (2007). Complex-valued matrix differentiation: Techniques and key results. *IEEE Trans. Signal Process.*, 55: 2740–2746.

Hoctor, R. T. and Kassam, S. A. (1990). The unifying role of the coarray in aperture synthesis for coherent and incoherent imaging. *Proc. IEEE*, 78(4): 735–752.

Hoerl, A. E. and Kennard, R. W. (1970). Ridge regression: Biased estimation for nonorthogonal problems. *Technometrics*, 12(1): 55–67.

Horn, R. A. and Johnson, C. A. (1985). *Matrix analysis*. Cambridge University Press, Cambridge.

Hua, Y., Gershman, A., and Cheng, Q., editors (2004). *High-resolution and robust signal processing*. Marcel Dekker, New York.

Huber, P. J. (1964). Robust estimation of a location parameter. *Ann. Math. Stat.*, 35(1): 73–101.

Huber, P. J. (1965). A robust version of the probability ratio test. *Ann. Math. Stat.*, 36(6): 1753–1758.

Huber, P. J. (1972). The 1972 Wald lecture robust statistics: A review. *Ann. Math. Stat.*, 43(4): 1041–1067.

Huber, P. J. and Ronchetti, E. M. (2009). *Robust statistics*. John Wiley & Sons, Hoboken, NJ.

Huber, P. J. and Strassen, V. (1973). Minimax tests and the Neyman–Pearson lemma for capacities. *Ann. Statist.*, 1(2): 251–263.

Hubert, M., Van Kerckhoven, J., and Verdonck, T. (2012). Robust PARAFAC for incomplete data. *J. Chemometrics*, 26(6): 290–298.

Hunter, D. R. and Lange, K. (2004). A tutorial on MM algorithms. *Am. Stat.*, 58(1): 30–37.

Jaeckel, L. A. (1972). Estimating regression coefficients by minimizing the dispersion of the residuals. *Ann. Math. Stat.*, 43(5): 1449–1458.

Jansson, M., Swindlehurst, A. L., and Ottersten, B. (1998). Weighted subspace fitting for general array error models. *IEEE Trans. Signal Process.*, 46(9): 2484–2498.

Jones, R. H. (1980). Maximum likelihood fitting of ARMA models to time series with missing observations. *Technometrics*, 22(3): 389–395.

Kalluri, S. and Arce, G. R. (2000). Fast algorithms for weighted myriad computation by fixed-point search. *IEEE Trans. Signal Process.*, 48(1): 159–171.

Kalman, R. E. (1960). A new approach to linear filtering and prediction problems. *Trans. ASME J. Basic Eng.*, 82: 35–45.

Kassam, S. A. (1981). Robust hypothesis testing for bounded classes of probability densities (Corresp.). *IEEE Trans. Inf. Theory*.

Kassam, S. A. and Poor, H. V. (1985). Robust techniques for signal processing: A survey. *Proc. IEEE*, 73(3): 433–481.

Kassam, S. A. and Thomas, J. B. (1975). A class of nonparametric detectors for dependent input data. *IEEE Trans. Inf. Theory*, 21(4): 431–437.

Katkovnik, V. (1998). Robust M-periodogram. *IEEE Trans. Signal Process.*, 46(11): 3104–3109.

Kelava, A., Muma, M., Deja, M., Dagdagan, J. Y., and Zoubir, A. M. (2014). A new approach for the quantification of synchrony of multivariate non-stationary psychophysiological variables during emotion eliciting stimuli. *Front. Psychol.*, 5.

Kent, J. T. (1997). Data analysis for shapes and images. *J. Stat. Plan. Inference*, 57(2): 181–193.

Kent, J. T. and Tyler, D. E. (1988). Maximum likelihood estimation for the wrapped Cauchy distribution. *J. Appl. Stat.*, 15(2): 247–254.

Kent, J. T. and Tyler, D. E. (1991). Redescending M-estimates of multivariate location and scatter. *Ann. Stat.*, 19(4): 2102–2119.

Kim, H.-J., Ollila, E., and Koivunen, V. (2013a). Sparse regularization of tensor decompositions. In *Proc. IEEE Int. Conf. Acoust. Speech Signal Process.*, 3836–3840, Vancouver, BC.

Kim, H.-J., Ollila, E., and Koivunen, V. (2015). New robust Lasso method based on ranks. In *Proc. 23rd Eur. Signal Process. Conf.*, 699–703, Nice, France.

Kim, H.-J., Ollila, E., Koivunen, V., and Croux, C. (2013b). Robust and sparse estimation of tensor decompositions. In *Proc. IEEE Global Conf. Signal Inform. Process.*, 965–968, Austin, TX.

Kim, H.-J., Ollila, E., Koivunen, V., and Poor, H. V. (2014). Robust iteratively reweighted Lasso for sparse tensor factorizations. In *Proc. IEEE Workshop Statist. Signal Process.*, 420–423, Gold Coast, Australia.

Kim, K. and Shevlyakov, G. (2008). Why Gaussianity? *IEEE Signal Process. Mag.*, 25(2): 102–113.

Kleiner, A., Talwalkar, A., Sarkar, P., and Jordan, M. I. (2014). A scalable bootstrap for massive data. *J. R. Statist. Soc. B*, 76(4): 795–816.

Koenker, R. (2005). *Quantile regression*. Cambridge University Press, New York.

Koivunen, V. (1996). Nonlinear filtering of multivariate images under robust error criterion. *IEEE Trans. Image Process.*, 5(6): 1054–1060.

Koivunen, V. and Ollila, E. (2014). *Model order selection, signal processing reference (handbook of signal processing)*, ch. 2. Vol. 3 of Zoubir et al., *Array and statistical processing* (2014).

Kolda, T. G. and Bader, B. W. (2009). Tensor decompositions and applications. *SIAM Rev.*, 51(3): 455–500.

Kozick, R. J. and Sadler, B. M. (2000). Maximum-likelihood array processing in non-Gaussian noise with Gaussian mixtures. *IEEE Trans. Signal Process.*, 48(12): 3520–3535.

Kraut, S., Scharf, L. L., and Butler, R. W. (2005). The adaptive coherence estimator: A uniformly most-powerful-invariant adaptive detection statistic. *IEEE Trans. Signal Process.*, 53(2): 427–438.

Kraut, S., Scharf, L. L., and McWhorter, L. T. (2001). Adaptive subspace detectors. *IEEE Trans. Signal Process.*, 49(1): 1–16.

Kreutz-Delgado, K. (2007). The complex gradient operator and the CR-calculus. Lecture notes supplement [online].

Krim, H. and Viberg, M. (1996). Two decades of array signal processing research: The parametric approach. *IEEE Signal Process. Mag.*, 13(4): 67–94.

Krishnaiah, P. R. and Lin, J. (1986). Complex elliptically symmetric distributions. *Commun. Stat. Theory Methods*, 15(12): 3693–3718.

Kumar, T. A. and Rao, K. D. (2009). A new M-estimator based robust multiuser detection in flat-fading non-Gaussian channels. *IEEE Trans. Commun.*, 57(7): 1908–1913.

Künsch, H. (1984). Infinitesimal robustness for autoregressive processes. *Ann. Stat.*, 12(3): 843–863.

Le, N. D., Raftery, A. E., and Martin, R. D. (1996). Robust Bayesian model selection for autoregressive processes with additive outliers. *J. Am. Stat. Assoc.*, 91(433): 123–131.

Ledoit, O. and Wolf, M. (2004). A well-conditioned estimator for large-dimensional covariance matrices. *J. Multivariate Anal.*, 88(2): 365–411.

Lee, Y. and Kassam, S. A. (1985). Generalized median filtering and related nonlinear filtering techniques. *IEEE Trans. Acoust. Speech Signal Process.*, 33(3): 672–683.

Lehmann, E. L. and D'Abrera, H. J. (1975). *Nonparametrics: Statistical methods based on ranks*. Holden-Day, Oxford.

Levy, B. C. (2009). Robust hypothesis testing with a relative entropy tolerance. *IEEE Trans. Inf. Theory*, 55(1): 413–421.

Li, J., Stoica, P., and Wang, Z. (2003). On robust Capon beamforming and diagonal loading. *IEEE Trans. Signal Process.*, 51(7): 1702 – 1715.

Li, T.-H. (2008). Laplace periodogram for time series analysis. *J. Am. Stat. Assoc.*, 103(482): 757–768.

Li, T.-H. (2010). A nonlinear method for robust spectral analysis. *IEEE Trans. Signal Process.*, 58(5): 2466–2474.

Li, Y. and Arce, G. R. (2004). A maximum likelihood approach to least absolute deviation regression. *EURASIP J. Adv. Signal Process.*, 2004(12): 1762–1769.

Ljung, G. M. (1993). On outlier detection in time series. *J. R. Statist. Soc. B*, 55(2): 559–567.

Lopuhaa, H. P. and Rousseeuw, P. J. (1991). Breakdown points of affine equivariant estimators of multivariate location and covariance matrices. *Ann. Stat.*, 19(1): 229–248.

Louni, H. (2008). Outlier detection in ARMA models. *J. Time Ser. Anal.*, 29(6): 1057–1065.

Madsen, K., Nielsen, H. B., and Tingleff, O. (2004). Methods for non-linear least squares problems. Lecture note.

Mandic, D. P. and Goh, V. S. L. (2009). *Complex valued nonlinear adaptive filters: Noncircularity, widely linear and neural models*, vol. 59. John Wiley & Sons, Hoboken, NJ.

Markets and Markets. (2016). Location-based services (LBS) and real time location systems (RTLS) market by location (indoor and outdoor), technology (context aware, UWB, BT/BLE, beacons, A-GPS), software, hardware, service and application area – Global forecast to 2021. www.marketsandmarkets.com.

Maronna, R. A. (1976). Robust M-estimators of multivariate location and scatter. *Ann. Stat.*, 5(1): 51–67.

Maronna, R. A. and Yohai, V. J. (2008). Robust low-rank approximation of data matrices with elementwise contamination. *Technometrics*, 50(3): 295–304.

Maronna, R. A., Martin, R. D., and Yohai, V. J. (2006). *Robust statistics theory and methods*. John Wiley & Sons, Ltd, Chichester, UK.

Marple, S. L. (1987). *Digital spectral analysis with applications*, vol. 5. Prentice-Hall, Englewood Cliffs, NJ.

Martin, R. D. and Thomson, D. J. (1982). Robust-resistant spectrum estimation. *Proc. IEEE*, 70(9): 1097–1115.

Martin, R. D. and Yohai, V. J. (1984). Influence function for time series. Technical Report 51, Department of Statistics, University of Washington, Seattle, WA.

Martin, R. D. and Yohai, V. J. (1986). Influence functionals for time series. *Ann. Stat.*, 14(3): 781–818.

Martin, R. D., Samarov, A., and Vandaele, W. (1982). Robust methods for ARIMA models. Technical Report 21, Department of Statistics, University of Washington, Seattle, WA.

Martinez-Camara, M., Béjar Haro, B., Stohl, A., and Vetterli, M. (2014). A robust method for inverse transport modelling of atmospheric emissions using blind outlier detection. *Geosci. Model Dev.*, 7: 2303–2311.

Masreliez, C. (1975). Approximate non-Gaussian filtering with linear state and observation relations. *IEEE Trans. Autom. Control*, 20(1): 107–110.

Mayo, M. S. and Gray, J. B. (1997). Elemental subsets: The building blocks of regression. *Am. Stat.*, 51(2): 122–129.

McQuarrie, A. D. and Tsai, C.-L. (2003). Outlier detections in autoregressive models. *J. Comput. Graph. Stat.*, 12(2): 450–471.

Middleton, D. (1996). *An introduction to statistical communication theory: An IEEE Press classic reissue*, vol. 1. Wiley-IEEE Press, Hoboken, NJ.

Middleton, D. (1999). Non-Gaussian noise models in signal processing for telecommunications: New methods and results for class A and class B noise models. *IEEE Trans. Inf. Theory*, 45(4): 1129–1149.

Mili, L., Cheniae, M. G., and Rousseeuw, P. J. (2002). Robust state estimation of electric power systems. *IEEE Trans. Circuits Syst. I*, 41(5): 349–358.

Moses, R. L., Simonypté, V., Stoica, P., and Söderström, T. (1994). An efficient linear method for ARMA spectral estimation. *Int. J. Control*, 59(2): 337–356.

Muler, N., Peña, D., and Yohai, V. J. (2009). Robust estimation for ARMA models. *Ann. Stat.*, 37(2): 816–840.

Müller, S. and Welsh, A. H. (2005). Outlier robust model selection in linear regression. *J. Am. Stat. Assoc.*, 100(472): 1297–1310.

Muma, M. (2014). Robust model order selection for ARMA models based on the bounded innovation propagation τ-estimator. In *Proc. IEEE Workshop Statist. Signal Process.*, 428–431, Gold Coast, Australia.

Muma, M. and Zoubir, A. M. (2017). Bounded influence propagation τ-estimation: A new method for ARMA model estimation. *IEEE Trans. Signal Process.*, 65: 1712–1727.

Nassar, M., Gulati, K., Sujeeth, A. K., Aghasadeghi, N., Evans, B. L., and Tinsley, K. R. (2008). Mitigating near-field interference in laptop embedded wireless transceivers. In *Proc. IEEE Int. Conf. Acoust. Speech Signal Process.*, 1405–1408, Las Vegas, NV.

Oja, H. (2013). *Multivariate median*, ch. 1. In Becker et al., *Robustness and complex data structures* (2013).

Ollila, E. (2010). *Contributions to independent component analysis, sensor array and complex-valued signal processing.* PhD thesis, Aalto University, Espoo, Finland.

Ollila, E. (2015a). *Modern nonparametric, robust and multivariate methods: A festschrift in honor of Professor Hannu Oja.* Springer, Basel, Switzerland.

Ollila, E. (2015b). Multichannel sparse recovery of complex-valued signals using Huber's criterion. In *Proc. Compressed Sens. Theory Appl. Radar, Sonar Remote Sens.*, 32–36, Pisa, Italy.

Ollila, E. (2016a). Adaptive Lasso based on joint *M*-estimation of regression and scale. In *Proc. 24th Eur. Signal Process. Conf.*, 2191–2195, Budapest, Hungary.

Ollila, E. (2016b). Direction of arrival estimation using robust complex Lasso. In *Proc. 10th Eur. Conf. Antennas Propag.*, 1–5, Davos, Switzerland.

Ollila, E. (2017). Optimal high-dimensional shrinkage covariance estimation for elliptical distributions. In *Proc. 25th Eur. Signal Process. Conf.*, 1689–1693, Kos, Greece.

Ollila, E. and Koivunen, V. (2003a). Influence functions for array covariance matrix estimators. In *Proc. IEEE Workshop Statist. Signal Process.*, 462–465, St. Louis, MS.

Ollila, E. and Koivunen, V. (2003b). Robust antenna array processing using *M*-estimators of pseudo-covariance. In *14th IEEE Proc. Pers. Indoor Mobile Radio Commun.*, vol. 3, 2659–2663, Beijing, China.

Ollila, E. and Koivunen, V. (2004). Generalized complex elliptical distributions. In *Proc. IEEE Sensor Array Multichannel Signal Process. Workshop*, 460–464, Barcelona, Spain.

Ollila, E. and Tyler, D. E. (2012). Distribution-free detection under complex elliptically symmetric clutter distribution. In *Proc. IEEE 7th Sensor Array Multichannel Signal Process. Workshop*, 413–416, Wiley, Hoboken, NJ.

Ollila, E. and Tyler, D. E. (2014). Regularized *M*-estimators of scatter matrix. *IEEE Trans. Signal Process.*, 62(22): 6059–6070.

Ollila, E., Croux, C., and Oja, H. (2004). Influence function and asymptotic efficiency of the affine equivariant rank covariance matrix. *Stat. Sin.*, 14(1): 297–316.

Ollila, E., Eriksson, J., and Koivunen, V. (2011a). Complex elliptically symmetric random variables – generation, characterization, and circularity tests. *IEEE Trans. Signal Process.*, 59(1): 58–69.

Ollila, E., Koivunen, V., and Poor, H. V. (2011b). Complex-valued signal processing – Essential models, tools and statistics. In *Inf. Theory Appl. Workshop*, 1–10. IEEE.

Ollila, E., Oja, H., and Croux, C. (2003a). The affine equivariant sign covariance matrix: Asymptotic behavior and efficiencies. *J. Multivariate Anal.*, 87(2): 328–355.

Ollila, E., Oja, H., and Koivunen, V. (2003b). Estimates of regression coefficients based on lift rank covariance matrix. *J. Am. Stat. Assoc.*, 98(461): 90–98.

Ollila, E., Quattropani, L., and Koivunen, V. (2003c). Robust space-time scatter matrix estimator for broadband antenna arrays. In *Proc. IEEE 58th Veh. Technol. Conf.*, vol. 1, 55–59, Orlando, FL.

Ollila, E., Soloveychik, I., Tyler, D. E., and Wiesel, A. (2016). Simultaneous penalized *M*-estimation of covariance matrices using geodesically convex optimization. *Arxiv:1608.08126v1*.

Ollila, E., Tyler, D. E., Koivunen, V., and Poor, H. V. (2012). Complex elliptically symmetric distributions: Survey, new results and applications. *IEEE Trans. Signal Process.*, 60(11): 5597–5625.

Österreicher, F. (1978). On the construction of least favourable pairs of distributions. *Zeitschrift für Wahrscheinlichkeitstheorie und Verwandte Gebiete*, 43(1): 49–55.

Owen, A. B. (2007). A robust hybrid of Lasso and ridge regression. *Contemp. Math.*, 443: 59–72.

Pan, J. and Tompkins, W. J. (1985). A real-time QRS detection algorithm. *IEEE Trans. Biomed. Eng.*, 32(3): 230–236.

Pang, Y., Li, X., and Yuan, Y. (2010). Robust tensor analysis with L1-norm. *IEEE Trans. Circuits Syst. Video Technol.*, 20(2): 172–178.

Papalexakis, E. E. and Sidiropoulos, N. D. (2011). Co-clustering as multilinear decomposition with sparse latent factors. In *Proc. IEEE Int. Conf. Acoust. Speech Signal Process.*, 2064–2067, Prague, Czech Republic.

Pascal, F., Chitour, Y., and Quek, Y. (2014). Generalized robust shrinkage estimator and its application to STAP detection problem. *IEEE Trans. Signal Process.*, 62(21): 5640–5651.

Pascal, F., Chitour, Y., Ovarlez, J.-P., Forster, P., and Larzabal, P. (2008). Covariance structure maximum-likelihood estimates in compound Gaussian noise: Existence and algorithm analysis. *IEEE Trans. Signal Process.*, 56(1): 34–48.

Patwari, N., Hero, A. O., Perkins, M., Correal, N. S., and O'dea, R. J. (2003). Relative location estimation in wireless sensor networks. *IEEE Trans. Signal Process.*, 51(8): 2137–2148.

Pedersen, K. I., Mogensen, P. E., and Fleury, B. H. (2000). A stochastic model of the temporal and azimuthal dispersion seen at the base station in outdoor propagation environments. *IEEE Trans. Veh. Technol.*, 49(2): 437–447.

Pesavento, M., Gershman, A. B., and Wong, K. M. (2002). Direction finding in partly calibrated sensor arrays composed of multiple subarrays. *IEEE Trans. Signal Process.*, 50(9): 2103–2115.

Picinbono, B. (1994). On circularity. *IEEE Trans. Signal Process.*, 42(12): 3473–3482.

Pillai, S. U. and Kwon, B. H. (1989). Forward/backward spatial smoothing techniques for coherent signal identification. *IEEE Trans. Acoust. Speech Signal Process.*, 37(1): 8–15.

Pitas, I. and Venetsanopoulos, A. N. (2013). *Nonlinear digital filters: Principles and applications*, vol. 84. Springer Science & Business Media, Berlin, Germany.

Poincaré, H. (1904). *L'état actuel et l'avenir de la physique mathématique*.

Poor, H. V. and Thomas, J. B. (1993). *Advances in statistical signal processing, signal detection (a research series)*, vol. 2. Jai Press, Greenwich, CT.

Priestley, M. B. (1981). *Spectral analysis and time series*. Academic Press, New York.

Prieto, J. C., Croux, C., and Jiménez, A. R. (2009). RoPEUS: A new robust algorithm for static positioning in ultrasonic systems. *Sensors*, 9(6): 4211–4229.

Prokhorov, Y. V. (1956). Convergence of random processes and limit theorems in probability theory. *Theory Probab. Appl.*, 1(2): 157–214.

Rao, B. D. and Hari, K. V. S. (1988). Performance analysis of subspace based methods (plane wave direction of arrival estimation). In *Proc. 4th Annu. ASSP Workshop Spectr. Estimation Modeling*, 92–97, Minneapolis, MN.

Razaviyayn, M., Hong, M., and Luo, Z. (2013). A unified convergence analysis of block successive minimization methods for nonsmooth optimization. *SIAM J. Optim.*, 23(2): 1126–1153.

Reeds, J. A. (1985). Asymptotic number of roots of Cauchy location likelihood equations. *Ann. Stat.*, 13(2): 775–784.

Remmert, R. (1991). *Theory of complex functions*. Springer-Verlag, New York.

Ribeiro, C. B., Ollila, E., and Koivunen, V. (2005). Propagation parameter estimation in MIMO systems using mixture of angular distributions model. In *Proc. IEEE Int. Conf. Acoust. Speech Signal Process.*, vol. 4, 885–888, Philadelphia, PA.

Ronchetti, E. (1997). Robustness aspects of model choice. *Stat. Sin.*, 7: 327–338.

Rousseeuw, P. J. (1984). Least median of squares regression. *J. Am. Statist. Assoc.*, 79(388): 871–880.

Rousseeuw, P. J. (1985). Multivariate estimation with high breakdown point. *Mathematical statistics and applications*. Reidel, Dordrecht, the Netherlands.

Rousseeuw, P. J. and Leroy, A. M. (2005). *Robust regression and outlier detection*, vol. 589. John Wiley & Sons, Hoboken, NJ.

Rousseeuw, P. and Yohai, V. J. (1984). Robust regression by means of S-estimators. In *Robust and nonlinear time series analysis*, 256–272. Springer, Berlin, Germany.

Rousseeuw, P. J. and Van Driessen, K. (2006). Computing LTS regression for large data sets. *Data Min. Knowl. Discov.*, 12(1): 29–45.

Ruckdeschel, P., Spangl, B., and Pupashenko, D. (2014). Robust Kalman tracking and smoothing with propagating and non-propagating outliers. *Statist. Papers*, 55(1): 93–123.

Rudin, L. I., Osher, S., and Fatemi, E. (1992). Nonlinear total variation based noise removal algorithms. *Physica D: Nonlinear Phenomena*, 60: 259–268.

Salibián-Barrera, M. and Yohai, V. J. (2012). A fast algorithm for S-regression estimates. *J. Comput. Graph. Stat.*

Salibián-Barrera, M. and Zamar, R. H. (2002). Bootstrapping robust estimates of regression. *Ann. Stat.*, 30(2): 556–582.

Salibián-Barrera, M., Van Aelst, S., and Willems, G. (2008a). Fast and robust bootstrap. *Statist. Methods Appl.*, 17(1): 41–71.

Salibián-Barrera, M., Willems, G., and Zamar, R. (2008b). The fast-τ estimator for regression. *J. Comput. Graph. Stat.*, 17(3): 659–682.

Salmi, J., Richter, A., and Koivunen, V. (2009). Sequential unfolding SVD for tensors with applications in array signal processing. *IEEE Trans. Signal Process.*, 57(12): 4719–4733.

Schäck, T., Harb, Y. S., Muma, M., and Zoubir, A. M. (2017). Computationally efficient algorithm for photoplethysmography-based atrial fibrillation detection using smartphones. In *Proc. 39th Annu. Int. Conf. IEEE Eng. Med. Biol. Soc.*, 104–108, Seogwipo, South Korea.

Schäck, T., Sledz, C., Muma, M., and Zoubir, A. M. (2015). A new method for heart rate monitoring during physical exercise using photoplethysmographic signals. In *Proc. 23rd Eur. Signal Process. Conf.*, 2666–2670, Nice, France.

Scharf, L. L. and Friedlander, B. (1994). Matched subspace detectors. *IEEE Trans. Signal Process.*, 42(8): 2146–2157.

Scharf, L. L. and McWhorter, L. T. (1996). Adaptive matched subspace detectors and adaptive coherence estimators. In *Proc. 30th Asilomar Conf. Signals Syst. Comput.*, 1114–1117, Pacific Grove, CA.

Schick, I. C. and Mitter, S. K. (1994). Robust recursive estimation in the presence of heavy-tailed observation noise. *Ann. Stat.*, 22(2): 1045–1080.

Schreier, P. J. and Scharf, L. L. (2010). *Statistical signal processing of complex-valued data: The theory of improper and noncircular signals*. Cambridge University Press, New York.

Schwarz, G. (1978). Estimating the dimension of a model. *Ann. Stat.*, 6(2): 461–464.

Shariati, N., Shahriari, H., and Shafaei, R. (2014). Parameter estimation of autoregressive models using the iteratively robust filtered fast-τ method. *Commun. Stat. Theory Methods*, 43(21): 4445–4470.

Sharif, W., Muma, M., and Zoubir, A. M. (2013). Robustness analysis of spatial time-frequency distributions based on the influence function. *IEEE Trans. Signal Process.*, 61(8): 1958–1971.

Shi, P. and Tsai, C. L. (1998). A note on the unification of the Akaike information criterion. *J. R. Statist. Soc. B*, 60(3): 551–558.

Sidiropoulos, N. D., De Lathauwer, L., Fu, X., Huang, K., Papalexakis, E. E., and Faloutsos, C. (2017). Tensor decomposition for signal processing and machine learning. *IEEE Trans. Signal Process.*, 65(13): 3551–3582.

Siegel, A. F. (1982). Robust regression using repeated medians. *Biometrika*, 69(1): 242–244.

Singh, K. (1998). Breakdown theory for bootstrap quantiles. *Ann. Stat.*, 26(5): 1719–1732.

Sion, M. (1958). On general minimax theorems. *Pac. J. Math.*, 8: 171–176.

Soloveychik, I. and Wiesel, A. (2015). Performance analysis of Tyler's covariance estimator. *IEEE Trans. Signal Process.*, 63(2): 418–426.

Song, I., Bae, J., and Kim, S. Y. (2002). *Advanced theory of signal detection: Weak signal detection in generalized observations*. Springer, Berlin, Germany.

Spangl, B. and Dutter, R. (2007). Estimating spectral density functions robustly. *REVSTAT Statist. J.*, 5(1): 41–61.

Stahl, S. (2006). The evolution of the normal distribution. *Math. Mag.*, 79(2): 96–113.

Stewart, C. V. (1999). Robust parameter estimation in computer vision. *SIAM Rev.*, 41(3): 513–537.

Stigler, S. M. (1973). Simon Newcomb, Percy Daniell, and the history of robust estimation, 1885–1920. *J. Am. Statist. Assoc.*, 68(344): 872–879.

Stockinger, N. and Dutter, R. (1987). Robust time series analysis: A survey. *Kybernetika*, 23(1): 3–88.

Stoica, P. and Selen, Y. (2004). Model-order selection: A review of information criterion rules. *IEEE Signal Process. Mag.*, 21(4): 36–47.

Stranger, B. E., Nica, A. C., Forrest, M. S., Dimas, A., Bird, C. P., Beazley, C., Ingle, C. E., Dunning, M., Flicek, P., and Koller, D. (2007). Population genomics of human gene expression. *Nat. Genet.*, 39(10): 1217–1224.

Strasser, F., Muma, M., and Zoubir, A. M. (2012). Motion artifact removal in ECG signals using multi-resolution thresholding. In *Proc. Eur. Signal Process. Conf.*, 899–903, Bucharest, Romania.

Stromberg, A. J. (1997). Robust covariance estimates based on resampling. *J. Stat. Plan. Inference*, 57(2): 321–334.

Sun, Y., Babu, P., and Palomar, D. P. (2014). Regularized Tyler's scatter estimator: Existence, uniqueness, and algorithms. *IEEE Trans. Signal Process.*, 62(19): 5143–5156.

Sun, Y., Babu, P., and Palomar, D. P. (2017). Majorization-minimization algorithms in signal processing, communications, and machine learning. *IEEE Trans. Signal Process.*, 65(3): 794–816.

Swami, A. and Sadler, B. M. (2002). On some detection and estimation problems in heavy-tailed noise. *Signal Process.*, 82(12): 1829–1846.

Swindlehurst, A. L. and Kailath, T. (1992). A performance analysis of subspace-based methods in the presence of model errors, part I: The MUSIC algorithm. *IEEE Trans. Signal Process.*, 40(7): 1758–1774.

Swindlehurst, A. L. and Kailath, T. (1993). A performance analysis of subspace-based methods in the presence of model errors: Part II – Multidimensional algorithms. *IEEE Trans. Signal Process.*, 41(9): 2882–2890.

Tabassum, M. N. and Ollila, E. (2016). Single-snapshot DOA estimation using adaptive elastic net in the complex domain. In *Proc. 4th Int. Workshop Compressed Sens. Theory Appl. Radar Sonar Remote Sens.*, 197–201, Aachen, Germany.

Tamada, J. A., Lesho, M., and Tierney, M. J. (2002). Keeping watch on glucose. *IEEE Spectr.*, 39(4): 52–57.

Tatsuoka, K. S. and Tyler, D. E. (2000). On the uniqueness of *S*-functionals and *M*-functionals under nonelliptical distributions. *Ann. Stat.*, 28(4): 1219–1243.

Tatum, L. G. and Hurvich, C. M. (1993). High breakdown methods of time series analysis. *J. R. Statist. Soc. B*, 55(4): 881–896.

Theil, H. (1950). A rank-invariant method of linear and polynomial regression analysis. *Proc. R. Neth. Acad. Sci.*, 53: 1397–1412.

Thomas, L. and Mili, L. (2007). A robust *GM*-estimator for the automated detection of external defects on barked hardwood logs and stems. *IEEE Trans. Signal Process.*, 55(7): 3568–3576.

Tibshirani, R. (1996). Regression shrinkage and selection via the Lasso. *J. R. Statist. Soc. B*, 58: 267–288.

Tibshirani, R., Saunders, M., Rosset, S., Zhu, J., and Knight, K. (2005). Sparsity and smoothness via the fused Lasso. *J. R. Statist. Soc. B*, 67(1): 91–108.

Tsay, R. S. (1988). Outliers, level shifts, and variance changes in time series. *J. Forecasting*, 7(1): 1–20.

Tseng, P. (2001). Convergence of a block coordinate descent method for nondifferentiable minimization. *J. Optim. Theory Appl.*, 109(3): 475–494.

Tukey, J. W. (1960). A survey of sampling from contaminated distributions. *Contrib Prob. Stat.*, 2: 448–485.

Tukey, J. W. (1977). *Exploratory data analysis*, vol. 1. Addison-Wesley Publishing Company, Boston, MA.

Tyler, D. E. (1987). A distribution-free *M*-estimator of multivariate scatter. *Ann. Stat.*, 15(1): 234–251.

van den Bos, A. (1994). Complex gradient and Hessian. *IEE Proc. Vis. Image Signal Process.*, 141(6): 380–382.

van den Bos, A. (1995). The multivariate complex normal distribution – A generalization. *IEEE Trans. Inf. Theory*, 41(2): 537–539.

Vardi, Y. and Zhang, C. H. (2000). The multivariate L1-median and associated data depth. *Proc. Natl. Acad. Sci.*, 97(4): 1423–1426.

Vastola, K. and Poor, H. V. (1984). Robust Wiener-Kolmogorov theory. *IEEE Trans. Inf. Theory*, 30(2): 316–327.

Visuri, S., Koivunen, V., and Oja, H. (2000a). Sign and rank covariance matrices. *J. Stat. Plan. Inference*, 91: 557–575.

Visuri, S., Oja, H., and Koivunen, V. (2000b). Nonparametric method for subspace based frequency estimation. In *Proc. 10th Eur. Signal Process. Conf.*, 1261–1264, Tampere, Finland.

Visuri, S., Oja, H., and Koivunen, V. (2001). Subspace-based direction-of-arrival estimation using nonparametric statistics. *IEEE Trans. Signal Process.*, 49(9): 2060–2073.

Vlaski, S. and Zoubir, A. M. (2014). Robust bootstrap based observation classification for Kalman filtering in harsh LOS/NLOS environments. In *Proc. IEEE Workshop Statist. Signal Process.*, 332–335, Gold Coast, Australia.

Vlaski, S., Muma, M., and Zoubir, A. M. (2014). Robust bootstrap methods with an application to geolocation in harsh LOS/NLOS environments. In *Proc. IEEE Int. Conf. Acoust. Speech Signal Process.*, 7988–7992, Florence, Italy.

Vorobyov, S., Gershman, A., and Luo, Z.-Q. (2003). Robust adaptive beamforming using worst-case performance optimization: A solution to the signal mismatch problem. *IEEE Trans. Signal Process.*, 51(2): 313–324.

Vorobyov, S. A., Rong, Y., Sidiropoulos, N. D., and Gershman, A. B. (2005). Robust iterative fitting of multilinear models. *IEEE Trans. Signal Process.*, 53(8): 2678–2689.

Wang, H., Li, G., and Jiang, G. (2007). Robust regression shrinkage and consistent variable selection through the LAD-Lasso. *J. Bus. Econ. Stat.*, 25(3): 347–355.

Wang, X. and Poor, H. V. (1999). Robust multiuser detection in non-Gaussian channels. *IEEE Trans. Signal Process.*, 47(2): 289–305.

Weiss, A. J. and Friedlander, B. (1989). Array shape calibration using sources in unknown locations – A maximum likelihood approach. *IEEE Trans. Acoust. Speech Signal Process.*, 37(12): 1958–1966.

Weiszfeld, E. (1937). Sur le point pour lequel la somme des distances de *n* points donnés est minimum. *Tohoku Math. J.*, 43: 355–386.

Werner, S., With, M., and Koivunen, V. (2007). Householder multistage Wiener filter for space-time navigation receivers. *IEEE Trans Aerosp. Electron. Syst.*, 43(3).

Wiesel, A. (2012). Geodesic convexity and covariance estimation. *IEEE Trans. Signal Process.*, 60(12): 6182–6189.

Wiesel, A. and Zhang, T. (2015). Structured robust covariance estimation. *Found. Trends Signal Process.*, 8(3): 127–216.

Williams, D. B. and Johnson, D. H. (1993). Robust estimation of structured covariance matrices. *IEEE Trans. Signal Process.*, 41(9): 2891–2906.

Wu, T. T. and Lange, K. (2008). Coordinate descent algorithms for Lasso penalized regression. *Ann. Appl. Stat.*, 2(1): 224–244.

Wu, T. T., Chen, Y. F., Hastie, T., Sobel, E., and Lange, K. (2009). Genome-wide association analysis by Lasso penalized logistic regression. *Bioinformatics*, 25(6): 714–721.

Ye, M., Haralick, R. M., and Shapiro, L. G. (2003). Estimating piecewise-smooth optical flow with global matching and graduated optimization. *IEEE Trans. Pattern Anal. Mach. Intell.*, 25(12): 1625–1630.

Yohai, V. J. (1987). High breakdown-point and high efficiency robust estimates for regression. *Ann. Stat.*, 15(2): 642–656.

Yohai, V. J. and Zamar, R. H. (1988). High breakdown-point estimates of regression by means of the minimization of an efficient scale. *J. Am. Statist. Assoc.*, 83(402): 406–413.

Zhang, T. and Wiesel, A. (2016). Automatic diagonal loading for Tyler's robust covariance estimator. In *Proc. IEEE Workshop Statist. Signal Process.*, 1–5.

Zhang, T., Wiesel, A., and Greco, M. S. (2013). Multivariate generalized Gaussian distribution: Convexity and graphical models. *IEEE Trans. Signal Process.*, 61(16): 4141–4148.

Zoltowski, M. D., Kautz, G. M., and Silverstein, S. D. (1993). Beamspace root-MUSIC. *IEEE Trans. Signal Process.*, 41(1): 344–364.

Zou, H. (2006). The adaptive Lasso and its oracle properties. *J. Am. Stat. Assoc.*, 101: 1418–1429.

Zou, H. and Hastie, T. (2005). Regularization and variable selection via the elastic net. *J. R. Statist. Soc. B*, 67(2): 301–320.

Zoubir, A. M. (2014). *Introduction to Statistical Signal Processing*, ch. 1. Vol. 3 of Zoubir et al., *Array and statistical signal processing* (2014).

Zoubir, A. M. and Boashash, B. (1998). The bootstrap and its application in signal processing. *IEEE Signal Process. Mag.*, 15(1): 56–76.

Zoubir, A. M. and Brcich, R. F. (2002). Multiuser detection in heavy tailed noise. *Digit. Signal Process.*, 12(2–3): 262–273.

Zoubir, A. M. and Iskander, D. R. (2004). *Bootstrap techniques for signal processing*. Cambridge University Press, Cambridge.

Zoubir, A. M. and Iskander, D. R. (2007). Bootstrap methods and applications. *IEEE Signal Process. Mag.*, 24(4): 10–19.

Zoubir, A. M., Koivunen, V., Chakhchoukh, Y., and Muma, M. (2012). Robust estimation in signal processing: A tutorial-style treatment of fundamental concepts. *IEEE Signal Process. Mag.*, 29(4): 61–80.

Zoubir, A. M., Viberg, M., Chellappa, R., and Theodoridis, S., editors (2014). *Array and statistical signal processing*, vol. 3. Academic Press Library in Signal Processing.

Index